An Introduction to Human–Environment Geography

An Introduction to Human–Environment Geography

Local Dynamics and Global Processes

William G. Moseley, Eric Perramond,
Holly M. Hapke, & Paul Laris

WILEY Blackwell

This edition first published 2014

Blackwell Publishing was acquired by John Wiley & Sons in February 2007. Blackwell's publishing program has been merged with Wiley's global Scientific, Technical, and Medical business to form Wiley-Blackwell.

Registered Office
John Wiley & Sons, Ltd, The Atrium, Southern Gate, Chichester, West Sussex, PO19 8SQ, UK

Editorial Offices
350 Main Street, Malden, MA 02148-5020, USA
9600 Garsington Road, Oxford, OX4 2DQ, UK
The Atrium, Southern Gate, Chichester, West Sussex, PO19 8SQ, UK

For details of our global editorial offices, for customer services, and for information about how to apply for permission to reuse the copyright material in this book please see our website at www.wiley.com/wiley-blackwell.

Library of Congress Cataloging-in-Publication Data

Moseley, William G.
An introduction to human-environment geography : local dynamics and global processes / William G. Moseley, Eric Perramond, Holly M. Hapke, Paul Laris.
pages cm
Includes bibliographical references and index.
ISBN 978-1-4051-8932-3 (hardback : alk. paper) – ISBN 978-1-4051-8931-6 (pbk. : alk. paper)
1. Human ecology–Textbooks. 2. Human geography–Textbooks. I. Title.
GF43.M67 2013
304.2–dc23

2013006409

A catalogue record for this book is available from the British Library.

Cover image: A man planting rice in paddy fields on the outskirts of Seoul, South Korea.
© Philippe Lissac/Godong/Panos.
Cover design by www.cyandesign.com

Set in 11/13pt Dante by SPi Publisher Services, Pondicherry, India

1 2014

For B. Ikubolajeh Logan (WM)
For Marshall Bowen (EP)
For my daughter, Syona, and to John Agnew (HH)
For B.L. Turner (PL)

Contents

Notes on the Authors

William G. Moseley is a professor and chair of geography at Macalester College, where he teaches courses on environment, development, and Africa. His research interests include political ecology, tropical agriculture, environment and development policy, and livelihood security. His research and work experiences have led to extended stays in Mali, Zimbabwe, South Africa, Botswana, Malawi, Niger, and Lesotho. He is the author of over 60 peer-reviewed articles and book chapters that have appeared in such outlets as *Proceedings of the National Academy of Science*, *Ecological Economics*, the *Geographical Journal*, the *Geographical Review*, *Applied Geography*, the *Singapore Journal of Tropical Geography*, and *Geoforum*. His books include: four editions of *Taking Sides: Clashing Views on African Issues* (2004, 2006, 2008, 2011); (with David Lanegran and Kavita Pandit) *The Introductory Reader in Human Geography: Contemporary Debates and Classic Writings* (Blackwell, 2007); (with Leslie Gray) *Hanging by a Thread: Cotton, Globalization and Poverty in Africa* (2008); and (with B. Ikubolajeh Logan) *African Environment and Development: Rhetoric, Programs, Realities* (2004). His fieldwork has been funded by the National Science Foundation and the Fulbright-Hays Program. He has served as editor of the *African Geographical Review*, as a national councilor to the Association of American Geographers, and as chair of the cultural and political ecology specialty group.

Eric Perramond, a geographer, is an associate professor in both the Environmental Science and Southwest Studies programs at the Colorado College. His teaching and research interests include cultural-political ecology, environment and development issues, GIS and research methods, and agro-climate governance issues. He conducts human–environment research in the Greater Southwest, semi-arid Mexico, and the French and Spanish Pyrenees. He has published in the *Geographical Review*, *Area*, the *Journal of Latin American Geography*, and the *Journal of Political Ecology*. He serves on the editorial board for the *Journal of Political Ecology*, and was associate editor for the *Journal of Latin American Geography* and on the editorial board for *ACME: An International E-Journal for Critical Geographies*. He is the author of *Political Ecologies of Cattle Ranching in Northern Mexico* (2010) and is a former Fulbright-Garcia Robles Fellow to Mexico. He also served as the Chairman of the *Conference of Latin Americanist Geographers* (CLAG) during 2008–10.

Holly M. Hapke is an associate professor in the Department of Geography at East Carolina University, where she also teaches courses in the International Studies program. Her research and teaching interests include political economy and development; fisheries and coastal livelihoods; ecological conflict; gender; migration; and research methods. Her regional area of expertise is South Asia, and she has conducted research on rural development issues in the US South. Her field research on gender, fisheries development, and fisherfolk livelihoods in India has been funded by the National Science Foundation, the Association of American Geographers, and the US Department of Education Fulbright-Hays Program. She has published articles in journals such as *Economic Geography*, *Annals of the Association of American Geographers*, *Professional Geographer*, *Gender, Place & Culture*, and *Geographical Review*, and she is a contributing author to Pulsipher and Pulsipher, *World Regional Geography: Global Patterns, Local Lives*, third edition. She currently serves on the editorial board of *Gender, Place & Culture*.

Paul Laris is professor and chair of the Department of Geography at California State University, Long Beach, where he also teaches in the Environment, Science, and Policy Program. His teaching and research interests include biogeography, cultural and political ecology, fire ecology, global change, ecological restoration, and remote sensing. He has conducted research in the savanna of Mali, the grasslands of Tierra del Fuego, and the shrublands of California. He has published in such journals as *Human Ecology*, *The Annals of the Association of American Geographers*, *Remote Sensing of Environment*, *Geoforum*, and *Bois et Forêts des Tropiques*. His fieldwork has been funded by NASA and the National Geographic Society.

Preface and Acknowledgments

This book has been a long time in the making. Like any good text, it emerged from a series of conversations, many in bars and cafés, and frequently when we met at our annual professional meeting. We had a few concerns which motivated us to write this book, foremost of which was a text that would convey geography's theoretically rich tradition and unique approach to environmental issues. Our other concern was to have a text that would be accessible to introductory students, many from allied environmental fields who were encountering geography for the first time, and others in geography for whom this was their first course on human–environment themes. While there are other environmental geography texts on the market, none (in our view) did all that we wanted. We felt that the lower level texts didn't do enough to convey geography's unique approach to the subject matter, frequently differing little from more generic environmental studies or environmental science texts. Those books that did convey the theoretical richness of the human environment tradition tended to be pitched at too high a level of student, or too narrowly focused on a particular subtheme of human–environment geography. The text that follows is our attempt to fill this niche.

You will note that the book is divided into four parts. The first part is meant to be a broad overview of the basic information needed to understand human–environment geography, from the geographic perspective, to environmental politics, to some basic physical geography and ecology. The second section explores a sampling of geography's rich theoretical traditions in the realm of human–environment geography. The third part is more thematic in nature, most closely resembling the traditional textbook approach except for a concerted effort to make connections between this material and the theoretical approaches detailed in the second section. The final part is meant to connect the book's material to the real world by showing the student how geographers undertake fieldwork and collect and analyze data. The concluding chapter makes suggestions for using the concepts in this text to understand environment-related problems and bring about change. Each of the chapters in these four sections has a similar structure. Chapters begin with an icebreaker, or a meaningful vignette which brings out the major themes of the chapter. This is followed by a statement of chapter objectives, an introduction, and then the main text. All chapters end with a chapter summary, critical questions, key terms, and references.

While this book was very much a collective project, our varied regional and thematic expertise helped ensure that a range of material would be covered from some position of comfort and familiarity. We also hail from different types of institutions, private colleges in Minnesota and Colorado, and public universities in North Carolina and California, and thus have experience working with different types of students. All of us relied heavily on our own teaching, research, and work experience to inform this project. This book took longer to complete than originally anticipated. We particularly wish to thank Justin Vaughan and Ben Thatcher at Wiley-Blackwell for attempting to keep us on track and for showing endless patience and understanding when we fell behind. We also thank our families for their understanding and support while we labored at writing, for reading and re-reading drafts in some cases, and for patiently listening to us over meals as we shared our geographic revelations. We finally express our appreciation to the anonymous reviewers who provided feedback on various portions of this text, and to our students with whom our interactions in the classroom have informed the way we present this material.

Part I

Fundamentals of Human–Environment Geography

1

Introduction

A Geographic Perspective on Human–Environment Interactions

Icebreaker: Human–Environment Connections Across Time and Space

Before chemical fertilizers came into heavy use in the 20th century, guano (bird or bat droppings) was the leading internationally traded source of agricultural plant nutrients. It was valued because of its high levels of phosphorous and nitrogen and lack of odor. The Incas of South America understood the value of guano long before the

An Introduction to Human–Environment Geography: Local Dynamics and Global Processes, First Edition. William G. Moseley, Eric Perramond, Holly M. Hapke and Paul Laris.

Published 2014 by John Wiley & Sons, Ltd.

Europeans and regulated its extraction quite carefully. The Incan government divided up the guano-bearing islands off the coast of modern-day Peru between its different provinces. Guano had accumulated on these islands over centuries because of abundant bird life due to rich fish stocks, a uniquely dry climate which enhanced guano preservation, rocky shores for nesting, and protection for the birds from predators and humans. Rules were established concerning when and where guano could be harvested and disturbing the nesting birds which produced guano was an offense punishable by death.

The geographer and explorer Alexander von Humboldt was the first European to recognize the potential value of guano. He returned from his 1799–1804 voyage around South America with samples which he shared with two French chemists who subsequently confirmed the value of the substance. American farmers experimented with guano in the 1820s, and then British farmers in the 1840s. Despite the initial concerns of farmers that such a powerful fertilizer would upset the nutrient balance of agricultural soils, demand for guano soon surged. The United Kingdom imported over 2 million tons of guano between 1841 and 1857. The fury over the guano trade was intense. It led to the Guano War of 1865–66 between Spain and Peru. The US Navy fought with Peru to maintain access to guano. The US also colonized over 50 islands in the Pacific and the Caribbean (including Midway Island) because of their guano resources. By 1900, the world's guano resources were all but depleted.

Fast forward to the 21st century, when one of the authors of this text was traveling with a group of students along the Atlantic Coast of South Africa. Here he visited Lambert's Bay, a fishing village on the coast with a history as a source of guano which was exported as fertilizer to Britain in the 19th century. The small island in Lambert's Bay was now a bird sanctuary where nature lovers and tourists could come and observe the courting rituals and the nesting habits of the Cape gannet. The gannet was a prodigious producer of the guano that had once accumulated in vast quantities on rocky islands along this semi-arid coastline. The author had been to the island the previous year and seen large numbers of Cape gannets (see Figure 1.01). As he crossed over the bridge to the island, he noticed that something was quite different, there were no gannets. He came to learn that the entire colony had left because they were being attacked by seals. This was, in itself, highly unusual as the seals had long coexisted with the gannets and never bothered them. The problem was that the seals were competing with fishermen for the same food source and were losing. As such, it was hunger which led the seals to attack the gannets on the island and it was this atypical behavior which caused the colony of Cape gannets to leave. While some of the overfishing in this area was caused by South African commercial fishers, the bigger culprit was large international fishing fleets.

The twists and turns of this story raise a number of important issues for consideration. These include: the ability of some societies to manage their resources sustainably, the role of science in the use and management of resources, the seeming inability of the global capitalist system to limit consumption, the role that non-human actors may play in transmitting the impacts of one human action to another human group, and the limits of

Figure 1.01 A colony of Cape gannets, Lambert's Bay, Atlantic Coast of South Africa.
Source: Photo by W.G. Moseley. Used with permission.

preservation in open ecosystems and economies. All of these themes and more are central to the dynamic subfield of human–environment geography.

Chapter Objectives

The objectives of this chapter are:

1. To suggest that humans, like other animals, are able to sustainably interact with their environment.
2. To highlight the pressing nature of some contemporary environmental problems.
3. To articulate the relevance of the geographic perspective to environmental questions.
4. To outline broad elements of a human–environment geography approach to environmental questions.
5. To demonstrate what new insights may be gleaned by applying the human–environment geography approach to some basic natural resource management concepts and an example of this in US environmental history.
6. To share the general plan and logic of the book.

Introduction

The broad objective of this chapter is to introduce to students to the way that human–environment geographers look at the world. We begin by exploring how humans are similar to, and different from, other animals which manipulate the environment. We then review geography and its distinctive human–environment tradition, followed by an exploration of some broadly similar ways that human environment geographers often examine environmental questions. The chapter ends with a specific case of how the geographic lens yields new insights when trained on some common environmental management approaches, namely exploitation, conservation, and preservation.

Animals and Their Habitats

Beavers (*Castor canadensis* in North America, *Castor fiber* in Eurasia) are known for their ability to modify the landscape for their own benefit and that of other species. By damming streams, beavers raise the water level to form protective moats around their lodges. The resulting beaver ponds also create the deep water needed for winter food storage in northern climates. While other animals struggle with winter cold and hunger, beavers stay warm in their lodges with an underwater food cache of branches in close proximity (see Figure 1.02). Beavers also harvest trees and branches for food and construction purposes. This pruning stimulates willows, cottonwood, and aspen to regrow more thickly the next spring. While some beaver behavior is instinctive, they also learn by imitation and from experience. As such, we find some beavers who are very adept at building dams and others who are not. Older, more experienced beavers also tend to build better dams than younger ones. The beavers' habitat modifications also impact other species. The wetlands they create support other mammals, fish, turtles, frogs, birds, and ducks. These wetlands also provide a variety of ecological services, such as the catchment of floodwaters, the alleviation of droughts (because beaver dams keep water on the land longer), the reduction of erosion, the local raising of the water table, and the purification of water.

Figure 1.02 Sketch of beaver lodge and dam.

Humans, like other animals, also modify the landscape. We manipulate the land, for example, through burning, cutting, tilling, planting, harvesting, dam building, and home construction to meet our own objectives. Through a process of experimentation, success and failure, observation, and the sharing and stealing of ideas, humans have learned how to manipulate the environment for their own purposes. For example, through careful observation of local environmental feedback, humans often developed farming systems that were highly productive, and sustained over centuries (Figure 1.03). A case in point is shifting cultivators in Papua New Guinea who created farming systems that were over five times more efficient (in terms of a ratio of crop yield over energy inputs) than modern maize-cropping systems in the United States and supported much higher levels of agrobiodiversity (Pimentel and Pimentel 1979). Women in rural Mali (West Africa) routinely collect dead wood and coppice (trim) branches from existing trees for firewood, lessening the chances of unmanageable bush fires and encouraging regrowth. Up until recently, many American farmers planted shelter belts (or tree hedges) around their fields in order to reduce aeolian (wind) erosion and encourage the proliferation of white-tailed deer (*Odocoileus virginianus*) which they hunted for game meat.

Of course, some societies took up unsustainable practices which eventually led to environmental decline and their downfall. Sometimes, but not always, these were highly stratified societies in which those making the decisions and those working the land were separated by many layers. In other cases, new migrants failed to understand the ecology of an area and attempted management approaches

Figure 1.03 A farm in Papua New Guinea. *Source*: © WaterFrame/Alamy.

that were inappropriate for their new location. Still others developed intensive production systems which required significant amounts of human labor to maintain. When political instability or disease disrupted these labor flows, such systems quickly fell into decline and the productivity of the environment declined.

As humans societies grew and prospered, and people traveled greater and greater distances, they began to trade. While trade was initially in luxury items, food and raw materials eventually came to be traded in significant quantities. By the 20th century, even garbage was being shipped around the world. The significance of this trade, combined with urbanization, was that it gradually separated people from the sources of their food and goods and the byproducts of their consumption. We were losing our ability to productively and sustainably engage with ecosystems. Today we live in a world where many consumers in the most developed areas of the world have little to no idea where their provisions originate from and how they are produced. We also live on a planet where the consequences of such detachment from the biophysical world seem to be growing. Increasing carbon emissions, and resulting climate change, is probably one of the most disquieting, global-scale environmental challenges. Other challenges, like deforestation, ground water depletion, and the loss of biodiversity, are also of great concern.

Not all ecological challenges are a direct result of humans modifying the environment in a problematic manner. In some cases it may have more to do with how humans position themselves vis-à-vis the biophysical world. Hurricanes, for example, become more of an issue for humans when they build homes close to coastlines, or inundations are a problem when towns and cities are established in floodplains. Some biomes have naturally sparse or erratic rainfall, so trying to live in such areas without adapting to these patterns is destined to be problematic.

Clearly many of the challenges described above could be avoided if we better understood our place within, and relationship to, the biophysical world. This text helps the student explore that world and how we got to this particular point in human history. While many disciplines and fields of study examine these questions, this text helps students understand these issues from the perspective of human–environment geography. We begin this chapter with a brief introduction to geography and then a more thorough examination of some basic elements of human–environment geography.

What Is Geography and What Does It Have To Do with Studying the Environment?

Geography is so basic that we all seem to have some idea of what it is, yet curiously, many would have trouble describing the subject to another person in casual conversation. Geography comes from the Greek word meaning "earth writing" or "earth describing." Even though the emphasis in geography has changed over the years, this is still a fairly accurate statement.

While the Greeks were the first to organize geography as a coherent body of knowledge, the need for geographic knowledge is as old as humankind. For as long as people have been traveling, exploring, and migrating, they have been encountering different environments and other human societies. As such, the survival and success of human populations meant that they needed to understand other groups, faraway lands, where these were located spatially (if for no other reason than to know how to get there again), the processes that connect one human group to others, and ways in which each group is unique. In the process, such travelers, explorers, and migrants learned a lot about where they had come from, that is, it helped them to understand what was special about their own homes.

Geography is a broad discipline that essentially seeks to understand and study the spatial organization of human activity and of people's relationships with their environment. It is also about recognizing the interdependence among places and regions, without losing sight of the individuality and uniqueness of specific places. Geography is rather unique for a discipline in that it straddles the science–social science–humanities divide, using a broad arsenal of methods and perspectives to tackle questions. It is also not an armchair science (in which data is downloaded for analysis) but rather has a long tradition of fieldwork. Finally, many geographers do get excited about maps (some might call us map geeks) but it is important to remember that maps are a means to an end for most geographers. By displaying data spatially, it pushes us to ask why things are distributed the way they are, or it may reveal patterns or correlations which had not previously been seen.

While the general tenets of a geographic approach (i.e., attention to spatial patterns, human–environment dynamics, the uniqueness of place, and connections between regions and across scales) apply to all areas of geography, geography has grown over time to recognize sub-specialties within the discipline based on the subject matter addressed. At the broadest level, there is a commonly recognized divide between the study of biophysical phenomena (**physical geography**) and the examination of human or social phenomena (**human geography**). Physical geographers seek to understand long-term climate patterns and change (climatology), patterns of plant and animal distribution (biogeography), and the origin and evolution of landforms (geomorphology). Human geographers study the patterns and dynamics of human activity on the landscape, including settlement, urbanization, economic activity, culture, population, development, and disease.

Between physical and human geography, lies the vibrant arena of **human–environment geography**. The investigation of nature–society relationships lies at the heart of geography and has been one of the pillars of the discipline since the modern academic structure crystallized in 19th-century Germany. This realm of inquiry also has been an important bridge between geography and other fields. Figure 1.04 depicts the position of human–environment geography within the discipline of geography.

This textbook is focused on a dynamic and burgeoning subfield of geography known as human–environment geography. The book introduces you to the study of human–environment interactions from a geographic perspective, with

Human geography	Human-Environment	Physical geography
(e.g., urban geography, economic geography, population geography, cultural geography, development geography, political geography)	(e.g., cultural ecology, political ecology, agricultural geography, water resources, human-dimensions of global change, hazards geography)	(e.g., biogeography, climatology, geomorphology)
Techniques (e.g., geographic information systems (GIS), remote sensing, cartography, statistics)		

Figure 1.04 Human-environment geography within the discipline of geography.

a special emphasis on the role of humans in changing the face of the earth and how, in turn, this changed environment may influence humans. We will examine environmental issues in a variety of geographic contexts (developed and developing countries) and the connections between environmental problems in different locations. While we tend to think of the environment as "natural" and more prominent in areas with fewer people, we will argue that the built environment is of no less concern than many so-called natural areas, and that both are products of human action. For example, in terms of generic interactions with the environment, what makes a peasant farmer any different than a suburban homeowner? Both live in environments modified by human activity, both manipulate the landscape (the farmer tilling her field, and the suburban man tending his lawn), and both are influenced by environmental conditions (the farmer planting six different varieties of millet in her field because rainfall varies from year to year; the suburban man, driving to the grocery store because his neighborhood has no sidewalks, is removed from shopping areas and lacks access to public transportation).

Human–environment geographers working in various subfields often interact with other academics or professionals working on similar themes (e.g., political ecologists with anthropologists and development practitioners, hazards geographers with geologists and disaster relief specialists, or water resource geographers with hydrologists and watershed managers).

Geography has long been known for its techniques for presenting and manipulating spatial data, particularly **cartography** or mapping. What is important to remember is that most geographers use these techniques as a bridge to greater understanding. For example, human–environment geographers may use dot maps to present and understand population distributions, **geographic information systems** (GIS) to analyze the potential relationship between population density and soil fertility, or **remote sensing** (aerial photography and satellite imagery) to monitor change in surface biomass over time. Some geographers specialize in a particular technique, rather than a thematic area of geography. These geographers often focus on further developing such technologies, devising methods for interpreting the data produced by them, or reflecting on the social implications of their use. A few selections in this volume will focus on the use of these technologies by human–environment geographers.

A Geographic Perspective on Environmental Questions[1]

Students may wonder what differentiates an introductory-level human–environment geography course from its corollary in environmental studies or environmental science. Later in this book, we will explore a number of subdisciplines in geography that offer distinctive lenses through which to explore environmental issues (e.g., cultural ecology, political ecology, hazards geography, environmental history, and environmental justice). In this chapter we articulate more fundamental geographic perspectives that often characterize the discipline's approach to human–environment questions. While not offering an exhaustive list of such generalized approaches, here we investigate and apply four perspectives: scale-sensitive analysis, attention to spatial patterns of resource use, a conception of the human–environment system as a single unit (rather than two separate parts), and a cognizance of the connections between places and regions.

As a way of introducing these four perspectives, we apply them to three basic approaches to environmental management that you would encounter at the start of most environmental studies texts: exploitation, conservation, and preservation. We start by exploring the conventional understandings of these approaches and then show how they may be understood somewhat differently from a geographic angle. We then re-examine, using this geographic perspective, a famous case in US environmental history that has been used to illustrate the conventional understandings of conservation and preservation, not to mention an early rift in the US environmental movement.

Conventional Understandings of Exploitation, Conservation, and Preservation

The concepts of exploitation, conservation, and preservation are typically used to differentiate human management and use of renewable resources (e.g., forests, fisheries, many sources of water). **Exploitation** is the easiest of these three concepts to grasp. It refers to the use of a resource without regard to its long-term productivity, usually by over-harvesting in the short term. As such, an exploitative approach to forest management might entail clear cutting, and not replanting, large tracts of land.

While the terms conservation and preservation are sometimes used interchangeably in public discussions, environment-related fields carefully use these words to refer to particular management regimes for renewable resources. **Conservation** (sometimes also described as the utilitarian approach in environmental history, or as resource conservation in the UK) typically refers to use within certain biological limits, or within the annual growth increment of a particular resource. In the case of forests or fisheries, this annual growth increment is also referred to as the **sustainable yield**[2] or maximum sustainable yield.

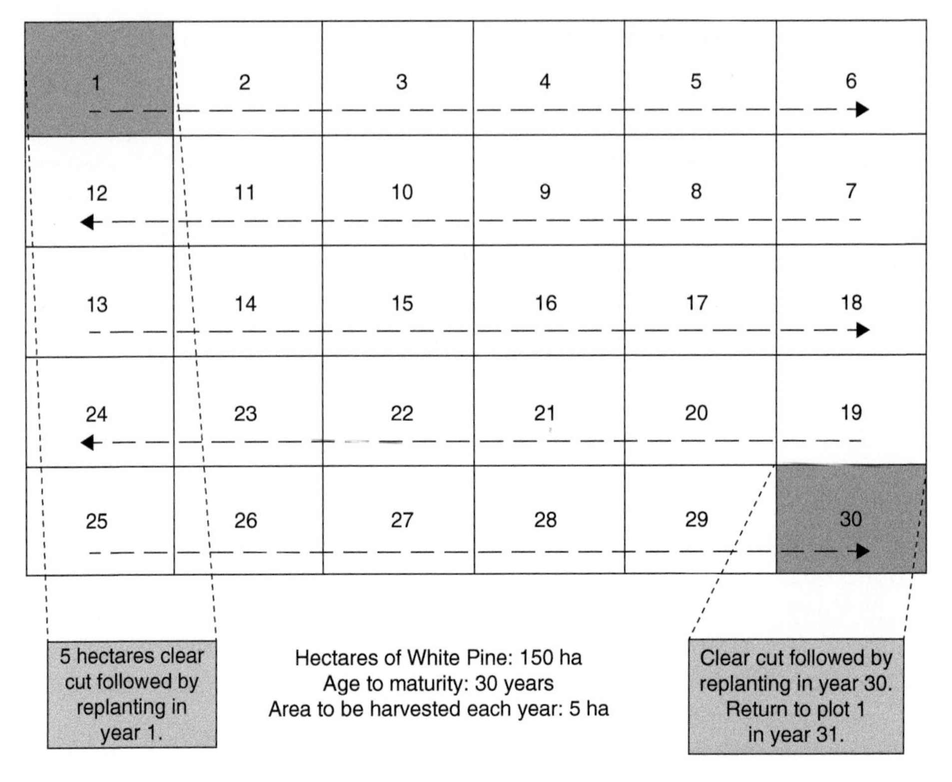

Figure 1.05 The principle of maximum sustainable yield as applied to an even-aged monoculture of white pine. *Source*: Macalester College cartographer Birgit Muhlenhaus/Moseley 2009.

The US government natural resource management agency most closely associated with the conservation approach is the US Forest Service (USFS), and the same approach is also applied by government forest agencies in many other parts of the world. Since the US Sustained-Yield Forest Management Act of 1944, the USFS has managed many of its forests under the principle of maximum sustainable yield. Typically the formula (forest area/age to maturity) is used to determine the percentage of the total forest area that may be harvested and replanted each year. Figure 1.05, for example, depicts spatially how an even-aged monoculture[3] of white pine that is 150 hectares in size, and for which the age to maturity is 30 years, would be harvested and replanted at the rate of 5 hectares per year (150 hectares/30 years).

In contrast to exploitation or conservation, **preservation** (also known as nature conservation in the UK) typically refers to the non-use or non-consumptive use of natural resources in an area. The practical expression of the preservationist approach in the North American context often comes in the form of a wilderness area or park. In some instances, an area is completely off limits to humans. More frequently, **non-consumptive uses** are allowed (e.g., hiking, camping). The rationale for preservation is that certain areas must be set aside for compelling aesthetic or biodiversity reasons. This approach to preservation has been described as the

"Yellowstone model,"[4] a model that emphasizes national parks which people may visit as tourists, but neither reside in nor exploit to support a resource-based livelihood. The US Park Service is the US government natural resource management agency most clearly identified with the preservationist approach (as is the case for government park services in many other parts of the world). The preservationist approach was introduced to developing countries during the colonial era when many parks and wilderness areas were established. Parks and preserves in the world's tropical regions have received considerable attention since the 1992 World Summit on Environment and Development in Rio de Janeiro. For example, over 10% of territory in some African countries is now managed by state and international organizations for preservation purposes.

Geographic Perspectives on Exploitation, Conservation, and Preservation

Geography's focus on scale, synergistic human–environment interactions, land-use patterns, and the connections between places and regions offers intriguing insights into the concepts of exploitation, preservation, and conservation. Attention to scale is a core geographic concern and a framing device that is profoundly implicated in any form of spatial analysis. The concept of **scale** may be used in somewhat different ways. In cartography (or the science of map-making), scale refers to the distance on the map in relation to the distance on the surface of Earth. As such, relatively small-scale maps show larger areas because the fraction of distance on the map over distance on the Earth's surface is small. In contrast, large-scale maps show smaller areas because the ratio or fraction of distance on the map over distance on the surface of the Earth is relatively large. Unlike the relatively specific idea of map scale, we can also think of this term more conceptually, e.g., local versus global scale (see Chapter 4 for further discussion of scale). As such, one might analyze a problem at the scale of a local community, or at the level of a park, or using data aggregated at the scale of a state or province (or some broader scale).

The geographer Stan Openshaw (1983) problematized a-scalar analysis in terms of the **modifiable areal unit problem**. In discussing this problem, Openshaw focused on two issues related to scale (the level at which data is aggregated and the boundaries of spatial units) to show how variation in these factors greatly affected findings.

Within geography, there is also a body of scholarship on the **politics of scale**. These studies examine the political implications of the choice of scale at which an environmental issue is articulated and conceptualized. Different groups frequently struggle over the scale at which an issue is framed. Mansfield and Haas (2006: 78), for example, discuss how "using scale as a framing device is a powerful political strategy ... because focusing on a particular scale presupposes certain kinds of solutions while foreclosing others." Similarly, attention to scale complicates conventional understandings of exploitation, conservation, and preservation. We may think about scale in at least three different ways in our case: the scale at which the approach is implemented, the scale at which the approach is analyzed, and the scale at which the approach is discussed, or discursive scale (the last point will be addressed in the subsequent section on Hetch Hetchy Valley). The scales at which an approach is implemented or analyzed are sometimes referred to as scale frames (e.g., Kurtz 2003).

In practical terms, the scale at which the preservationist approach may be implemented is limited by the need for humans to use natural resources. As such, unless an area is lightly populated, it is challenging to set aside extremely large tracts of land as preserves because people need to use some land to sustain themselves. With the possible exception of Antarctica (a continental example), most lands set aside for preservation are modest in scale. More specifically, most of the world's big IUCN (International Union for the Conservation of Nature) category Ia and Ib parks are in the high Arctic or sparsely populated tropic forests, whereas most preservation units (IUCN category IV) in densely populated Europe are relatively small. While parks appear as preservation (if the unit of analysis stops at the park boundary), this perception quickly changes at broader scales of analysis if surrounding areas are overexploited. For example, national parks in Costa Rica have been referred to as diamonds in a sea of devastation (Sanchez-Azofeifa et al. 2002).

In contrast to preservation, conservation could (at least in theory) be implemented at a much broader scale because it allows for human use of resources within biological limits. In such a situation, people in all places would be allowed to tend to their needs, yet would be required to operate within the biological limits of the environment. This is a very integrated and spatially broad vision of conservation that shares commonalities with certain (i.e., the strong or radical green) conceptions of sustainable development (see Chapter 2). In practice, such an approach would require a significant departure from current development patterns. This departure would be necessary because market economies (which tend to produce haves and have-nots) may not be able to coexist with an approach where the limits of all environments are respected. In other words, many would assert that capitalism itself promotes a patchwork of uses on the landscape, with capital accumulation in one area leading to capital depletion in another (Frank 1979; Wallerstein 1979; Harvey 1996).

In the real world, conservation (like preservation) is often implemented at a more local scale. As described earlier, foresters managing a wood lot under the principles of sustainable yield carve it up into equal-sized plots (derived via the formula: forest area/age to maturity) and then harvest and replant one such plot per year until eventually they return to the first plot that was cut and replanted. Here again, examining the situation at a variety of scale frames allows one to recognize that the management of the forest as a whole might be labeled as conservation (scale frame A in Figure 1.06). Conversely, when examined at the scale of the individual plot being harvested (often several hectares in size), the situation might more aptly be described as exploitation (scale frame B in Figure 1.06). Exploitation might be the more appropriate term at this scale because such plots are often

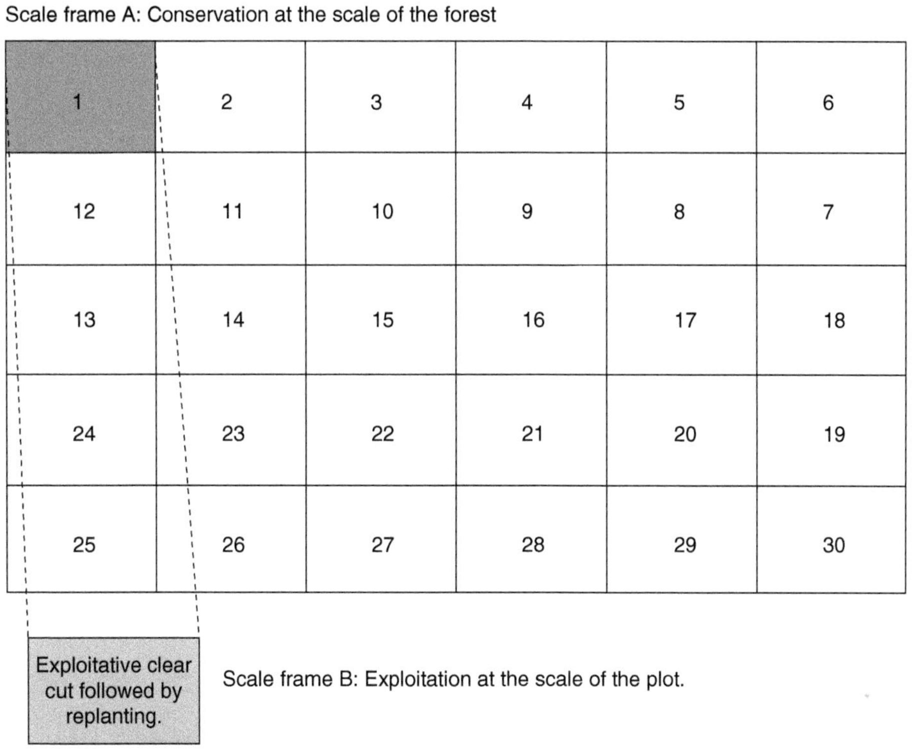

Scale frame A: Conservation at the scale of the forest

1	2	3	4	5	6
12	11	10	9	8	7
13	14	15	16	17	18
24	23	22	21	20	19
25	26	27	28	29	30

Exploitative clear cut followed by replanting.

Scale frame B: Exploitation at the scale of the plot.

Figure 1.06 Scale analysis of forest managed under principles of maximum sustainable yield. *Source*: Macalester College cartographer Birgit Muhlenhaus/Moseley 2009.

clear-cut, and lie barren for some time before they are replanted. While such a process incrementally impacts biodiversity, soil stability, and infiltration at the scale of the forest management unit, all three of these factors decline dramatically at the plot scale after a clear cut.

Analyzing preservation and conservation at the scale of a land management unit allows one to arrive at one set of conclusions about the nature of these approaches. However, as the scale frame is narrowed or broadened, the homogeneity or heterogeneity of land-use practices changes, and the characterization of what is happening changes as well. As such, the scale at which an approach is presented or analyzed is a choice with political and ideological implications.

If one pulls back from the land management unit and begins to analyze the situation at broader scales, at least three other geographic issues begin to become apparent: (1) patterns of land use (i.e., how the landscape is divided up into different land-use units); (2) the economic and ecological connections between different areal units – and the politics of these linkages; and (3) synergistic human–environment interactions.

In the first instance, for example, preservation at a limited scale means that humans must divide up the landscape into areas designated for preservation and those for other types of land use (ranging in use from overexploitation to

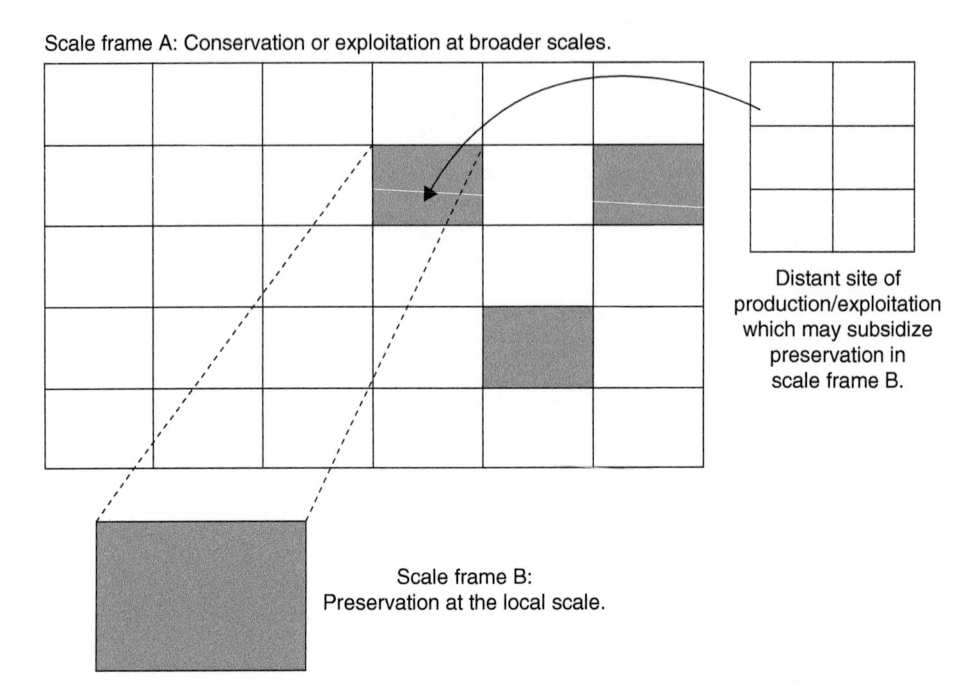

Figure 1.07 Scale/space analysis of preservation. *Source*: Macalester College cartographer Birgit Muhlenhaus/Moseley 2009.

conservation). As such, when viewed from a broader scale perspective (scale frame A in Figure 1.07), one sees a patchwork landscape of exploitation, conservation, and preservation, which (by definition) could not be considered preservation. Such patchwork landscapes, with preservation in some areas (scale frame B in Figure 1.07) and different uses in others, may represent conservation at best (use within biological limits) and (more likely) overexploitation in many instances. In other words, preservation at the local scale could violate conservation at a broader scale if it leads to overexploitation on other parcels.

Secondly, underpinning the land-use mosaic are a variety of economic and ecological linkages between preservation areas and other points on the landscape. Accounting for linkages, or chains of explanation, between local land-use strategies and the broader political economy has been standard practice in geography, especially in such subfields as political ecology, where there is an emphasis on the political economy of human–environment interactions (see Chapter 4).

At a very basic level, the non-use or non-consumptive use of resources in certain areas implies that uses that could have occurred in these areas likely have shifted elsewhere. While it is acknowledged that US national parks often were established on economically marginal lands, it is difficult to deny that these could have been sites of resource extraction. In other words, it is the "subsidy" provided by intensive use of "normal use areas" (both as sources of resources and sites of human habitation) that allows people to set aside areas for preservation. Another way to conceptualize one unit of land subsidizing preservation on another is to consider the **net primary productivity**[5] (NPP) of any ecosystem. If a human population was to consume a large portion of an area's NPP per year, and then half the land was turned into a preserve, then the human population would need to look elsewhere for the resources it needs to survive. These resources (garnered from unpreserved land) then support or subsidize the preserve. The subsidy provided to preserves from normal land-use areas may come from inside or outside the country (see Figure 1.07 showing the subsidy provided by a distant site of production or exploitation). For example, a reduction in US timber harvests from 1990 (because of stricter requirements to sustain biodiversity and other ecosystems functions) was possible in the face of growing US demand for wood products because of increasing imports from Canada (Martin and Darr 1997) and several tropical countries (Tucker 2002). The global economic system, with its increasingly global set of commodity chains, is sufficiently opaque to prohibit most people from seeing the impact of resources they may be drawing on from overseas (Princen 2002).

Beyond shifts in resource extraction, a second connection between preserves and other points on the landscape may be the dislocation of peoples. A large body of geographic scholarship on parks and peoples has documented how the creation of parks in developing countries often implies the relocation of peoples to other areas (see Chapter 11). For example, Guha (1997) describes an on-going controversy in Nagarhole National Park in Karnataka Province, India, where the Forest Department has been trying to relocate 6000 local people. Relocated peoples, while (arguably) lessening impact within the preserve, often augment impact elsewhere (another dimension of outside areas "subsidizing" preserves). Furthermore, displaced peoples often bear considerable costs for the creation of such preserves in terms of compromised livelihoods. While a lesser-known phenomenon, the establishment of national parks in the US (and other areas of the Global North) often involved the displacement of native peoples (Burnham 2000; Braun 2002). While most proposed US national parks were described as economically worthless and uninhabited by their proponents in the early 20th century in order to avoid conflicts with economic interests, many of these areas were far from uninhabited. The environmental historian Philip Burnham has described the role of public agencies in removing Native Americans from lands that were to become Glacier, Badlands, Mesa Verde, Grand Canyon, and Death Valley national parks (2000).

A third connection exists to outside areas when preserves are supported by the fees of ecotourists, increasingly the case in many developing countries. These fees represent real financial transfers that may offset the resources forgone when an area is set aside for a preserve. However, those bearing the costs of the park (often local people in terms of compromised livelihoods) and those benefiting from user fees and tourist revenues (e.g., national governments, tour companies) are often different. Globetrotting ecotourists also generate significant environmental externalities when they consume large amounts of resources to travel across the world to visit wildlife preserves and parks in Africa, Latin America, and Asia.

What the above examples suggest is that preservation is only preservation (non-use or non-consumptive use) at the scale of the preserve. When such a preserve is viewed at broader scale frames, we see that this is not really preservation because non-use/non-consumptive use in one area is almost always subsidized by use in surrounding areas (or even distant centers of production and consumption). In either instance (conservation or preservation), we suggest that the distinctions between exploitation, conservation, and preservation begins to become blurred when they are analyzed at multiple scales. Key to understanding this multi-scaler analysis is attention to how humans divide up space to apply either principle (conservation or preservation) and to socio-economic and ecological connections that exist between regions and places.

Another issue which becomes clear when discussing approaches to resource management (exploitation, conservation, and preservation) is a tendency to separate humans from nature, rather than viewing humans as part of nature (that is, just another animal among many). As the story about beavers at the beginning of this chapter suggests, humans are not really all that different from other animals who manipulate their environments to achieve certain ends. This is not to insinuate that, if humans are part of nature, then anything goes. Rather, the idea is to acknowledge that we are not all that different from other actors in the environment and that it is problematic to think of ourselves as being able to operate outside of such systems. If humans are a part of nature, then the preservation approach in particular becomes problematic as it operates from a position that humans are (or ought to be) outside of natural areas if these are to be considered natural. The approach also ignores that fact that preserves are a product of human action as these are areas where we have deliberately chosen not to pursue certain types of activities.

Gifford Pinchot, John Muir, and Conventional Interpretations of the Hetch Hetchy Valley Controversy

In the US, the conservationist and preservationist philosophies evolved quickly in the wake of large-scale deforestation (i.e., exploitation) in the post-Civil War period as American cities and industry boomed (Williams 1989). Along with this destruction came the realization by some that America's vast natural resource base was not inexhaustible.

Two iconic figures, Gifford Pinchot and John Muir, are used in many texts to help illustrate the two approaches to resource management (Miller 1990; Cunningham and Saigo 2001; Holechek et al. 2003; Chiras and Reganold 2005; Righter 2005). **Gifford Pinchot**, a German-trained forester who established the Yale School of Forestry (the first school of forestry in the US) and the founding head of the US Forest Service, was probably the most visible early 20th-century proponent of the conservationist approach in the US (Miller 2001). Pinchot did not see conservation and development as incompatible. In fact, Pinchot saw the wise use of natural resources as the key to sustained development and production over time (not all that dissimilar from the sustainable development discourse that would rise to prominence in the late 1980s and 1990s (see Chapter 2)). Pinchot wanted to maximize human benefit from the resource base over time (instead of just maximizing human benefit in the short run). According to Pinchot, resources should be used "for the greatest good, for the greatest number for the longest time" (quoted in Cunningham and Saigo 2001: 18). Pinchot's portrayal of forestry practices (especially sustainable yield) as "scientific," allowed him to build the US Forest Service into a powerful and well-resourced agency (Clarke and McCool 1985).

John Muir played a key role in the establishment of some of the first national parks in the US and was one of a handful of American intellectuals who began to write about the aesthetic beauty of the American wilderness in the late 19th century (Runte 1987). Muir, a Scottish immigrant, spent his early years in Wisconsin and then traveled to California where he helped establish and became the first president of the Sierra Club. Muir and others (e.g., Ralph Waldo Emerson, Henry David Thoreau) often described the American wilderness in quasi-religious terms. This sentiment is reflected in Muir's description of the Hetch Hetchy Valley. He wrote that to dam the Hetch Hetchy one "may as well dam for water-tanks the people's cathedrals and churches, for no holier temple has ever been consecrated by the heart of man" (Muir 1908). As others have noted (most notably Cronon 1996), "nature" for these writers was a place without humans. As such, the preservation of nature not only entailed limits on the consumptive use of resources, but a prohibition on people living in these spaces. Wilderness eventually came to be defined as a place "where man himself is a visitor who does not remain" (US Wilderness Act 1964, section 2c).

The early 20th-century conflict between Gifford Pinchot and John Muir over the decision to dam or not dam the **Hetch Hetchy Valley** in California is used by many texts to starkly differentiate between conservation and preservation. Hetch Hetchy Valley was reported by many, including Muir, to be more beautiful than Yosemite Valley. Both of these valleys, Hetch Hetchy and Yosemite, are found within Yosemite Park, which was established as a California state park in 1864 and then became a US national park in 1906. Figure 1.08 shows Yosemite National Park and its position within the state of California. The conventional wisdom is that Muir favored preserving the valley for aesthetic beauty whereas Pinchot advocated damming the valley to provide water and hydroelectric power for the city of San Francisco.

Muir and the preservationists eventually lost the battle with Pinchot and the conservationists when the US Congress passed the Raker Bill in 1913 allowing flooding of Hetch Hetchy Valley. The Hetch Hetchy Valley was submerged in 1923 when construction of O'Shaughnessy Dam was completed. Figure 1.09 features historical photos of Hetch Hetchy Valley before and after the dam was built. As demonstrated by the two quotes below, readers of texts using the Hetch Hetchy case in this context are led to conclude that conservation is a highly problematic approach that ends in the destruction of beautiful sites.

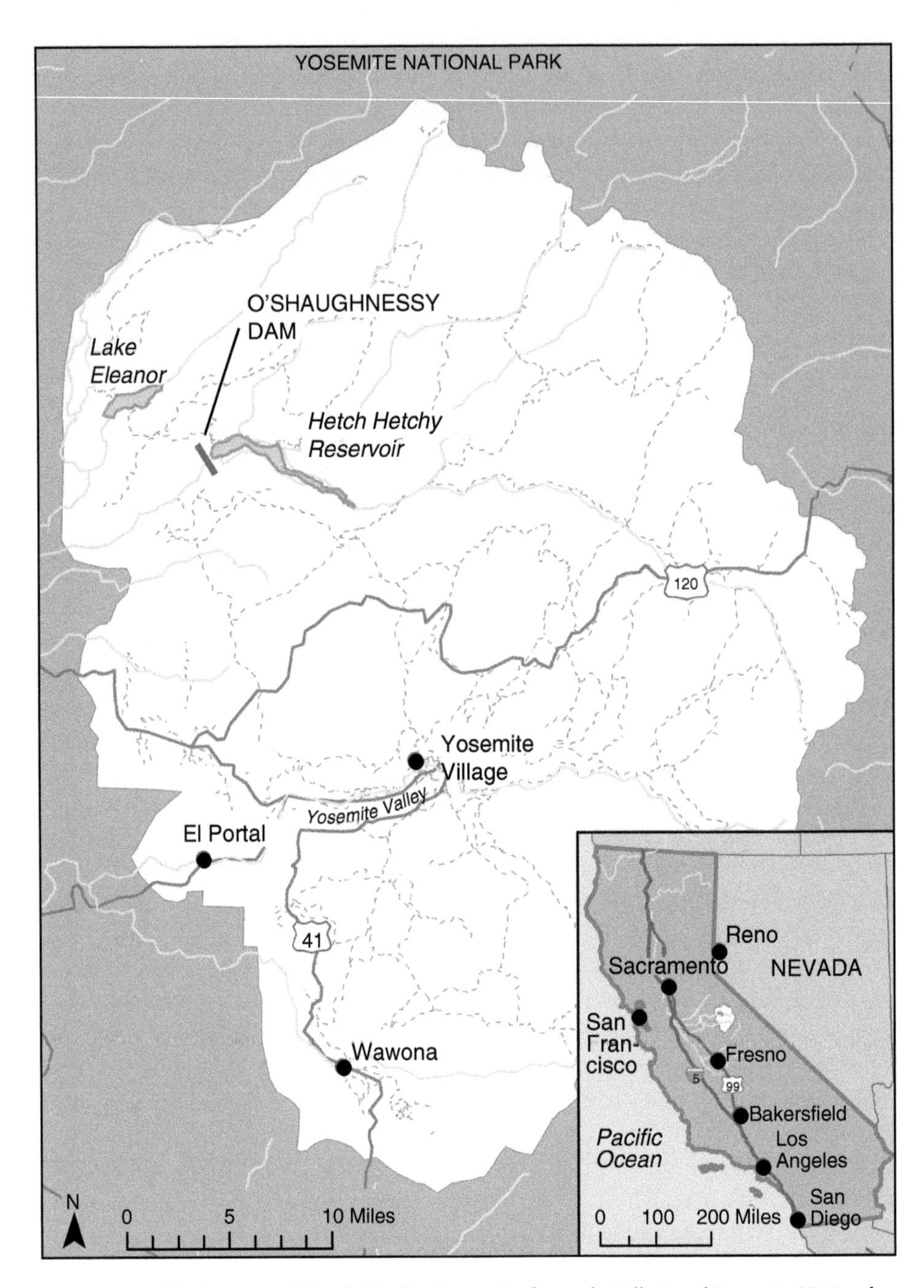

Figure 1.08 The location of Hetch Hetchy Reservoir (formerly Valley) and Yosemite National Park. *Source*: Macalester College cartographer Birgit Muhlenhaus/Moseley 2009.

Figure 1.09a Hetch Hetchy Valley before the O'Shaughnessy Dam. *Source*: F.E. Matthes/United States Geological Survey Photographic Archive.

Figure 1.09b Hetch Hetchy Valley after the O'Shaughnessy Dam. *Source*: © Anthony Dunn/Alamy.

> Scientific conservationists, led by Gifford Pinchot, wanted to build a dam and flood the valley to create a reservoir to supply drinking water for San Francisco. Preservationists, led by John Muir, wanted to keep the *beautiful spot* from being flooded. After a long and highly publicized battle, a dam was built and the valley was flooded. (Miller 1990: 39; emphasis added)

> One of the first and most divisive of these battles was over the flooding of the Hetch Hetchy Valley in Yosemite National Park. In the early 1900s, San Francisco wanted to dam the Tuolumne River to produce hydroelectric power and provide water for the city water system ... Muir said that Hetch Hetchy Valley rivaled Yosemite itself in beauty and grandeur and should be protected. After a prolonged and bitter fight, the developers won and the dam was built. (Cunningham and Saigo 2001: 439)

Hetch Hetchy Valley Reinterpreted

So how does an alternative, geographic perspective change our understanding of how this controversy is often presented? Many writers, and the mainstream environmental movement more broadly, depict this narrative from a certain scale perspective in that they tell a story focused on changes in the valley (rather than at another scale such as the river basin).

> Early in this century conservationists disagreed over how *the beautiful Hetch Hetchy Valley* in what is now Yosemite National Park was to be used. This controversy split the American conservation movement into two schools of thought, the preservationists and the scientific conservationists. (Miller 1990: 41; emphasis added)

Given that the story is presented at the scale of the valley (scale frame A in Figure 1.10), is what happened in the valley after the O'Shaughnessy Dam was built a reasonable and fair example of conservation? We would argue that this story misrepresents the true meaning of conservation (use within biological limits) by focusing on a particular scale frame where the result is really one of total destruction or exploitation (as the valley was completely flooded). In other words, the story, as conventionally told, is focused on the area submerged following construction of the O'Shaugnessy Dam, i.e., the destroyed riparian zone and upper reaches of the valley floor. For example, Philp (2002) notes that the "Hetch Hetchy is submerged under 117 billion gallons of water. In one of the great contradictions of all time, Congress decided ... in 1913 to drown Hetch Hetchy to supply water for San Francisco." Interestingly, and further proof that the environmentally minded public is most concerned about the impact of the dam on the valley itself (rather than the larger river basin) are contemporary efforts to remove the dam (see Restore Hetch Hetchy 2006). By manipulating the discursive scale, conservation (operating at the scale of the watershed) is presented as exploitation (operating at the scale of the valley). As such, the case study as conventionally told conflates conservation and exploitation (making it a poor example of the difference between conservation and

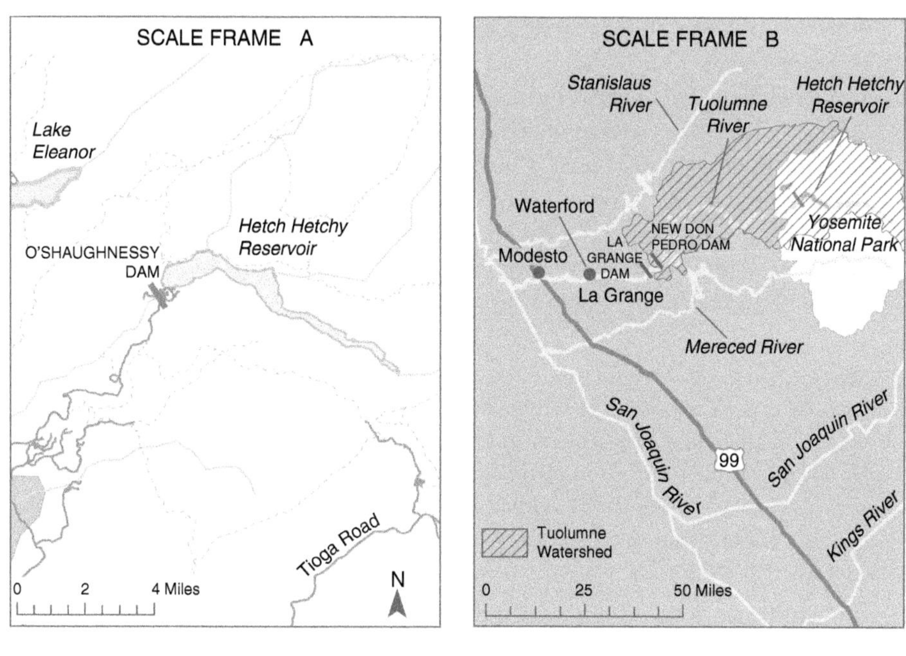

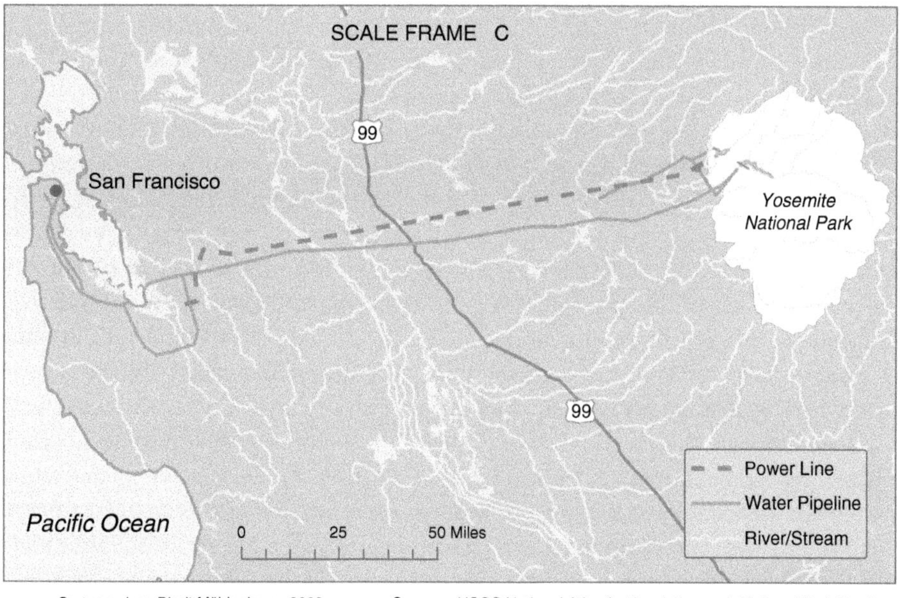

Cartographer: Birgit Mühlenhaus, 2008
Macalester College, Geography Department

Sources: USGS National Atlas (nationalatlas.gov), National Park Service, and the San Francisco Public Utilities Commission (SFPUC).

Figure 1.10 Hetch Hetchy Reservoir and Yosemite National Park as viewed from three different scale-frame perspectives. *Source*: Macalester College cartographer Birgit Muhlenhaus/Moseley 2009.

preservation) and leaves most environmentally minded students thinking that conservation is a bad approach because it led to the destruction of the "beautiful Hetch Hetchy Valley." Preservation, in contrast, is held up as the better (if not more ethical) method.

Furthermore, the full story of the preservation efforts in Yosemite National Park is not discussed in conventional presentations of the Hetch Hetchy Valley controversy. As noted previously, when a preserve is created there are often knock-on effects (or connections) in terms of impacts on local livelihoods and resource extraction on nearby and faraway lands. Interestingly, in contrast to the situation in other national parks, Native Americans were allowed to continue living in Yosemite National Park (in a central Indian village) because they provided valuable labor and served as a tourist attraction. However, as Burnham describes, John Muir found the local Yosemite people "distasteful."

> Muir ... found the Yosemite people dirty and indolent during a park visit in 1869, remarking with distaste their fondness for imbibing ant and fly larvae. In his search for the pristine, there was little room for hunter gatherers who "seemed sadly unlike Nature's neat well-dressed animals." ... Muir decided that the worst thing about them was their uncleanliness, adding that, to his way of thinking, "nothing truly wild is unclean." (Burnham 2000: 21)

Unfortunately, John Muir's and the National Park Service's passive-aggressive attitude towards Native Americans (fueled in part by a conception of nature as a place without humans) undermined their existence in Yosemite. Eventually the Yosemite people left the park as the Park Service progressively raised rents over time. By 1970, the Indian village had been razed.

While dams are clearly problematic in most instances, the Hetch Hetchy story would take on a different tenor if it were presented at the scale of the Tuolumne River or the Tuolumne River Basin (scale frame B in Figure 1.10). At this scale, the dam might be viewed only as impacting a portion of a much larger river or river basin. Of course, it is known that dams can significantly impact upstream and downstream environments (see Chapter 12), that is, an area that extends well beyond the flooded zone. However, unlike the flooded Hetch Hetchy Valley, such a system does continue to function, albeit in a different manner. While contentious, it must also be said that if one accepts that "nature" is socially constructed (see Chapter 2), then the "impacted" Tuolume River system may be no more shaped by human action than the idealized "pristine" wilderness we so cherish.

Finally, we could also tell the story at an even broader scale (scale frame C in Figure 1.10) which encompasses the Hetch Hetchy Valley, Yosemite National Park, the Tuolumne River Basin, as well as a distant site of consumption, the San Francisco metropolitan area where the power and water from the Hetch Hetchy Reservoir is delivered. A discussion at this final and broadest scale frame challenges the reader to confront the relationship between urban America and dams and parks in the surrounding hinterlands. It brings into sharp relief the tension between urban Americans' desire for wilderness amenities and their consumption of water and power resources. As such, the pedagogical value of Hetch Hetchy Valley controversy does not lie in its ability to illustrate the difference between conservation and preservation (it is a poor example of this if we focus on the story at the scale of the valley), but rather as one of the first cases where the tension between urban America's dichotomous impulses to consume resources (see Chapter 9) and to preserve natural areas occurred in the same place (making this an inherently spatial story). In most cases, we have continued to "have our cake and eat it too" by spatially disaggregating sites of exploitation and preservation across the landscape, rather than opting for the much more challenging option of living within biological limits (read conservation). In the end, one might still think that the dam was a bad idea, but the point is that by presenting the story at the scale of the Hetch Hetchy Valley, writers: (1) obscure the complicated social history and economic linkages involved in creating and sustaining Yosemite National Park and the O'Shaugnessy Dam; and (2) discursively equate conservation with the complete destruction of an area.

Plan for the Rest of the Book

The explanations and story above have, we hope, illustrated how geographers might look at a human–environment question in a slightly different manner than other environmental scientists. The main goal for the rest of this book is to present human–environment geography's rich tapestry of theoretical approaches and then to demonstrate how these may be productively engaged to understand human–environment interactions. After explaining some fundamental concepts in Part I of the book, Part II details the major theoretical traditions within human–environment geography. Part III reviews major thematic issues within the field (e.g., population, food and agriculture, water resources). Part IV aims to connect the book's material to the real world by showing the student how geographers undertake fieldwork and collect and analyze data. It also makes suggestions for using the concepts in this text to understand environment-related problems and bring about change.

While the text deliberately seeks to explicate the theoretical underpinnings of human–environment geography, it also seeks to relate these insights to real-world policy questions. To make these connections as explicit as possible, many chapters contain one or more text boxes featuring op-eds from the world's major newspapers that have been published by nature-society geographers. We also emphasize geography's strong tradition of fieldwork via guest field notes and critical case studies focused on pressing human–environment issues and challenges. These field notes are deployed as examples to illustrate how geographers have gone about conducting fieldwork to answer key human–environment questions.

Part I: Fundamentals of Human–Environment Geography

Following this introductory chapter, this section has two more chapters focused on broad introductory themes. Chapter 2 is about the politics of nature. It seeks to explore human conceptions of nature to examine the contemporary environmental movement in historical and global context. While the question "What is nature?"

may seem simple, most geographers would tell you it is not. As such, this chapter begins by exploring the social construction of nature and resources – and the implications of this for policy and praxis. The chapter then examines the history of contemporary environmentalism in the Global North and Global South. Particular emphasis will be given to the rise of the conservation, preservation, and environmental movements in the Global North (from the mid-19th century onward), the different phases of these movements, and their spread to other world regions. We will also explore environmental ethics from different cultural perspectives. Throughout the discussion, we will attempt to create a healthy tension between need for conservation and the often problematic way in which it has been implemented. Chapter 3 is about the biophysical environment. It reviews key concepts in ecology and physical geography. Its purpose is to impart some basic knowledge of biophysical processes and to introduce the spatial perspective that physical geographers often bring to these issues.

Part II: Contemporary Perspectives in Human–Environment Geography

Each of the chapters in this section explores a different conceptual perspective within geography on human–environment interactions. The goal is to get readers to think about the environment in terms of its "relationship" to people and to examine how different social science and humanities perspectives are useful for explaining human environmental phenomenon and processes. Chapter 4 explores the lens of cultural and political ecology. The interdisciplinary field of cultural ecology came to prominence in the 1960s and 1970s largely in reaction to modernist interpretations of development in the Global South. The field played an important role in the decolonization process by documenting and explaining local agricultural and resource management practices. In reaction to a series of critiques leveled against the field in the 1980s, it eventually came to view local human–environment interactions within the context of a broader political economy. This new field is now known as political ecology. Chapter 5 examines environmental history. This highly interdisciplinary field is a key reference point for many human–environment geographers. The chapter examines important environmental histories from the post-industrial era as well as the growing literature on paleoclimatology. Chapter 6 explores the evolving subfield of hazards geography, which has shifted over time from a technocratic concern with risk to the political economy of natural hazards. Geography's hazards tradition will be presented, as well as an analysis of various natural hazards (e.g., drought, tsunamis, earthquakes, floods, etc.) and related societal vulnerability around the world. The final chapter of this section examines environmental justice, or the uneven distribution of pollution and environmental opportunity. This tradition bears some resemblance to political ecology, but has a separate history and an increasing number of practitioners who are geographers. While historically focused on the disproportionate burden of pollution borne by minority communities in the USA, environmental justice is now being used as a lens to understand problems in the Global South.

Part III: Thematic Issues in Human–Environment Geography

The goal of this section is to explore major environmental themes and apply some of the theoretical perspectives described in Part II to these issues. Chapter 8 examines climate, atmosphere, and energy. Here, energy resources will be discussed in relation to their impact on climate and the atmosphere. Local atmospheric issues will be reviewed, followed by an examination of the big three global climate issues (acid deposition, stratospheric ozone depletion, and global warming). Chapter 9 examines the nexus of human population, consumption, and technology in terms of their environmental impacts. Chapter 10 looks at agriculture and food systems. This topic has a long history of study within human–environment geography. This chapter explores agricultural systems around the world as well, and the impact of global markets for various products. Chapter 11 examines biodiversity, wildlife, and protected areas. Building on discussions in Chapters 2, 3, and 4, this chapter introduces more sophisticated biodiversity concepts and examines conservation efforts in this realm. Water resources and fishing livelihoods are explored in Chapter 12. The chapter examines basic water concepts, freshwater and marine resources, as well as fishing livelihoods.

Part IV: Bridging Theory and Practice

The final section articulates the connections between theory and practice. Human–environment geographers' approach to research is detailed in Chapter 13, as many introductory students have a limited understanding of how geographers actually conceptualize research projects and collect and analyze data to arrive at certain conclusions. This chapter details a number of approaches to collecting and analyzing information as examples of how some human–environment geographers practice their craft. In the concluding chapter (14), we examine how the many ideas in this book may be employed in the real world to understand environment-related problems and bring about change. It is suggested that the way we understand the world has important implications for how we act in the world. Furthermore, human–environment geographic research need not necessarily be an activity isolated from "real-world" policy and change. There are many examples wherein the actual process of research (particularly action-research) is a vital component of the change process. Finally, there are strains of human–environment geography which are quite applied in nature (meaning used in the everyday work environment, be it consulting firms or government offices) and there are a number of geographers who have worked directly on policy issues. As such, 21st-century geography students are well positioned to bridge the worlds of theory and practice by being thoughtful practitioners and engaged scholars.

Chapter Summary

This chapter began by giving students a sense of how humans, like other animals, have manipulated the environment in order to encourage the production of certain types of resources. Sometimes people have been successful at doing this over time and on other occasions we have exhausted the resource base with unwelcome consequences. We then introduced the field of geography and its relevance to environmental questions. We outlined some generalized perspectives that human–environment geographers often take when examining resource questions. These perspectives included attention to: scale, spatial patterns of environmental use, a conception of the human–environment system as a single unit (rather than two separate parts), and the connections between places and regions. In order to illustrate such perspectives, the second half of the chapter applied these to develop a revised understanding of three resource-management approaches as well as to a famous case in US environmental history, the Hetch Hetchy Valley controversy. The last part of the chapter simply described the plan for the rest of the book.

Critical Questions

1. Is human manipulation of the landscape necessarily a bad thing? Explain.
2. Are humans unique in their ability to manipulate the physical environment? If not, what may distinguish humans' ability to manage the environment from that of other animals?
3. How might you describe geography to a friend? If this is your first geography class, how has your understanding of geography changed since you started this course and read this chapter?
4. What are some fairly general, yet distinctive, elements of the way human–environment geographers approach natural resource questions?
5. Why, according to geographers, is the Hetch Hetchy controversy in US environmental history not a very good example of the difference between conservation and preservation?

Key Vocabulary

cartography
conservation
exploitation
geographic information systems (GIS)
Gifford Pinchot
Hetch Hetchy Valley
human geography
human–environment geography
John Muir
modifiable areal unit problem
net primary productivity
non-consumptive use
physical geography
politics of scale
preservation
remote sensing
scale
sustainable yield

Notes

1 Much of the material in this section is based on: Moseley 2009.
2 This is the amount of a renewable resource that could be safely harvested each year without reducing its long-term productive potential.
3 Monoculture refers to a farm field or forest stand that is planted with one crop or tree as opposed to a variety of such plants.
4 As Yellowstone National Park in the USA was the first expression of a preserve in the modern era.
5 For terrestrial ecosystems, net primary productivity (NPP) refers to the total amount of biomass generated for a given area (usually per square meter) for a given time period (usually a year).

References

Braun, B. (2002) *The Intemperate Rainforest: Nature, Culture and Power on Canada's West Coast* (Minneapolis: University of Minnesota Press).

Burnham, P. (2000) *Indian Country, God's Country: Native Americans and the National Parks* (Washington, DC: Island Press).

Chiras, D.D. and Reganold, J.P. (2005) *Natural Resource Conservation: Management for a Sustainable Future* (Upper Saddle River, NJ: Prentice Hall).

Clarke J.N. and McCool, D. (1985) *Staking Out the Terrain: Power Differentials among Natural Resource Management Agencies* (Albany, NY: State University of New York Press).

Cronon, W. (1996) The trouble with nature or, getting back to the wrong wilderness. *Environmental History*, 1(1), pp. 7–28.

Cunningham, W.P. and Saigo, B.W. (2001) *Environmental Science: A Global Concern*. 6th edn. (New York: McGraw-Hill).

Frank, A. (1979). *Dependent Accumulation and Underdevelopment* (New York: Monthly Review Press).

Guha, R. (1997) The authoritarian biologist and the arrogance of anti-humanism: wildlife conservation in the Third World. *The Ecologist*, 27(1), pp. 14–20.

Harvey, D. (1996) *Justice Nature and the Geography of Differences* (Oxford: Blackwell).

Holechek, J.L., Cole, R.A., Fisher, J.T., and Valdez, R. (2003) *Natural Resources: Ecology, Economics and Policy* (Upper Saddle River, NJ: Prentice Hall).

Kurtz, H.E. (2003) Scale frames and counter-scale frames: constructing the problem of environmental justice. *Political Geography*, 22, pp. 887–916.

Mansfield, B. and Haas, J. (2006) Scale framing of scientific uncertainty in controversy over the endangered Steller sea lion. *Environmental Politics*, 15(1), pp. 78–94.

Martin, R.M. and Darr, D.R. (1997) Market responses to the US timber demand-supply situation of the 1990s: implications for sustainable forest management. *Forest Products Journal*, 47(11–12), pp. 27–32.

Miller, C. (2001) *Gifford Pinchot and the Making of Modern Environmentalism* (Washington, DC: Island Press).

Miller, G.T. (1990) *Living in the Environment: An Introduction to Environmental Science*. 6th edn. (Belmont, CA: Wadsworth).

Moseley, W.G. (2009) Beyond knee-jerk environmental thinking: teaching geographic perspectives on conservation, preservation and the Hetch Hetchy Valley Controversy. *Journal of Geography in Higher Education*, 33(3), pp. 433–51.

Muir, J. (1908) The Hetch Hetchy Valley. *Sierra Club Bulletin*. January.

Openshaw, S. (1983) *The Modifiable Areal Unit Problem* (Norwich, UK: Geo Books).

Philp, T. (2002) Water: bring back Hetch Hetchy? *Sacramento Bee*, April 21.

Pimentel, D. and Pimentel, M. (1979) *Food, Energy and Society* (London: Edwin Arnold).

Princen, T. (2002) Distancing: consumption and the severing of feedback. In T. Princen, M. Maniates, and K. Conca (eds.), *Confronting Consumption*, pp. 103–31 (Cambridge, MA: MIT Press).

Restore Hetch Hetchy (2006) Available at http:www.hetchhetchy.com (accessed January 19, 2013).

Righter, R. (2005) *The Battle over Hetch Hetchy: America's Most Controversial Dam and the Birth of Modern Environmentalism* (New York: Oxford University Press).

Runte, A. (1987) *National Parks: The American Experience*, 2nd edn. (Lincoln: University of Nebraska Press).

Sanchez-Azofeifa, G.A., Rivard, B., and Calvo, J. (2002) Dynamics of tropical deforestation around national parks: remote sensing of forest change on the Osa Peninsula of Costa Rica. *Mountain Research and Development*, 22(4), pp. 352–8.

Tucker, R. (2002) Environmentally damaging consumption: the impact of American markets on tropical ecosystems in the twentieth century. In T. Princen, M. Maniates, and K. Conca (eds.), *Confronting Consumption*, pp. 177–95 (Cambridge, MA: MIT Press).

Wallerstein, I.M. (1979) *The Capitalist World Economy: Essays* (New York: Cambridge University Press).

Williams, M. (1989) *Americans and Their Forests: A Historical Geography* (New York: Cambridge University Press).

2

The Politics of Nature

Icebreaker: Evolving Environmentalism

In the 1960s, environmental issues became a visible and controversial part of daily life in many parts of the world. The impacts of pesticides on the environment, concerns over population growth, and the constant fear of radiation and nuclear warfare were all at the forefront at that time. Since then, the prime environmental issues have shifted. From population, to the decline of biological diversity, to the latest work on global climate change, these all involve a real politics of nature. But what is nature? And why does your particular conception of nature influence real life in your community?

An Introduction to Human–Environment Geography: Local Dynamics and Global Processes, First Edition. William G. Moseley, Eric Perramond, Holly M. Hapke and Paul Laris.

Published 2014 by John Wiley & Sons, Ltd.

Chapter Objectives

The objectives of this chapter are:

1. To understand the different perspectives on "nature" within geography.
2. To consider how different perspectives on nature might influence one's environmental politics.
3. To understand how environmentalism and environmental politics have evolved over the past 150 years.
4. To situate the student's own region and environmental movement with respect to the global push to recognize humanity's increasing footprint on the planet.

Introduction

This chapter has two broad goals: (1) to explore human conceptions of nature; and (2) to examine the contemporary environmental movement in historical and global context. While the question "What is nature?" may seem simple, most geographers would tell you it is not. As such, this chapter begins by exploring the social construction of nature and resources – and the implications of this for policy and praxis. The chapter then examines the history of contemporary environmentalism in the Global North and Global South. Particular emphasis will be given to the rise of the conservation, preservation, and environmental movements in the Global North (from the mid-19th century onward), the different phases of these movements, and their spread to other world regions. We will also explore environmental ethics from different cultural perspectives. Throughout the discussion, we will attempt to create a healthy tension between the need for conservation or preservation and the often problematic ways these approaches have been implemented.

Human Conceptions of Nature

In contemporary geography, there are arguably three major ways of conceptualizing the relationship between humans and their environment.

- Human–environment geography: Humans play a role in modifying environments, and these environments – in turn – influence (but do not determine) human actions or behaviors. It is a two-way, synergistic relationship.
- Ecocentric (non-human) perspective: Non-human nature not only matters, but organisms are agents of change, change humans, and deserve consideration and special focus.
- Social construction of nature: Nature or the environment is constructed (mentally) by humans and human concepts of nature are culturally, temporally, and politically changing.

Since the rest of this book is largely about the human–environment perspective, here we give some attention to the other two major perspectives in geography: the ecocentric, and the social constructivist. These are somewhat artificial boundaries as many human–environment geographers actually sympathize with and deploy ecocentric or constructivist perspectives in their work. The aim here is to at least explain some different human–environment perspectives in geography.

The **ecocentric approach** treats non-human nature as equally important as people (if not more so). This means acknowledging that other animals, insects, microbes, and plants have not only an obvious role in our mutual relations on Earth, but they should be given a more important consideration and focus. The argument is twofold. First, "consideration" refers to an ethical stance that would have us humans recognize the rights of other beings on the planet. Second, by "focus" ecocentric geographers mean that we should be paying greater attention in our research to the agency of non-human nature (see Wolch and Emel 1998 for a ground-breaking example).

Let's briefly explore two examples which illustrate this approach. For the ethical dimension, ecocentrists might ask "What are the rights of domesticated animals, like cats and dogs, in our society?" (Haraway 2003). On the second question, they might ask all of us "How do the ants in the soils you study shape the soils themselves, and how can you account for that biological agent role when you write about soils?" A newer, more subtle version of this ecocentric perspective is taking shape in what some are calling "**post-humanism**," a philosophy that pushes beyond a focus on the meaning of the world for humans, and includes non-human nature in it. It should be noted that in most cases, ecocentrism and the constructivist perspective are hard to tell apart.

The social constructivists in geography take an explicitly social stance on the concept of nature. Or, as Noel Castree (2001: 5) has put it: nature has never been simply "natural." From the constructivist notion of environment, or nature, they speak of **socionature**. This is a combination of the social and the natural that, from this perspective, removes the separating lens between the social and the natural. For these authors and philosophers, then, they are one and the same and you cannot talk about "the natural world" as if it was separate from humans.

Take, for example, the seemingly natural condition of a drought. Human experience of such a situation is deeply conditioned by the way a society is structured. One farming society might anticipate drought as a form of natural variability and store surplus grain in good years to be redistributed to households in years of poor rainfall. Another society might be enmeshed in a global economic system which extracts all food and fiber from a region at minimal prices. As such, the experience of these two groups with the same drought in a given year may be very different, with one group surviving with little hardship and the other facing extreme hunger and possible famine. Clearly the experience of this natural event is as much a product of social factors as it is of biophysical ones (based on the work of Watts 1983).

We can also illustrate these differences, between human–environment, ecocentric, and constructivist stances, with a set of diagrams (see Figures 2.01, 2.02, and 2.03).

Box 2.01 What is your conception of the human–environment relationship?

Out of the three brief portraits above, can you easily identify which stance or approach appeals to you most? Can you then identify why one or two of those perspectives match your idea of the relationship between humans, society, and nature?

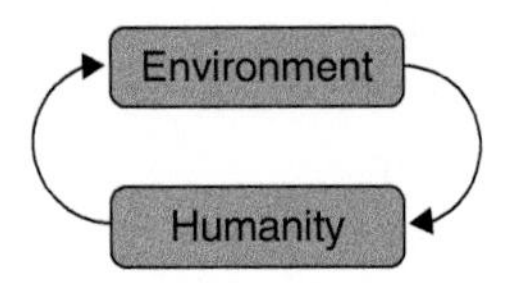

Figure 2.01 The human–environment perspective, simplified in a diagram. *Source*: https://www.e-education.psu.edu/geog030/book/export/html/111.

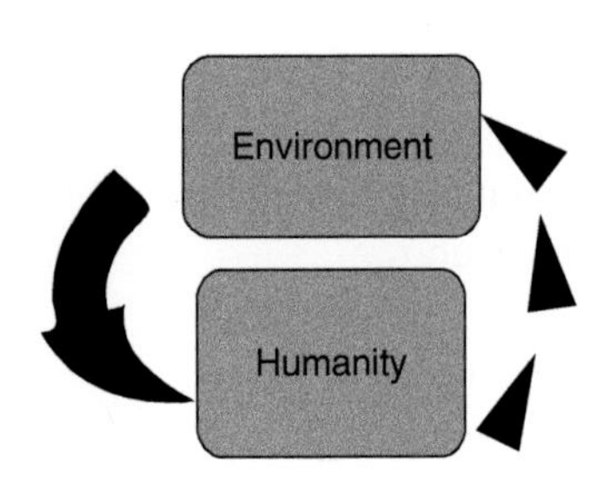

Figure 2.02 The ecocentric point of view, where humans and environments are not just in a reciprocal relationship, but where nature informs and is the source of inspiration and future human action. *Source*: Drafted by Eric Perramond.

Figure 2.03 The socionature perspective: the view that all of nature is really a social construction. One cannot separate society from nature; therefore it is viewed as one entity: socionature. *Source*: Drafted by Eric Perramond.

In both the human–environment and ecocentric perspectives, there is a clear flow and dialectic between the environmental and human sides of the relationship. The separate boxes are a convenient way to think with and through this relationship, and should not necessarily be taken as a sure-fire sign that thinkers view these components as separate or discrete. In the third perspective (Figure 2.03), proponents claim that this holistic, socially informed view is more representative of how society goes about working in nature, producing socionature that is difficult to disentangle from economics, politics, and daily life (see Braun 2002; Mansfield 2003). These figures are gross simplifications, but should at least spark some discussion in your class (see Box 2.01 for an activity idea).

The Contemporary Environmental Movement in Historical and Global Context

From Limitless Resources to Conservation

Europe in the 17th century was a region of turmoil and change. Changes in maritime technology and European thirst for gold driven by mercantilism[1] had opened up the New World a little over a century before. A combination of changes in European cities and rural areas was simultaneously creating a surplus of people and a thirst for more and more raw materials. Formerly feudalist landscapes had given rise to a new mercantile class in the cities who were vying with the rural, landed aristocracy for power. Rural landlords were responding by enclosing much of the land that had previously been available to commoners (or villagers) for the grazing of their animals or hunting and gathering (see a discussion of the enclosure movement in Chapter 9). The elimination of such safety nets was putting pressure on the poor, forcing many of them to leave the countryside. This led to a great exodus of Europe's poor to North America, South Africa, and Australia/New Zealand in the 18th and 19th centuries.

Many of these poor were excited by the prospect of having land they could call their own in their new-found homes. As in many frontier societies, resources in the USA were essentially viewed as limitless prior to the 1860s. Once resources in one area were exhausted, settlers just moved on to a new area. This approach, also known as exploitation (discussed in Chapter 1) or **cowboy economics**,[2] worked as long as there were new areas to which to move. This frontier mentality was not unique to North America but was repeated in other regions where European settlers moved, such as Australia and South Africa (Beinart and Coates 1995).

Of course, local people inhabited many of the areas into which European settlers moved and they typically had a different approach to resource management given their longer-term residence in the area. This is not to romanticize local people's approach to resource management (see Denevan 1992) but simply to acknowledge important differences. In many instances, Europeans had brought exotic species and diseases with them that proved deadly to environments and local people with no previous exposure. Such biological warfare (intentional or not) was less detrimental in regions or with peoples who had previous exposure to such diseases (much of Africa) or in areas with diseases (such as malaria) which were equally problematic for Europeans (Crosby 2004). The combination of disease, theft, and warfare meant that European newcomers often displaced local peoples.

The beginnings of modern environmentalism began after roughly a century of industrialization in Europe and some of its settler colonies.[3] **Environmentalism** in the modern sense[4] refers to a concern for human impacts on the environment. Up until this point, most scholarship had focused on the effects of various environments on human cultures rather than the other way around. Industrialization fueled two phenomena to which environmentalism might, in some sense, be considered a reaction: urbanization and resource exploitation. First, urbanization produced a longing for rural life. In Europe, this produced a generation of painters (late 18th to mid-19th century) known as the Romantics who, among other themes, idealized rural life. In the United States, this gave rise to several intellectuals, essayists, and writers in the 19th century (e.g., Ralph Waldo Emerson, Henry David Thoreau, and John Muir), who went beyond idealizing working rural landscapes to celebrating areas without people – or what we would now call wilderness. In some sense, urbanization had separated people from the rural landscape and, in their attempts to reconnect with it, they came to view it differently. It was a place to be visited and appreciated for its aesthetic beauty rather than a landscape to be cultivated or managed to produce food, fiber, or fuel. Furthermore, some writers were concerned that young boys in particular needed to have experiences in underdeveloped, rural, or wilderness areas in order to become truly male (Cronon 1996). This new understanding of nature led to a desire to set aside areas as preserves (also known as preservation, discussed in Chapter 1) or nature reserves as it became clear that less and less undeveloped land would remain.

The other problem giving rise to early environmentalism was the over-exploitation of resources. Raw materials were needed to supply the factories and build the cities. In North America, previously a continent of endless resources for European immigrants, it was becoming clear by the second half of the 19th century that such resources might run out if something were not done. As discussed in Chapter 1, conservation is a resource management philosophy which seeks to use renewable resources within their biological limits. Gifford Pinchot most famously advocated this approach by applying it to forestry as per his training in Germany (a place where timber supplies had come under stress much earlier). What Pinchot understood was a basic relationship between natural capital and man-made capital (subsequently written about by ecological economists like Herman Daly (1991). In the case of forestry, were the US to run out of timber, this would constrain all of the industries that depended on wood and cripple the country's economy. An investment in a timber mill would be rendered useless without adequate supplies of raw trees to hew into lumber. As the first head of the US Forest Service (1905–10), Pinchot attempted to rationalize the use of the nation's forests in order to produce "the greatest good for the greatest number of people for the greatest amount of time."

A second environmental crisis would bring about another wave of conservation in the 1930s, except this time soil resources would be the target. From 1927 to 1932, there was an extended drought in the central United States which caused massive crop failure. This drought and it consequences were the result of both natural events (decreased rainfall) and human action. Farmers had plowed under large areas of the central US grasslands, leaving topsoil exposed. With neither ground cover nor rainfall of any significance, this region was racked by terrible windborne erosion or, as it came to be known, the **Dust Bowl** (Worster 1979). The government responded with aggressive soil erosion control plans, promoting contour plowing, wind breaks, shelter belts, and the like (see Chapter 10 for a longer discussion of these measures). At the same time, a Depression-era work program was created, known as the Civilian Conservation Corps (CCC). While the short-term goal was jobs, the long-term goal was building infrastructure to support the nation's conservation and preservation efforts. The Dust Bowl experience and the way it was addressed would reverberate outside of the US For example, Hugh Bennett, then Chief of the US Soil Conservation Service, toured South Africa in the early 1940s. He identified soil erosion in South Africa's wheat belt and recommended a series of soil conservation measures (Meadows 2003). Similarly, ideas from the Dust Bowl experience reinforced a concern that **desertification**[5] was primarily a human-caused phenomenon which influenced concerns about this in West Africa (Swift 1996; Moseley and Laris 2008).

The Rise of Modern Environmentalism in the 1960s and 1970s

In contrast to environmental movements in the late 19th century and 1930s, environmentalism in the 1960s and 1970s was fundamentally different in a few ways. First, it largely was focused on **brown issues** (pollution related to industrialization and urbanization) rather than **green issues** (forestry, parks, wildlife).

Second, the support for this movement reached deeper into the middle class than in previous eras, when the majority of support came from the upper and upper-middle classes.

The environmental movement of the 1960s was largely the result of an increasing awareness of the environmental downside of economic growth. This is the first time that economic growth, and modernization more generally, were seriously questioned. Rachel Carson's *Silent Spring* (1962), which raised awareness regarding the problematic side of pesticide use, was a seminal work in this regard. In this era, we also see other volumes published which raised the profile of the environment in the public imagination. These include Paul Ehrlich's *The Population Bomb* (1968), which – rightly or wrongly – makes population growth an environmental issue (see a discussion of this in Chapter 9). There was also Donella Meadows et al.'s *Limits to Growth* (1972), which argued that there were real environmental limits to future economic growth. A high-water mark for this movement was the first **Earth Day** celebration on April 22, 1970, a day intended to inspire awareness and understanding of the Earth's natural environment (see Figure 2.04).

Figure 2.04 New York; photo of the first Earth Day, April 22, 1970. *Source*: © Associated Press/Press Association Images.

The environmental issues of the Global North described above dominated the agenda at the 1972 Conference on the Human Environment in Stockholm, Sweden, the first international UN conference on the environment. At this conference, environmentalism was largely framed as antagonistic to economic development. The environmentalism of this era was also exported to many developing countries through technical assistance programs that sought to improve the natural resource management expertise of local people. Forestry programs in particular were promoted by many donor agencies. Increasingly, environmental problems were blamed on human ignorance, lack of technical expertise, and excessive population growth.

The Era of Sustainable Development (1980s–1990s)

The concept of **sustainable development** was first articulated in a report by the International Union for the Conservation of Nature (IUCN) in 1980. The term, however, would not really take hold in policy circles until it was used in a 1987 report by the World Commission on Environment and Development entitled *Our Common Future* (also known as the **Brundtland Report**). This report defined sustainable development as: "Development that meets the needs of the present without compromising the ability of future generations to meet their own needs" (WCED 1987).

Sustainable development represented a significant departure from the environmentalism of the 1960s and 1970s because it suggested that environmentalism and economic development were not necessarily antagonistic objectives, but could be complementary (this potential complementarity will be explored in the coming paragraphs). This was appealing not only to business interests but also to developing countries, which had been concerned that earlier environmentalism would lead to international agreements which constrained their economic development aspirations. It is for this reason that sustainable development was a powerful organizing principle at the second UN Conference on Environment and Development in 1992 in Rio de Janeiro, Brazil (also known as the **Earth Summit**). Unlike the first UN Conference on Environment and Development 20 years earlier, this conference had significant buy-in from both developed and developing countries. In addition to **Agenda 21** (this was the main product of the Earth Summit, an action plan for sustainable development), an important sub-agreement was the **UN Convention on Biological Diversity**. This convention would keep biodiversity at the top of the international environmental agenda throughout the 1990s and beyond (see Box 2.02).

The reality is that sustainable development, as articulated in the 1987 Brundtland Report, was very vaguely defined. This vagueness was either a blessing or a curse, depending on your perspective. It was a blessing because it allowed a broad group of actors to rally behind the concept. It was a curse because, when people discussed sustainable development, they could mean very different things. While there are multiple takes on sustainable development, below we will discuss two distinct definitions (or two basic schools of thought). Before getting to these two very different understandings, we will first review the meanings of the terms "development" and "sustainable."

While one could write an entire book on the meaning of **development**, it generally refers to improvement in the living standards of a society. Such improvement has traditionally been measured in terms of economic indicators (such as per capita Gross National Product (GNP)). Others argue that economic growth and development are not synonymous (and that there is not always a tight correlation between the two). As such, development needs to be measured by a variety of indicators, not just economic ones. Such quality of life measures include: literacy rates, life expectancy, infant mortality, and female participation in national parliaments.

The concept of sustainability can be traced back to the discipline of forestry and the notion of maximum sustainable yield (see Chapter 1). **Sustainability** is the idea that an activity could be carried out indefinitely without resource depletion. A renewable resource that is managed sustainably allows for a constant rate of harvest indefinitely that is equivalent to the annual growth increment of the resource.

As indicated above, there are two basic schools of thought on the meaning of sustainable development. The first, and minority, view is that one needs to

Box 2.02 Biodiversity as a primary environmental concern in the 1990s

It was not until the late 1980s and early 1990s that biodiversity as a label and an issue attained popular prominence. While environmentalists had discussed forests and wildlife preservation in previous eras, biodiversity was now the operative term. In some ways, this was a triumph of the ecologists and biogeographers because the prominence of the term biodiversity forced people to consider many different species (as opposed to a few charismatic ones) and the interconnections between them.

Despite a new interest in biodiversity, international environmental NGOs (non-governmental organizations) and agencies turned to an old idea, the park, as one of the primary mechanisms for preserving biodiversity. Many of these international preservation efforts were focused on the tropics, where species diversity at the level of the tropical forest biome was highest. In the Global South, this spawned concern about "green imperialism" by those in the Global North. The main issue was that environmentalists were so concerned about biodiversity conservation that they often forgot about the livelihoods of local people who were adversely impacted by a park model and a conception of nature that did not include people (see Figure 2.05).

Figure 2.05 A park ranger in Kruger National Park, South Africa. *Source*: W.G. Moseley. Used with permission.

fundamentally change development so that it is more compatible with the environment. Under this definition, development is less about economic growth and more about addressing basic human needs (which may not require consuming more). Furthermore, it is development that respects real biological limits. As such, society cannot produce more pollution than can be assimilated by the environment, or use renewable resources more quickly than these can regenerate. Using fewer resources often implies consuming less, recycling more, relying on renewable energy, improving efficiency of energy use, etc. The ethic of thrift embedded in this version of sustainable development is similar to the approach of early 20th-century environmental movements, which also emphasized efficiency of use in line with Gifford Pinchot's maxim, "Conservation is the foresighted utilization, preservation and/or renewal of forests, waters, lands and minerals, for the greatest good of the greatest number for the longest time." Moving a society to this particular vision of sustainable development would likely imply regulations to insure that people and organizations do not surpass biological limits. This more "ecocentric" version of sustainable development is less popular because mainstream policymakers tend to dislike development that is not equivalent to economic growth.

The second, and more common, understanding of sustainable development suggests that development and environmental stewardship are complementary. The argument is that economic growth is good for natural resource management because: (1) you need wealth to invest in conservation (or the technology to promote this); and (2) the household wealth generated by economic growth allows individuals to have a longer-term vision and to more highly value environmental amenities. Furthermore, it is suggested that an enhanced natural resource base is also good for longer-term development. The quotes below exemplify the more conventional understanding of sustainable development.

> "Poverty is also a factor in accelerating environmental degradation, since the poor ... are unable and often unwilling to invest in natural resource management ..." (World Bank 1996)

> "Economic growth is key to environmental progress, because it is growth that provides the resources for investment in clean technologies." (former US President George W. Bush, in a speech announcing his Clear Skies Initiative, 2002)

> "Only when people are rich enough to feed themselves do they begin to think about the effect of their actions on the world around them and on future generations." (Bjorn Lomborg 2002)

> "Degradation of these [environmental] resources reduces the productivity of the poor – who most rely on them – and makes the poor even more susceptible to extreme events ..." (World Bank 1996)

One of the key theoretical models in support of the notion that increasing wealth leads to better environmental management (an idea that was central to the more common understanding of sustainable development) was the

Environmental Kuznets Curve. This curve hails from the original Kuznets Curve (developed by the macroeconomist Simon Kuznets in the 1950s), which suggested that there was an inverted U-shaped relationship between increasing national wealth and income inequality. He posited that the least developed countries have relative equality between household incomes. As a country develops, income inequality increases. Incomes then become more equal again as wealth increases even further. Others then hypothesized that there was a similar relationship between increasing wealth and environmental quality, which came to be known as the Environmental Kuznets Curve. Figure 2.06 is an example of such a curve for sulfur dioxide emissions.

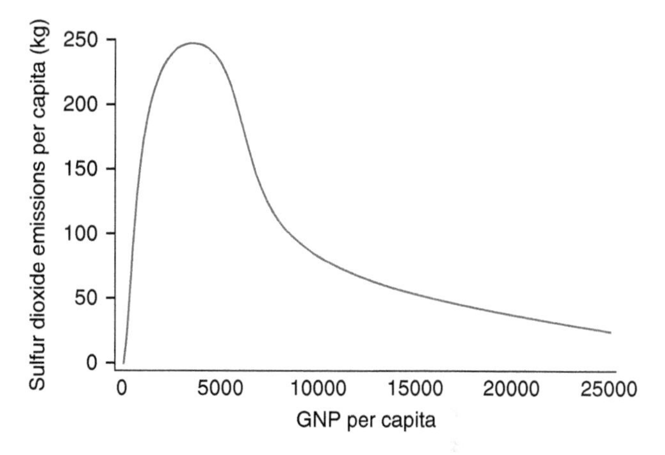

Figure 2.06 The Environmental Kuznets Curve for sulfur dioxide emissions. *Source*: T. Panayotou, "Empirical tests and policy analysis of environmental degradation at different levels of development" (1993), cited in *Encyclopedia of the Earth*. http://www.eoearth.org/article/Macroeconomics_and_ecological_sustainability. Reprinted by kind permission of the author.

The Environmental Kuznets Curve, and the broader notion that increasing wealth leads to better environmental management, have been critiqued on a number of fronts. First, while the relationship holds for some air pollutants (such as sulfur dioxide), it does not hold up for many others. Second, while wealthier economies often appear to be cleaner, they may just be exporting dirty industries to poorer countries (and hence there is no net decline in pollution occurring at the global scale). Third, while economic growth may create wealth that can be invested in environmental protection, the additional production and consumption associated with this economic growth also create waste and use resources. As such, it is not clear if the increased environmental protection afforded by this wealth fully compensates for the waste and resources used to generate it. Fourth, while wealth creates the possibility to invest in cleaner technology, this is not a sufficient guarantee that such investments will be made (absent regulations or other incentives to encourage this). A good example of this is the US in the 1990s, when purchases of fuel-inefficient sport utility vehicles (SUVs) soared during a period of economic expansion (Gray and Moseley 2005).

The idea that poor people are less likely to prioritize environmental stewardship (another idea embedded in the conventional understanding of sustainable development) is bolstered by ideas in economics about "time preference," as well as by beliefs in psychology and sociology literatures regarding the "hierarchy of needs." These theories suggest that once a person's basic needs are addressed, they may consider higher-order needs and wants, including environmental amenities. However, there is evidence that many poor households in developing countries do not treat the environment as a higher-order amenity. Rather, they often manage the environment as essential working capital that should be preserved whenever possible (Moseley 2001). Or they are unlikely to engage in deleterious environmental management practices because they are less connected to the global capitalist economy (Moseley 2005).

In sum, many have suggested that the mainstream version of sustainable development represented a co-optation of the concept by multilateral donors and status quo capitalism. The World Bank, for example, devoted its annual report to sustainable development in 1992 (World Bank 1992). These entities essentially argued that if a country pursued development as usual, it would lead to improved environmental management (Lele 1991).

Environmentalism in the New Millennium (2000–Present)

While there was a follow-up conference to the Rio meeting in 2002 in Johannesburg, South Africa (known as the Earth Summit 2002 or Rio +10), most of the key international environmental discussions in the new millennium have focused on the narrower topic of climate change. These climate change negotiations emerged from the 1992 Earth Summit in a sub-agreement known as the **UN Framework Convention on Climate Change** (UNFCCC). This framework led to the Kyoto Protocol in 1997 and subsequent meetings in the 2000s (discussed in Chapter 8). While other types of environmental concerns have received consideration in the 2000s, climate change has arguably dominated the agenda. Concerns about climate change have also influenced the way we think about other environmental issues. Take food and agriculture for example. Among other motivating factors, the **local food movement**[6] in the Global North has in part been fueled by a concern that shipping food around the world produces an inordinate amount of greenhouse gas emissions (see Box 2.03). Concerns about climate change have also reinvigorated a concern about energy efficiency that was central to the environmental movement

Box 2.03 Op-ed in the *San Francisco Chronicle* (Sunday, November 18, 2007)

Farmers in developing world hurt by "eat local" philosophy in US
By William G. Moseley

Increasing awareness of climate change has transformed the way Americans think about organic food. While organic consumers used to focus on how food was produced, such as whether pesticides were used, they now are also concerned about how far food has traveled to arrive at their plate. The issue is that greater distances often equate to more energy use and greenhouse gas emissions.

The preference for eating local has been popularized, among others, by UC Berkeley journalism professor Michael Pollan in the "Omnivore's Dilemma" and by Barbara Kingsolver in "Animal, Vegetable, Miracle." This "eating local" philosophy has a huge following among those consumers who buy organic food. But what about the consequences of the local food craze for farmers in the developing world who have joined the organic and fair trade movements?

We're getting a glimpse of the future of this debate in the United Kingdom, where the tension between the local food and fair trade movements is acute. Just recently, the UK Soil Association, a nonprofit group that promotes sustainable and organic farming, called on the British government to restrict imports of organic produce brought in by air. In a concession to the fair trade movement, this group would allow for imports from countries actively seeking to promote organic and fair trade markets within their own borders. Despite this concession, British fair trade activists are worried.

Whether the British government would ever adopt such a ban is questionable, but labeling schemes and use of concepts such as "food miles" (the distance a product has traveled to reach the store) are likely to increase consumer awareness and influence purchasing habits.

The suggestion that developing countries should promote local markets for organic produce in order to wean themselves off of export markets is a false alternative. These markets often already exist in everything but name.

Many farmers in the poorest of African nations – where I do my research – already supply local markets with their grains and produce. While not formally recognized as such, these markets are virtually organic because most poor African farmers restrict pesticide use to traditional export crops such as cotton, cacao and coffee, while local foodstuffs are grown with few or no chemical inputs.

Traditional export-oriented agriculture is problematic in many ways, but the organic and fair trade movements are beginning to diversify opportunities for African farmers in this sector. Just as Mexico and South America supply large amounts of organic produce to California, European demand for organic and fair trade products from Africa is surging. These are not just niche markets where developing world farmers can potentially gain a higher return, but these channels also promote better working conditions and the reduced use of chemicals.

If the local food movements in Europe and North America reduce their demand for organic and fair trade products from afar, the most likely consequence is that African farmers who have entered these niche markets will return to producing their export crops in the conventional, pesticide-intensive manner. While local food markets can provide some income for these farmers, they still are reliant on export opportunities for the bulk of their cash income.

Although our decisions as consumers have the power to influence how our food is produced, this approach is limited. What we really need are changes in the basic rules that govern the global marketplace.

If international bodies, such as the World Trade Organization, set and enforced rules about basic working conditions and environmental standards, then we would not be relegated to trying to promote organic farming and fair labor practices via labeling schemes and informed consumption. If African countries were allowed to protect nascent industries, then they might not be so reliant on agricultural exports.

But until these changes are made, it is a cruel joke to condemn developing-world farmers to commodity crop production and then remove the only hope they have for higher returns – organic and fair trade crops and products.

While the local food craze is all well and good, we should not be so quick to denounce organic and fair trade foods that are imported from the developing world. By shunning these products, we do not encourage local markets to flourish in these countries, but we condemn these farmers to the ills of conventional production for the global market (the only other real alternative at this time). We should remain open to such products in the short term, but also work for broad scale changes in the rules of the global market place to ensure that even conventional agricultural production is safe and fairly compensated.

Source: Used by permission of the author.

in the 1970s, but then fell out of favor for two decades. In the 2000s, alternative energy and energy efficiency became a central concern again because of climate change concerns and – more recently – rising energy prices.

Chapter Summary

This chapter began by reviewing different conceptions of nature within human–environment geography. It then examined the rise and evolution of the modern environmental movement since the 19th century. This review highlighted the conservation and preservation movements in the late 19th century, soil conservation in the 1930–1940s, a concern with brown environmental issues in the 1960s and 1970s, sustainable development in the 1980s–1990s, and finally, a focus on climate change in the new millennium.

Critical Questions

1. What are the three major perspectives regarding nature within human–environment geography?
2. In what ways did industrialization and urbanization lead to the rise of environmentalism in the latter half of the 19th century?
3. How did sustainable development represent a departure from the environmentalism of the 1960s and 1970s? What are the strengths and weaknesses of this concept?
4. What has been the focus of environmentalism, environmental politics, and environmental activism in your area over the past five years? How do you envision this evolving over the next five years?

Key Vocabulary

Agenda 21
brown issues
Brundtland Report
cowboy economics
development
desertification
Dust Bowl
Earth Day
Earth Summit
environmental determinism
Environmental Kuznets Curve
environmentalism
green issues
local food movement
post-humanism
socionature
sustainability
sustainable development
UN Convention on Biological Diversity
UN Framework Convention on Climate Change

Notes

1. A proto-capitalist philosophy which emphasized running a trade surplus in order to amass gold reserves at the level of the nation-state.
2. Cowboy economics refers to an approach to resource management and consumption in which resources are believed to be limited. This is slightly different than exploitation, in which resources are managed/consumed with no regard for their future.
3. Settler colonies are those colonies where Europeans settled and persisted in large numbers (e.g., the United States, Australia, South Africa).
4. There is an older understanding of environmentalism which equates this term with **environmental determinism** (see Chapter 5 for a discussion of environmental determinism).
5. Desertification refers to the degradation of surface biomass and soils in the world's dryland areas (most often in semi-arid regions rather than in true deserts). The degree to which this process is driven by human mismanagement or climatic factors is hotly contested (Moseley and Jerme 2010).
6. A movement concerned with producing and consuming food locally. It was inspired by works such as Michael Pollan's *Omnivore's Dilemma* (2006).

References

Beinart, W. and Coates, P. (1995) *Environment and History: The Taming of Nature in the USA and South Africa* (New York: Routledge).

Braun, B. (2002) *The Intemperate Rainforest: Nature, Culture and Power on Canada's West Coast* (Minneapolis: University of Minnesota Press).

Carson, R. (1962) *Silent Spring* (New York: Houghton Mifflin).

Castree, N. (2001) Socializing nature: theory, practice, and politics. In N. Castree and B. Braun (eds.), *Social Nature: Theory, Practice and Politics*, pp. 1–21 (New York: Wiley-Blackwell).

Cronon, W. (1996) The trouble with nature or, getting back to the wrong wilderness. *Environmental History*, 1(1), pp. 7–28.

Crosby, A.W. (2004) *Ecological Imperialism: The Biological Expansion of Europe. 900–1900* (Cambridge: Cambridge University Press).

Daly, H. (1991) *Steady-State Economics*, 2nd edn. (Washington, DC: Island Press).

Denevan, W.M. (1992) The pristine myth: the landscape of the Americas in 1492. *Annals of the Association of American Geographers*, 82(3), pp. 369–85.

Ehrlich, P. (1968) *The Population Bomb* (New York: Ballantine Books).

Gray, L.C. and Moseley, W.G. (2005) A geographical perspective on poverty–environment interactions. *Geographical Journal*, 171(1), pp. 9–23.

Haraway, D. (2003) *The Companion Species Manifesto: Dogs, People, and Significant Otherness* (Chicago: Prickly Paradigm Press).

Lele, S.M. (1991) Sustainable development: a critical review. *World Development*, 19(6), pp. 607–21.

Lomborg, B. (2002) The environmentalists are wrong. *New York Times*, August 26.

Mansfield, B. (2003) From catfish to organic fish: making distinctions about nature as cultural economic practice. *Geoforum*, 34(3), pp. 329–42.

Meadows, D.H., Meadows, D.L., Randers, J., and Behrens, W.W. (1972) *Limits to Growth: A Report for the Club of Rome's Project on the Predicament of Mankind* (New York: Universe Books).

Meadows, M.E. (2003) Soil erosion in the Swartland, Western Cape Province, South Africa: implications of past and present policy and practice. *Environmental Science & Policy*, 6, pp. 17–28.

Moseley, W.G. (2001) African evidence on the relation of poverty, time preference and the environment. *Ecological Economics*, 38(3), pp. 317–26.

Moseley, W.G. (2005) Global cotton and local environmental management: the political ecology of rich and poor small-hold farmers in southern Mali. *Geographical Journal*, 171(1), pp. 36–55.

Moseley, W.G. and Jerme E. (2010) Desertification. In B. Warf (ed.), *Encyclopedia of Geography*, vol. 2, pp. 715–19. (New York: Sage Publications).

Moseley, W.G. and P. Laris (2008) West African environmental narratives and development-volunteer praxis. *Geographical Review*, 98(1), pp. 59–81.

Pollan, M. (2006) *The Omnivore's Dilemma: A Natural History of Four Meals* (New York: Penguin Press).

Swift, J. (1996) Desertification: narratives, winners and losers. In M. Leach and R. Mearns (eds.), *The Lie of the Land*, pp. 73–90 (Oxford: James Curry).

Watts, M. (1983) On the poverty of theory: natural hazards research in context. In K. Hewitt (ed.), *Interpretations of Calamity*, pp. 231–62 (Boston: Allen & Unwin).

WCED (World Council on Environment and Development) (1987) *Our Common Future* (New York: Oxford University Press).

Wolch, J. and Emel, J. (eds.) (1998) *Animal Geographies: Place, Politics, and Identity in the Nature-Culture Borderlands* (New York: Verso Press).

World Bank (1992) *World Development Report 1992. Development and the Environment* (New York: Oxford University Press).

World Bank (1996) *Toward Environmentally Sustainable in Sub-Saharan Africa: A World Bank Agenda.* Development in Practice Series (Washington, DC: World Bank).

Worster, D. (1979) *Dust Bowl: The Southern Plains in the 1930s* (New York: Oxford University Press).

3

The Biophysical Environment

An Introduction to Human–Environment Geography: Local Dynamics and Global Processes,
First Edition. William G. Moseley, Eric Perramond, Holly M. Hapke and Paul Laris.

Published 2014 by John Wiley & Sons, Ltd.

Icebreaker: Amazonian Black Earths (*Terra Preta de Indio*)

Perhaps no place on earth receives more attention concerning the negative impacts of humans on the natural environment than the Amazon rainforest. The statistics are mind-boggling: each year thousands of hectares of rainforest are cut. This may threaten the predictability of local climates but it certainly endangers and extinguishes rare forms of biological life. We often think of the Amazon Basin as the epitome of tropical nature, somehow biologically blessed with natural resources that have nothing to do with humans. But did you know that some of the most biologically diverse patches of the Amazonian rainforest are found on lands that people once burned, farmed, and abandoned? These fertile areas, with soils known as "black earths" or **terra preta**, *are the outcome of a specific form of burning and farming practiced by native Amazonians a thousand years ago. Today these nutrient-rich soils support some of the highest levels of biodiversity on Earth.*

Residents of Amazonia, and visitors to the region, have long known about the so-called "dark earths." These rich soils, valued by farmers, are the result of long-term human settlement, use, and perhaps improvement. If they first drew some attention in the late 1800s, due to observations of Smith and later Hartt, their origins and their areal extent have only recently attracted more scholarly focus. Previous and current scholars proposed a wide variety of theories for why such rich, dark soil would occur in otherwise poorly developed tropical soils. But key points for their occurrence are previous human occupation, high organic matter, and the clear use of low-temperature fires in patches of terra preta. The net result is that these are not just productive soils, but they seem to play host to micro-organisms that continue to create dark soils. Several scientists have argued that this "regenerative" capacity is a distinctive marker and indicator for a deeper human role in making these soils what they are. In many areas of Amazonia, nearby farmers will come to "harvest" dark earth soils, to then be spread on their farms for cultivation.

What on Earth is going on here? Did previous residents of Amazonia actually figure out how to improve soils with fire? Although contemporary efforts to duplicate the processes behind terra preta have been mixed, there are promising signs that a lighter touch with fire, charring instead of complete combustion, enhances the soil's capacity to retain black carbon and improve soil fertility. The implication of terra preta soils is that humans, using the physical properties of fire and old organic matter, transformed their own local soils in a way that activated new biological processes to create a soil that can regenerate itself. This is a remarkable story and anecdote, but it is also much-needed good news in a world seemingly filled with mostly depressing news about the human–environment experience. Humans can not only thrive in seemingly difficult environments, but can also improve on those aspects that support human and non-human life. Not all human–environment interaction is doom and gloom.

Sources: Glaser and Woods (2004); Balée and Erickson (2006)

Does the realization that the Amazon rainforest has been modified by the "human touch" detract from how we value it? Just because the Amazon rainforest can no longer be considered entirely pristine, or untouched by humans, does not change the fact that it is a major harbor of biodiversity and provides us a critical service by converting carbon dioxide to oxygen. But the Amazon Basin is only one example, out of thousands possible, of how biological and physical environments guide and respond to human actions. Humanized or not, the Amazon rainforest is still being logged, burned, or converted to agricultural fields at an alarming rate. Here, we spend some time discussing and reviewing the important biophysical parameters, useful knowledge for understanding what aspects of environmental change are natural, and which ones are affected or accelerated by the human hand.

Chapter Objectives

The goals of this chapter are for the student to be able to:

1. Describe the major physical and biological factors underlying human–environment relationships.
2. Understand the principles of biophysical relationships and why they are important to understand.
3. Explain why physical setting, hill slope, aspect, climate, soils, and vegetation may or may not condition certain human uses of these physical settings.
4. Illustrate and relate how different ecosystems respond differently to human use and why this matters for human–environment geography.
5. Critically distinguish between factors that are largely biophysical, largely anthropogenic, or a complex mix of both in human–environment relationships.

Introduction

The point of this chapter is to better understand the biophysical world. In doing so, we will have to attempt to separate out the influences of natural and human factors. This is not an encyclopedic inventory of biophysical geography, as you probably already know, and we urge you to consult further readings to gain a full appreciation of Earth's biological, chemical, and physical complexity. The Amazonian example illustrates a key point of relevance to this chapter; although at first glance the human impact on the environment may seem an obvious one, the deeper one digs, the more complex the relationship between humans and nature becomes (see also Goudie 2005). Those of us who live in developed, urban areas find the human imprint unmistakable. Yet as we travel to less-developed and more rural environments, the line between what is natural and what is human-created can become blurred.

For policymakers trying to regulate negative impacts on the environment, such as biodiversity loss, distinguishing between the human and natural is imperative. The same is true for organizations trying to restore degraded landscapes to a more natural state, and for the climate scientist who seeks to understand how human impacts are altering the Earth's climate. The Amazonian case serves as a cautionary tale reminding us to be critical thinkers as we seek to understand the biophysical world in all of its diversity, and our impact on it.

Life on our planet is incredibly diverse. The diversity of the environments we find on Earth is a function of many factors including climate, geologic history, biophysical factors, and, more recently, key human impacts. Before we can develop a deep understanding of people's relationships with their environments, we need a basic understanding of the biophysical environment and specifically how biophysical factors have come to shape the diversity of life on Earth as we know it today.

Studying the biophysical environment requires that we pay special attention to **stability and change**. The Earth's environments may appear relatively stable but in fact they are constantly changing and evolving. We know, for example, the environment in which you find yourself reading this book is vastly different today than it was 10,000, 20,000, or even 100,000 years ago. As you may well know, human actions are a major reason for many of the more recent environmental changes, but certainly not all. Physical changes, such as subtle shifts in the Earth's rotation and orbit around the sun, have caused past climatic changes which have dramatically altered the **biotic** (living) and **abiotic** (non-living) world. Even without these broad-scale physical changes, species move about and compete with each other in constant struggle, resulting in extinction, evolution, and the modification of ecosystems.

Consider the concept of environmental change. We know, for example, that many *natural* (that is, not caused by people) changes have occurred over time. The key point then becomes how to distinguish between natural change and **anthropogenic** (human-induced) change. One goal of this chapter is to introduce and discuss different ways of distinguishing human from natural forces of change.

In our quest to understand the human role in the processes that cause environmental change we must begin with the most basic of questions: Are humans a part of the environment or not? The title of this book gives a hint of our perspective, for if we aim to investigate human–environmental *relations* then the two have been considered separate even if they are not. We accept this distinction only because it is useful as it helps to understand how the world works and our role in it.

The Science of Physical Geography

Our inquiry begins with **physical geography**, which we define as the study of the Earth. Physical geography is not just a catalogue of rivers, forests, mountains, and oceans, but a science that investigates why we find certain features on the Earth where we do and how and why the Earth's surface changes from day to day, year to year, and over millions of years (Strahler 2011).

At a basic level, physical geography is concerned with four realms – the hydrosphere, the atmosphere, the lithosphere, and the biosphere. The **hydrosphere** (*hydro* is Greek for water) is water in all its forms, including lakes, rivers, oceans, vapor, and ice. The **atmosphere** (*atmo* is Greek for air) is the gaseous layer that surrounds the Earth. It supplies vital elements of carbon, nitrogen, and oxygen to sustain life. The **lithosphere** (*litho* is Greek for stone) is the platform of the life layer. It includes solid rock and soil. This layer is sculpted into **landforms** such as hills and valleys (see Goudie and Viles 2010). Finally, the **biosphere** (*bio* is Greek for life) encompasses all living organisms on Earth which depend on the other three layers. Here we focus especially on the last three (atmosphere, lithosphere, and biosphere), and refer you to Chapter 12 on water for more on the hydrosphere and its connections to human–environment relationships.

The biosphere, which is the primary focus of this chapter, can be thought of as the **life layer**. It is a relatively thin layer that includes the surface of the lands and upper surface of the ocean. It is where the other realms come together and interact, and it is also the human habitat.

In this chapter we focus on some basic environmental concepts beginning with the two fields that encompass the study of the biosphere, ecology and biogeography. Although both perspectives (or disciplines) concern themselves with the biotic environment, all life is influenced by the abiotic, and the two are explored together.

Biogeography is the study of the patterns and relationships found in the biotic or living environment. Biogeographers use scientific methods to study past and present distributions of plants and animals and other organisms (McDonald 2003: 1). Why bother to study the past when we have so many present environmental problems? First, the study of past distribution of plants and animals can tell us a lot about past climates since each species is found only within a specific climate range. This is a major aid for scientists who seek to understand what future climates might be like and how different species might adapt to climate changes (see Chapter 8). Second, by studying past distributions, movements, evolution, and extinctions of species we can learn much about the basic mechanisms of life.

Ecology, biogeography's better-known and larger umbrella discipline, is the science of the interrelationships of organisms and their environment. A key ecological term is **ecosystem** which is defined as the set of biotic and abiotic components in a given environment. Simply put, ecology deals more with interrelationships among organisms while biogeography is most concerned with patterns or distributions of them. As we will see, however, you cannot understand one without the other. Indeed, one of the lessons of contemporary ecology and biogeography is that pattern influences process. Of course the opposite is also true, process influences pattern.

Global to Local Patterns

Biogeographers search for patterns. Have you ever wondered why the Earth appears to have ribbons of various shades of green and brown when viewed from space? Take a look at the image of Africa in Figure 3.01. When viewed from space

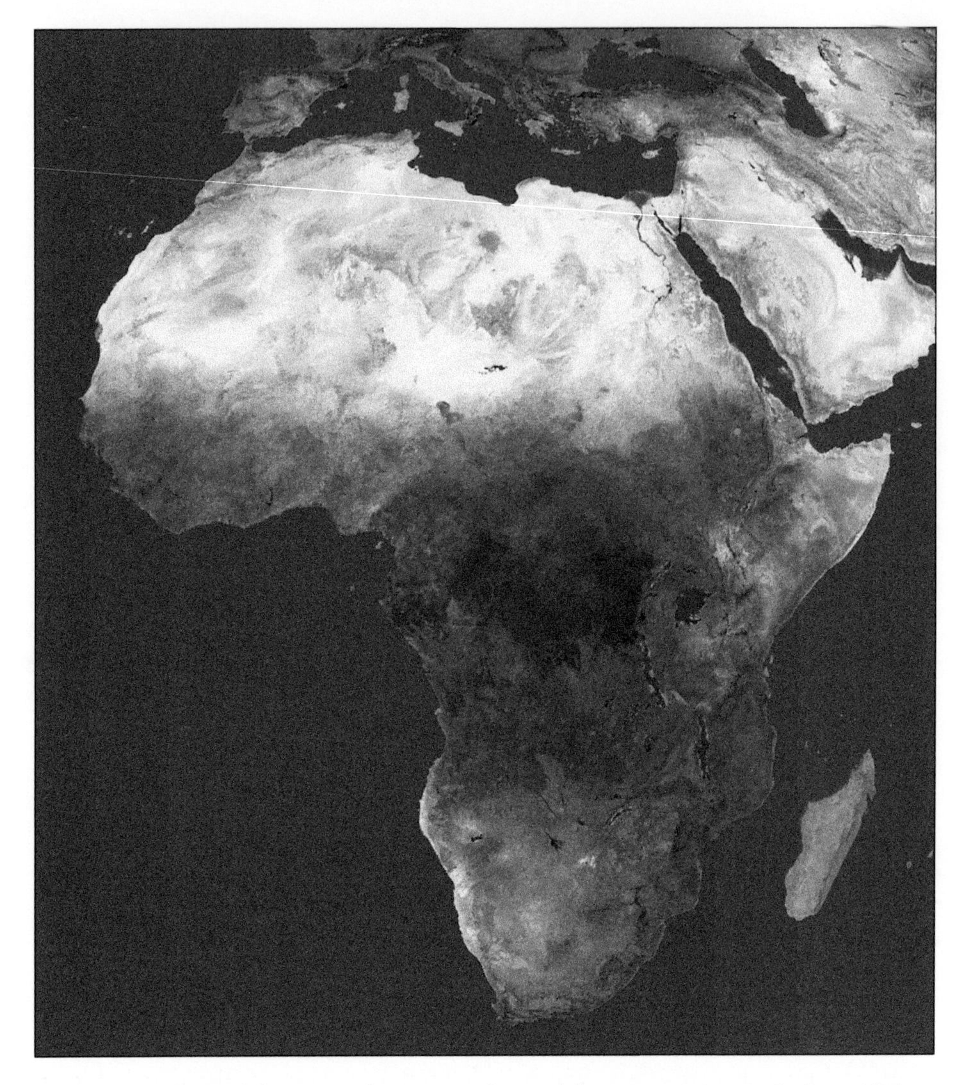

Figure 3.01 The Earth from space focusing on Africa and showing variations in shading with the general locations of biomes on the African continent. *Source*: © NASA.

using remote sensing technology and satellite imagery, African vegetation appears to form a series of bands. If one begins with the large green band of rainforest at the equator, the vegetation zones extend in either direction forming bands of different vegetation types which appear as different colors in the image. These patterns are largely the outcome of Earth–sun relations. They also hint at the close association between climates and vegetation.

In this chapter we will be examining the Earth at many different **spatial scales**. We will begin at the macroscale by looking at the Earth as a whole and its relationship to the sun.

Earth–Sun Relations

If you have ever travelled to another country in a different region on Earth, you probably learned that the intensity of the sun varies from place to place. Those that have left the temperate zone for the tropical sunshine (such as on Spring Break, for example) may have learned this the hard way. The tropical sun is more intense than in the temperate zones and can quickly cause sunburn.

Incoming **solar radiation** sustains life on Earth and drives the Earth's climate system and ocean currents. Variations in the sun's intensity are the key drivers of climate and an important reason for the diversity of environments on Earth. The sun not only warms the Earth's land and water surfaces, it is captured and utilized by plants which convert it into different forms energy.

The reason the sun is more intense near the equator has to do with the tilt of the Earth's axis and the **angle of incidence**. We are all familiar with the angle of incidence because it changes throughout the day. But the angle of incidence also varies by latitude and by season (see Figure 3.02). The first point to understand is that as the Earth rotates around the sun it spins on an axis that is tilted to a perpendicular

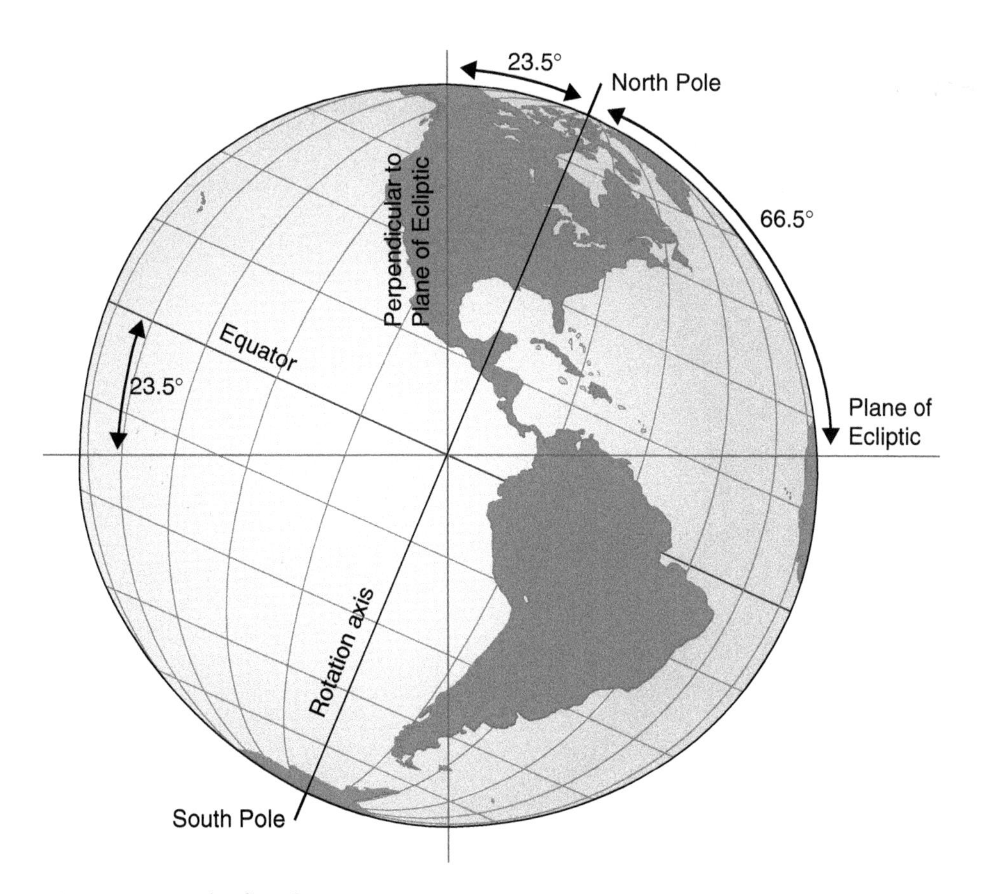

Figure 3.02 Angle of incidence.

from an orbital plane at an angle of 23.5 degrees from the vertical. In other words, as the Earth rotates, it is currently tilted 23.5°. In addition, this angle is relatively fixed (except over very long time periods); the axis always points in the direction of Polaris, the North Star.

This simple tilt affects our planet and our lives in important ways. As the Earth rotates around the sun the northern hemisphere is tilted toward the sun during one part of the annual cycle and tilted away from the sun during another. It is the tilt, and not our distance from the sun, which causes the seasons. When the tilt does vary over long spans of time, this can drive a completely different solar relationship, with dramatic climate consequences on Earth (see Chapter 8).

Solstice and Equinox

The Earth's 23.5° tilt causes an annual cycle whereby solar energy falls perpendicularly on different regions of the Earth during different periods. We know this cycle as the Earth's **seasons**. The seasons are not only characterized by different temperature and weather patterns, they have different lengths of day and night. This tilt is thus the reason why days are longer in summer and shorter in winter (see Figure 3.03). When our hemisphere is tilted away from the sun, the sun appears to us to be low in the sky and to the south (in the northern hemisphere).

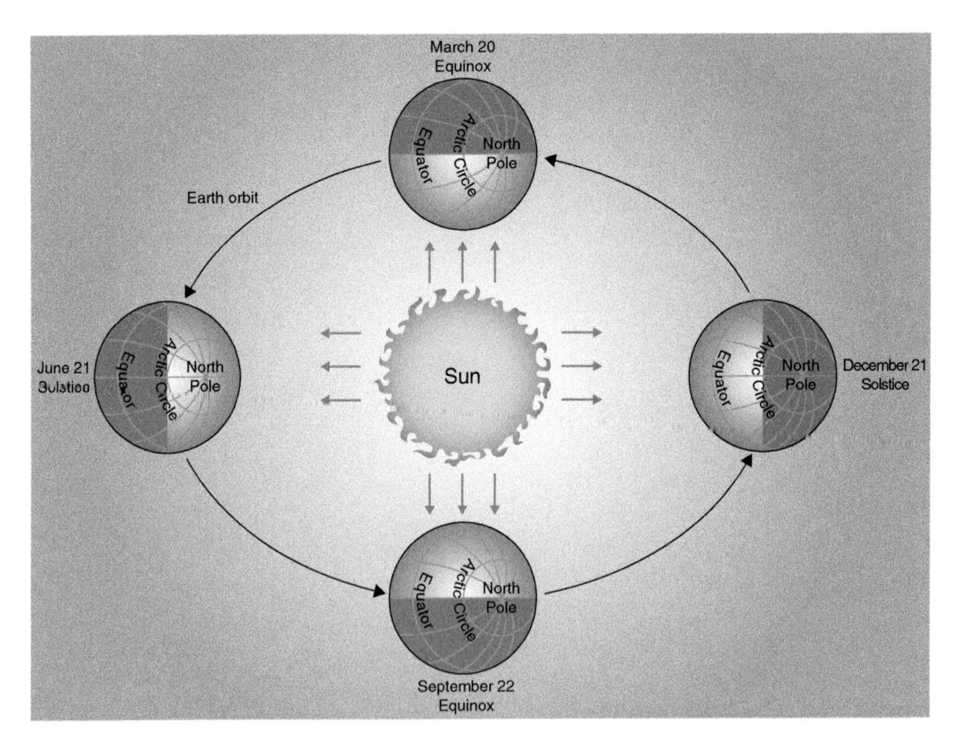

Figure 3.03 Tilt for equinox and solstice.

When it is tilted toward the sun, the days are longer and the sun passes nearly overhead at noon. Only at the equator are there exactly 12 hours of daylight every 24 hours throughout the year. Only on the **equinoxes** (on March 21 and September 22 each year), when the sun's rays fall perpendicularly on the equator, do all parts of the Earth experience 12 hours of sunlight. On these days the equatorial latitudes are heated more intensely while the poles receive the least amounts of energy. After the spring equinox the days get longer until the summer solstice (June 22 in the northern hemisphere). On June 22 the sun's rays fall perpendicularly on the Tropic of Cancer (23.5° N latitude) and the northern hemisphere is heated more intensely and experiences longer days than nights while the southern hemisphere experiences winter, with cooler days and longer nights. The effects of the Earth's tilt increase with increasing latitude. The higher the latitude, the longer the days in summer until one reaches the Arctic or Antarctic circles at 66.5°, where there is 24 hours of daylight on the summer solstice and the sun never sets. This "midnight sun" is matched by 24 hours of darkness on the winter solstice!

Unequal Heating of Land and Water

When solar radiation strikes a surface several things may happen: it may be partly reflected, absorbed, and/or transmitted. **Albedo** is a measure of the reflecting power of a surface. Specifically, it is the fraction of radiation received that is reflected by the surface. Albedo is important because it determines how quickly a surface heats up. A low albedo means most energy is absorbed and the surface heats quickly, as is the case with black top (asphalt). A high albedo means that more energy is reflected and heating is slow, as is the case with snow. Solar radiation that is not reflected by a surface is largely absorbed, and the absorbed energy heats the surface of a given mass. The simplest way to experience albedo is to walk barefoot across a black top parking lot on a hot summer day. The white stripes on the pavement are much cooler than the hot black top because of the albedo. The black top has absorbed most of the solar energy and its temperature rises, while the white stripes have reflected a good portion of the energy.

Solar energy is largely reflected or absorbed by land surfaces, but a percentage is transmitted through water. Water is translucent and light penetrates into the surface layers. Depending on the qualities of the water (its clarity), the water will absorb or transmit a different percentage of the sun's energy. We can experience the transmission of energy through water as light when we swim in a clear pool of water. Have you noticed how the upper surface is warmer than the deep water?

Zooming in Closer: Land/Water

The first reason for unequal heating of the Earth's surface is the tilt of the Earth's axis; the second reason is the unequal rate of heating of land and water (see Figure 3.04). At the global scale, the Earth's tilt has the greatest influence on

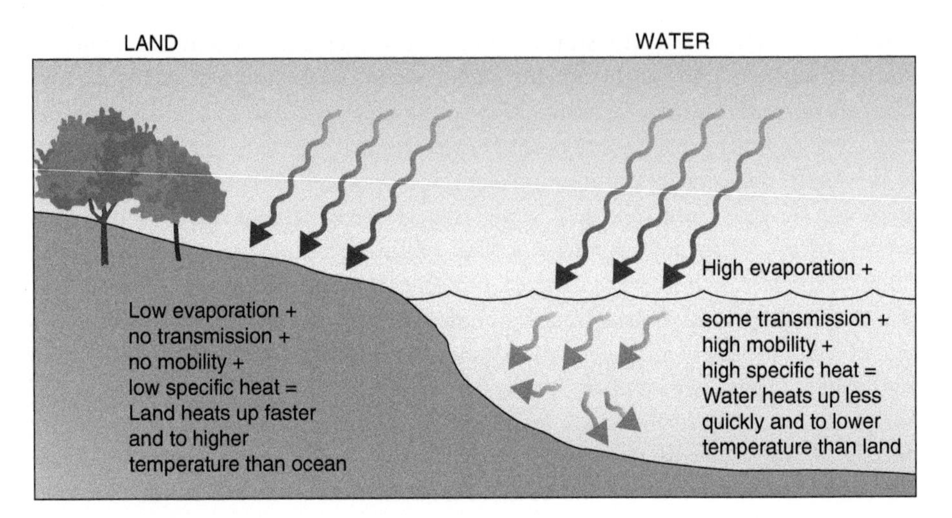

Figure 3.04 Unequal land/water heating and heat capacity.

temperature patterns. When we zoom in closer to the continental or regional scale, another important principle of physical geography comes into play – the surface of any large, deep water body heats more slowly than a land surface. Large water bodies also retain heat longer and thus cool more slowly. This causes land areas to heat more quickly during the day and cool more quickly at night than water bodies. Similarly land areas heat more quickly in summer and cool more quickly in winter, resulting in large temperature differences between land and water.

These differences are also easier to differentiate in tabular form:

Water	**Land**
Transparent	Opaque
Heat distributed over a great volume	Heat concentrated near surface (few inches)
The greater the body of H2O,the smaller the temperature contrast (i.e. Pacific)	The larger the landmass, the greater the contrast is in temperature (i.e. the Eurasian land mass)

There are four reasons for the different heating and cooling properties of land and water:

1 Land heats rapidly and reaches a high temperature (in particular dark surfaces).
2 Water heats slowly and reaches moderate temperature.
3 Land cools off rapidly and reaches low temperature.
4 Water cools off slowly and maintains a moderate temperature.

The most important difference between land and water is that solar radiation penetrates water, distributing some heat deep into the water layers. As noted, water heats much more slowly than a land mass composed of rock or soil. This is due to the different values of **specific heat** of the substances. In addition, water is subject to thermal mixing. Water moves, mixing the warm surface with cool water below. Finally, evaporation cools the water surface. These four factors drive not only wind patterns over land, but also wind and ocean currents.

Zooming in Closer: Elevation

As one zooms in still closer and focuses on a single region of the Earth, the effects of elevation become apparent. Elevation has an important cooling effect on the Earth's surface and air parcels. Just as temperatures decrease as in higher latitudes, they decrease as elevation increases. This may seem somewhat counterintuitive (why is it colder closer to the sun?), but is explained by the fact that the density and pressure of the air decrease with increasing elevation. The temperature change as a function of elevation is governed by the **environmental lapse rate**=6.4° C/km (3.5° F/1000 ft). This change in air temperature by elevation, however, is variable. The amount of water (humidity) in the air mass governs how quickly or slowly the air parcel will cool (as it rises) or warm (as it descends).

Pressure, Winds, and Precipitation

Differential heating of the Earth's surface not only causes temperature differences, but air pressure differences as well. Differences in air pressure in turn are key drivers of climate regimes. As noted, when the sun beats down on a surface, the area gradually absorbs heat and its temperature rises. As this occurs, the air near the surface is also warmed and begins to rise. Rising air masses are associated with lower pressure. Similarly, when an air mass cools it descends, causing the pressure to rise. Differences in pressure give rise to winds as flows from high to low pressure areas.

Take a moment to consider how the different aspects of what we have learned about the properties of solar energy, temperature and pressure, and land and water, can be combined to explain common weather patterns. Wind and ocean currents are caused in large part by the uneven heating of the Earth's surface. When you feel the breeze coming off of the ocean, think about the fact that it is occurring because there is unequal heating going on. For example, the land heats faster than the ocean. Anyone who has walked across a hot sandy beach and then plunged into cool ocean water has experienced this simple phenomenon.

The "land–sea" breeze is caused by the unequal rates of heating of land and water. During the day, the land heats more rapidly than the water, resulting in higher land temperatures and a relatively lower pressure over the land surface than over water. As the land heats up and the air rises, the drop in pressure relative to the pressure over water causes air to move from above the ocean to over land. This "onshore" breeze cools the land surface. In the evening the process reverses as the temperature over land falls faster than that of the ocean, often resulting in a mild "offshore" breeze. The differential heating properties of land and water and the breezes these differences help create are a key reason why coastal areas are cooler in summer and warmer in winter than the middle of continents. This difference is

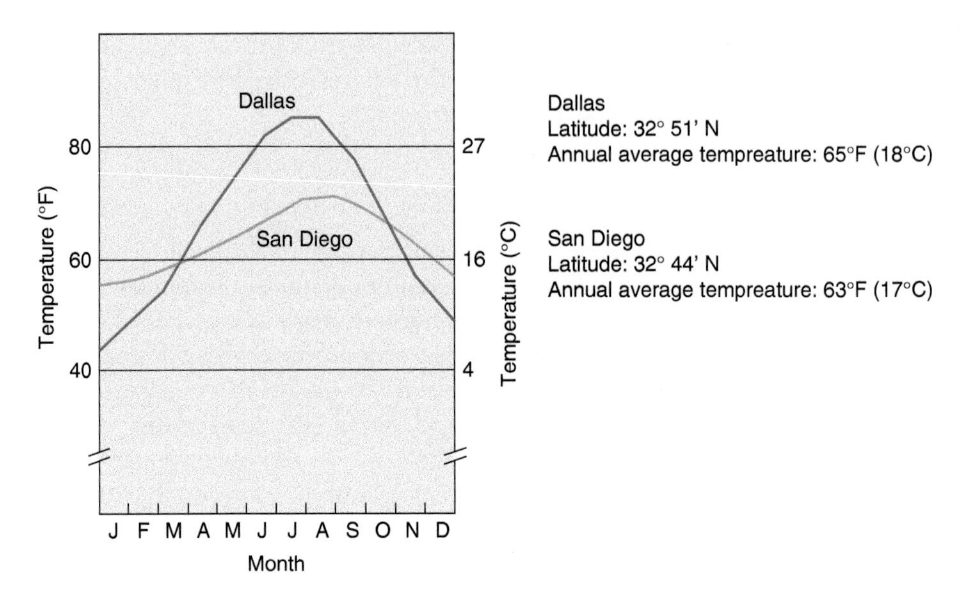

Figure 3.05 Average annual temperature fluctuation for the cities of Dallas and San Diego.

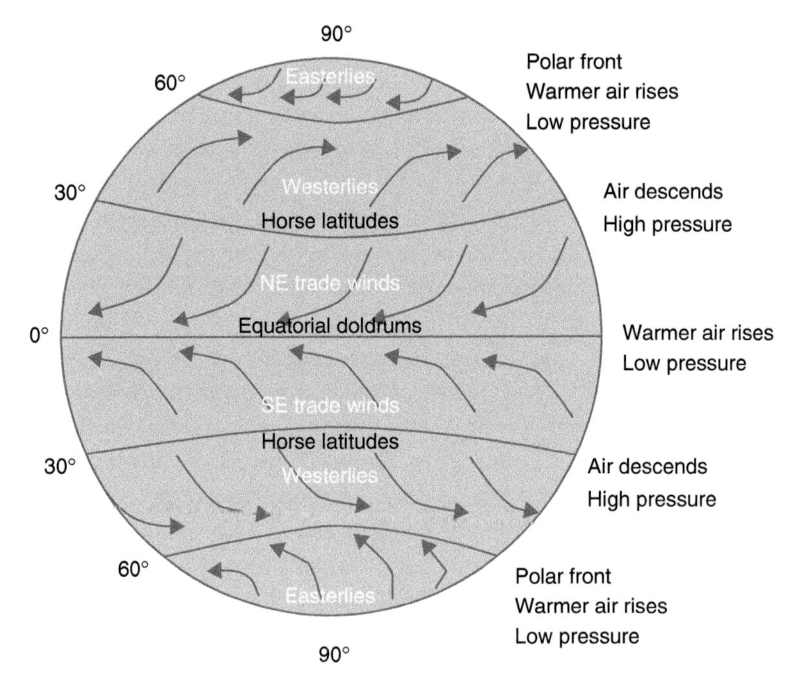

Figure 3.06a Global wind patterns and pressure systems.

visible when you compare the average temperatures over the course of a year between a coastal city, say San Diego, California, and the more interior land-locked city of Dallas, Texas (see Figure 3.05).

Global pressure differences are the main cause of wind patterns (see Figure 3.06a), which in turn are also a key cause of ocean currents (as seen in Figure 3.06b).

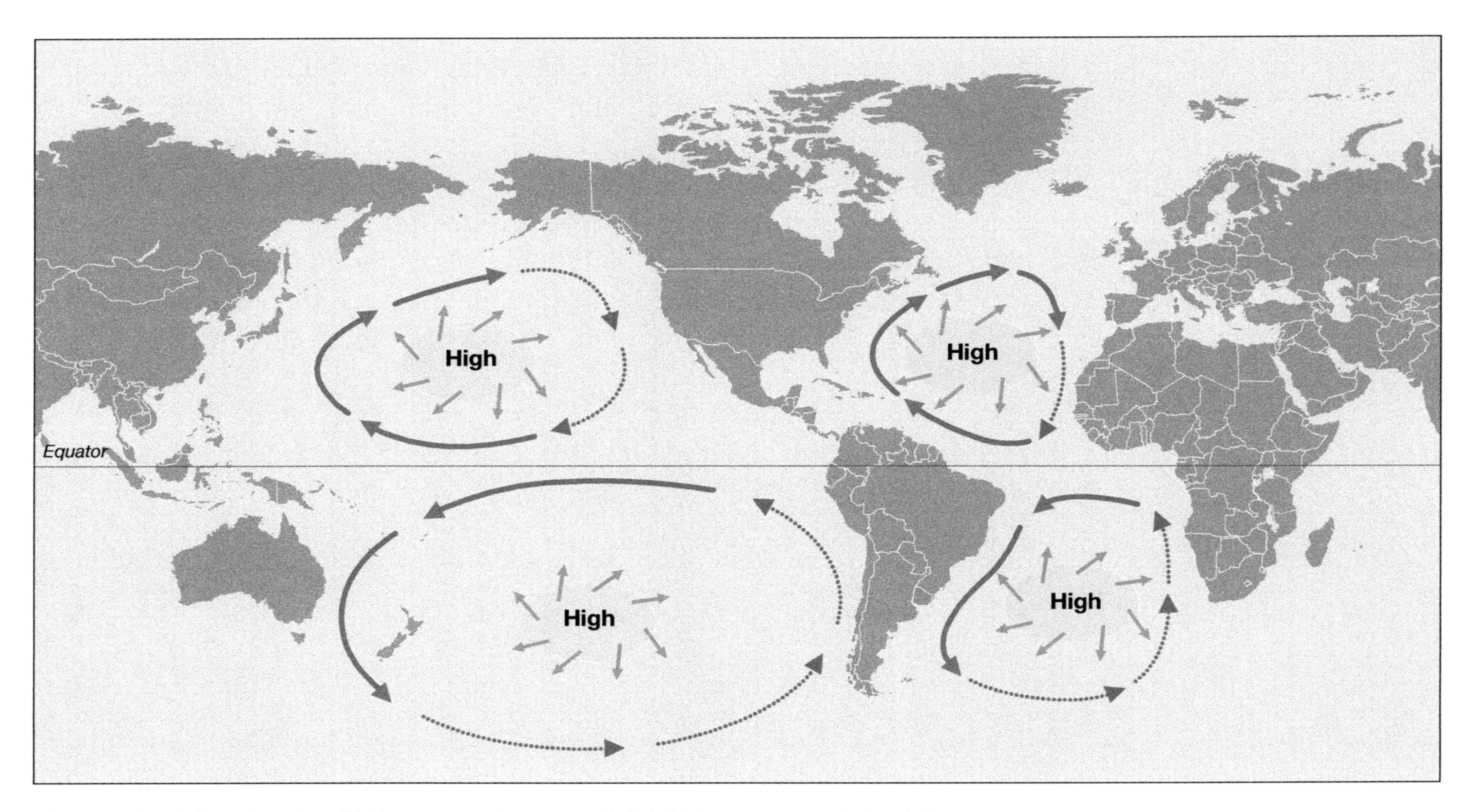

Figure 3.06b Relative location of high-pressure air masses and their influence on oceanic circulation.

Think of winds literally sweeping the surface water in one direction or another. Pressure differences also explain much about patterns of precipitation since winds often transfer air masses laden with moisture, and low-pressure systems can cause precipitation.

The Physics of Water and Precipitation

To better understand these patterns we need to know something about the physics of water. As air warms it can absorb increasing amounts of water. As it cools, it eventually reaches the **dew point**, at which point the air is saturated; further cooling results in **condensation** and the formation of clouds. If this process occurs rapidly enough, the particles of water or ice become too heavy to remain airborne and they fall as rain or snow.

Recall that warm air rises, but as air rises it also cools. The temperature of rising air cools as it expands due to reduced air pressure at higher elevation. As the air cools it can hold less moisture. Thus, when air rises fast enough it cools rapidly, causing air to condense, and under the right conditions precipitation will occur. A good example of this phenomenon is when a mass of air is forced to rise over a mountain. If the mountain is high enough the air will cool to the dew point and condensation will form (see Figure 3.07). This phenomenon, known as the **orographic effect**, is why we often see clouds and fog clinging to mountain tops and why the **windward** sides of mountains receive more rain than the **leeward** or downwind sides. When moist air is pushed rapidly up over a mountain, it rises and cools and precipitation occurs on the windward side. After the air mass passes over the mountain it begins to descend; as it does so, it warms and dries and precipitation halts, creating a **rain shadow** on the leeward side. This means that, within a few miles, the range of ecosystems can vary dramatically. It also means that the natural resources in any given area are structured by these relationships. And this can translate to very different human lifestyles and livelihoods. Climate does not determine human life, but different climates give us different options for using landscapes and making a living.

Zooming in Further: Urban/Rural Temperature Profile

If we zoom in even closer to the Earth to a scale familiar to us all – that of the urban-rural landscape – we can observe another crucial factor influencing temperature patterns, the impact of urbanization. Urban areas are often warmer than non-urban ones due to the **urban heat island effect**. Many urban areas are composed of large expanses of dark surfaces such as roads and rooftops. As observed above, dark surfaces, which have low albedo, absorb a relatively high percentage of solar radiation and thus heat to higher temperatures than lighter surfaces. In addition, urban areas typically have less vegetation than non-urban ones. **Transpiration** is when vegetation releases moisture into the air, which also

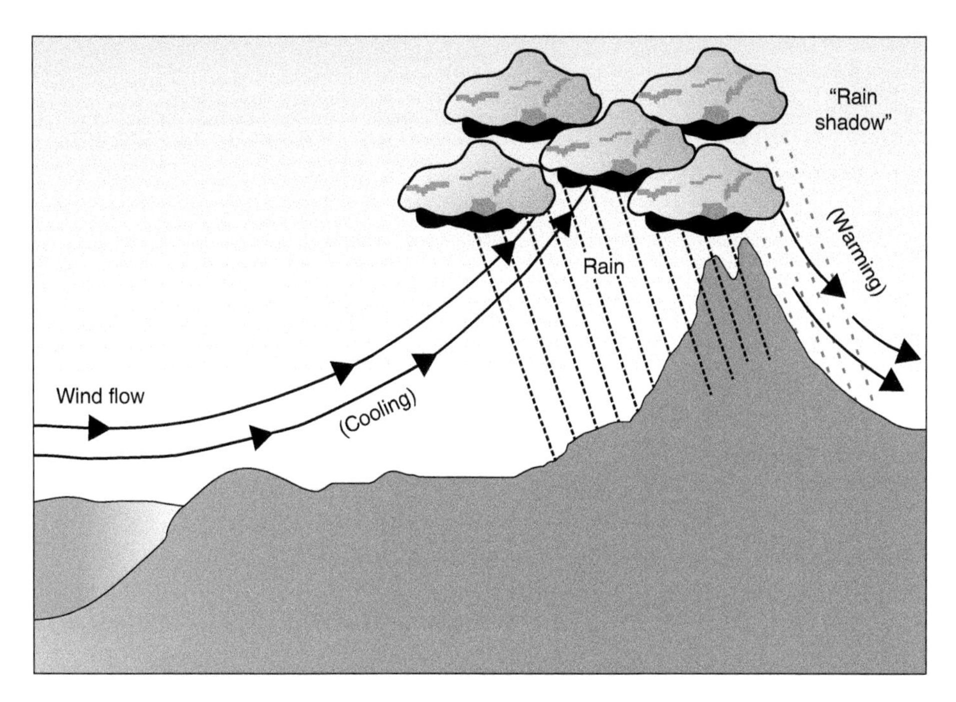

Figure 3.07 The orographic effect, where precipitation is created by a topographic barrier, and a rain shadow is created on the opposite side.

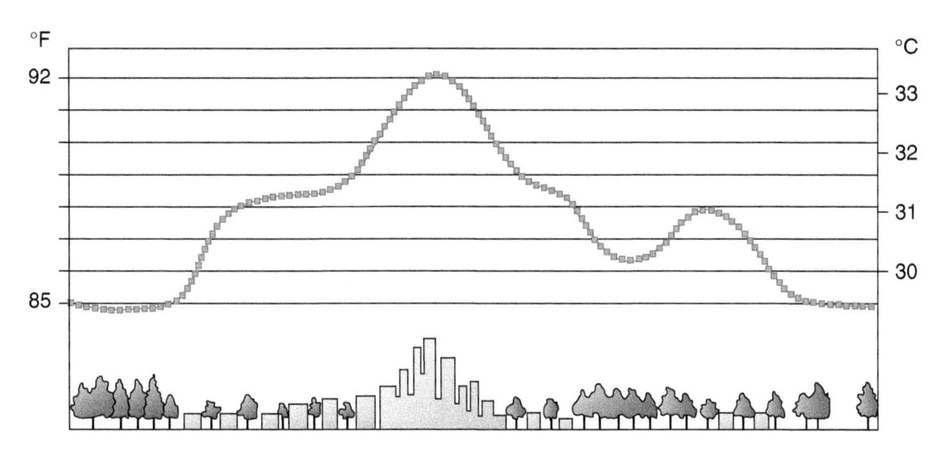

Figure 3.08 Illustration showing the urban heat island effect as it varies spatially across a hypothetical city. Note the hypothesized temperature "rise" with the hypothetical skyline of the city. What factors might change the appearance of the heat island effect and its geographic occurrence in large cities?

has a cooling effect. As a result, urban areas usually have higher daytime temperatures than rural and semi-urban ones (see Figure 3.08).

In summary there are four determining factors driving global temperature, precipitation, and wind patterns going from global to local: (1) latitude, (2) land/water differences, (3) elevation, and (4) urban heat islands.

Biomes and Major Climates

Climate is defined as the average condition of the weather over a long period of time. **Weather** we can define as the day-to-day state of the atmosphere. Weather involves the short-term change in moisture and air movement. Weather and climate result from processes that redistribute heat (solar energy) around the world. Although globally the sun is the major driver of the annual shifts in weather, as we will see, winds play a crucial role at the regional and local scale.

Climate is the most important factor determining the natural vegetation of a place, and it is no surprise that the major plant communities correspond to key climate regimes (see Figures 3.09 and 3.10 for climates and biomes). Precipitation and temperature are two key variables that have the greatest influence on the kinds of vegetation found in a given area. Four aspects of precipitation are important: (1) the total amount; (2) the form in which it arrives (snow, rain, hail etc.); (3) the uniformity or seasonality (e.g., wet/dry pattern); and (4) the variability or dependability (droughts or extreme events). **Temperature regimes**, or how temperatures vary over the course of a year, are also important. Tropical areas have warm temperatures with relatively little daily or seasonal variation, posing few limitations on plant growth. Areas near the poles experience long, cold winters with short summers that impact the ability of plants to grow and establish. Other regions of the world have temperature regimes that are more evenly divided between warm and cold periods, requiring plants to develop means for adjusting to changes.

Even a casual observer will notice that certain types of plants tend to group together in particular climates to create distinct vegetation communities. A **community** is defined as the assembly of all organisms living in a prescribed area. **Biomes** are plant and animal communities with distinct characteristics that cover large geographical areas. Biomes are usually designated according to a dominant or obvious vegetation form on the landscape. For example, the forest biome is dominated by trees, whereas the tundra lacks trees and is dominated by low-lying shrubs and grasses. Although biomes are named for a predominant type of vegetation, each includes a specialized group of animals adapted to the plants and climate.

The concept of biomes is useful for understanding how broad-scale patterns of climate shape the biotic environment. It should be stressed that one finds great differences among the species found in the same biome located in different regions of the world. Even within biomes, there is a good degree of variation to be found at finer scales of observation. However, in broad terms the basic structure of the plant communities is similar for a given biome.

Although temperature and precipitation patterns are the primary determinants of global vegetation patterns, several other factors may influence these patterns. Three key factors are soil type, disturbance, and elevation. The

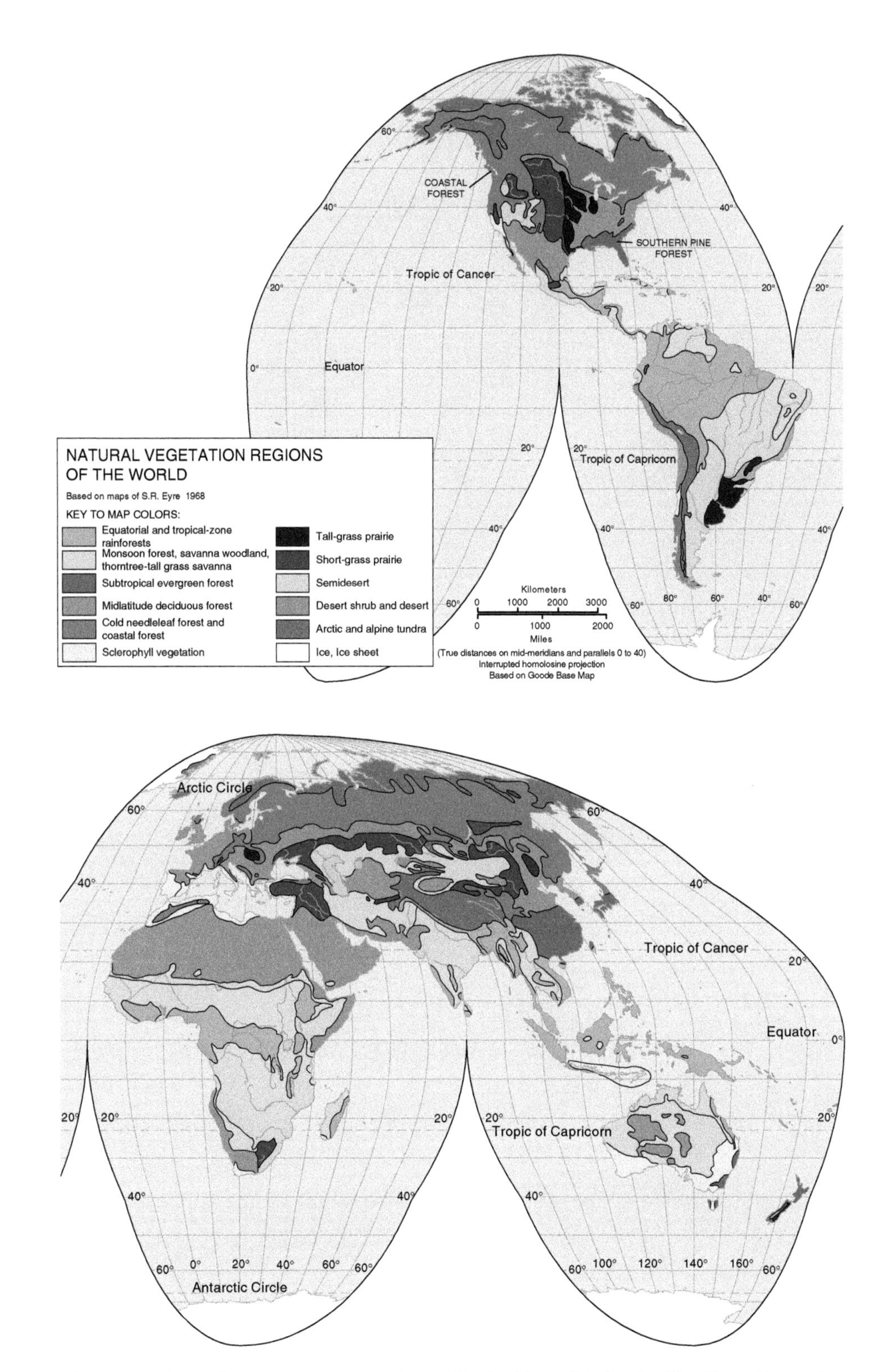

Figure 3.09 Natural vegetation zone map. *Source*: Alan Strahler and Arthur Strahler, *Physical Geography: Science and Systems of the Human Environment* (Hoboken: John Wiley & Sons, 2005), figure 24.3.

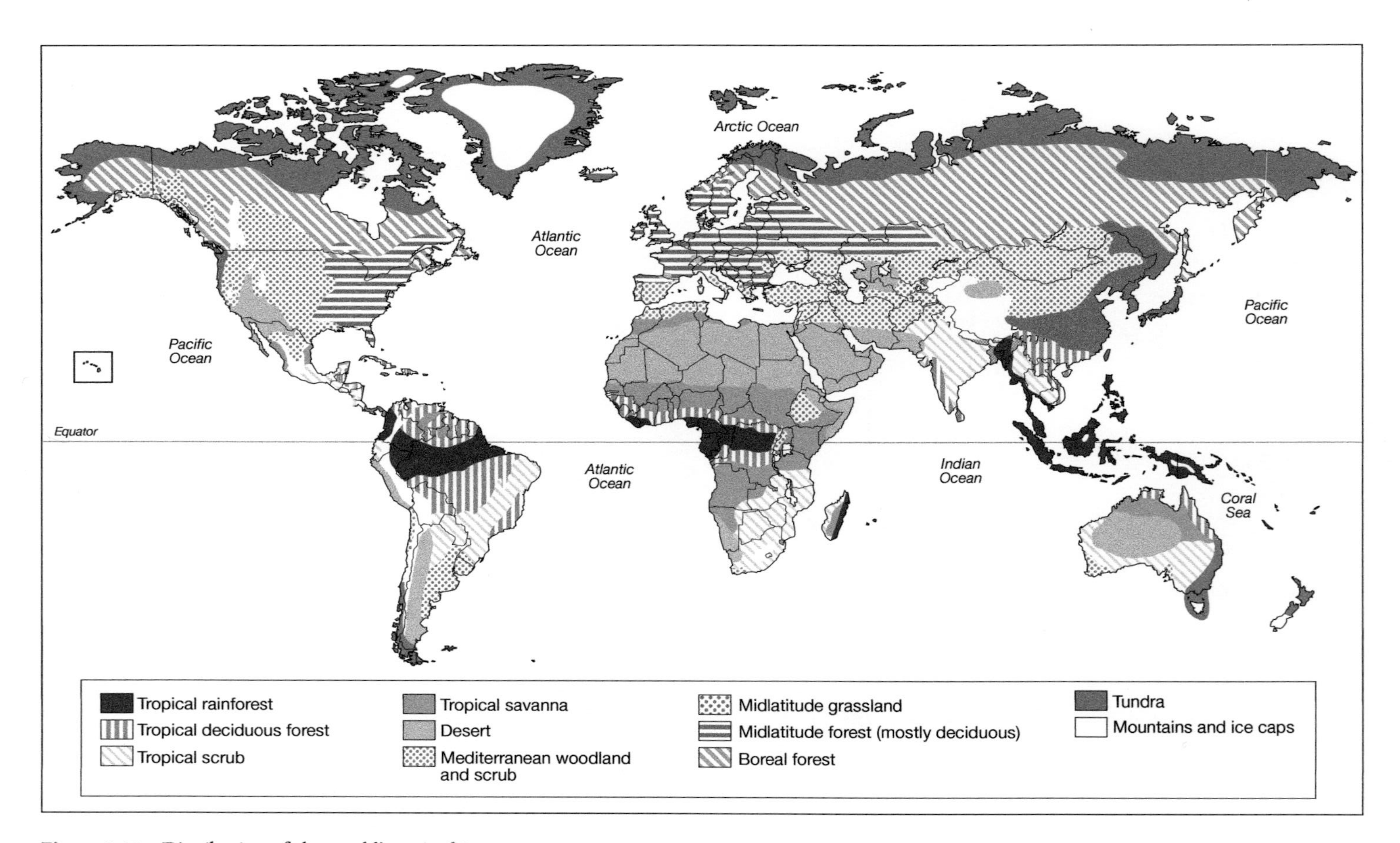

Figure 3.10 Distribution of the world's major biomes.

characteristics of soil can mediate the effects of climate. Some soils, such as those with high sand content, dry out rapidly following rain, while others, such as those with high clay content, will remain moist for long periods of time following a rainfall event. These different soil characteristics influence the kinds of plants that can exist in a given locale. For example, patches of savanna grassland exist in parts of the Congo rainforest basin as a result of patches of deep, sandy soil conditions in some areas.

A disturbance is an event that causes short-term changes to plant communities. Common disturbances include fires, floods, windstorms, or human activities such as logging or farming (see below). Globally, the most important disturbance is periodic fire. Geographers have long noted that some parts of the world have sufficient precipitation to support forest yet they are composed of grasslands or savanna. Periodic fires prevent trees from establishing, and maintain grass cover. Persistent high winds or periodic floods also prevent the establishment of trees in some areas. Conversely, large herds of grazing animals consume grasses and may promote the establishment of dense tree or shrub populations.

Finally, as noted above, elevation affects temperature as well as precipitation. High-elevation areas thus tend to be cooler and moister than lowlands, affecting the kinds of plants found in these areas. These three factors interact together with climate to create a patchwork of vegetation patterns on a landscape.

Although one might think that subdividing the geographic distribution of life into communities is a straightforward endeavor, it can be problematic and the source of much controversy. It is important to note that plant communities overlap with each other. There are few sharp boundaries between ecosystems or biomes, so it is best to think of these mapped categories as rough locations with much overlap. Biogeographers also recognize these blurred boundaries as **ecotones**, transition zones of competition in which the typical species of one biome intermingle with those of another (see Figure 3.11).

Although there are numerous ways to classify vegetation, biogeographers generally recognize six major terrestrial vegetation forms:

1. Forest: a tree-dominated community with a nearly closed canopy.
2. Savanna or woodland: a community where trees and grasses coexist, in which individual trees are more or less widely spaced with a grassy understory.
3. Shrubland: a fairly contiguous layer of shrubs, sometimes with a grassy understory.
4. Grassland: a community dominated by grasses and forbs.
5. Scrub: a mostly shrub-dominated community in which individuals are widely spaced or discrete.
6. Desert: a community with very sparse plant cover in which most of the ground is bare.

Figure 3.11 An example of California coastal sage scrub and grassland ecotone, where two different vegetation types transition and blend. *Source*: Paul Laris. Used with permission.

The Biomes

The above vegetation forms translate into eight well-recognized biomes. We begin by describing biomes at the equator and then gradually move toward the poles (Figure 3.10).

Lower Latitudes

Tropical rainforest

Tropical rainforest is characterized by continuous forest cover, creating a **closed canopy** with a shaded understory. There are notable aspects of the plant canopy that distinguish a tropical forest from, say, a temperate and deciduous forest. First, tropical forests have multiple vertical layers of vegetation, from soil to tree-top. Second, a tropical forest almost always has additional plant life and not just "trees." Lianas, vines, and epiphytes are all common in the tree canopies of the tropics. Third, and in contrast to most grasslands, the vast majority of biological life and organic matter occurs above the soil surface. This is largely due to the poor, highly drained, soils in most areas of the tropics. Key exceptions would be areas where soils have been humanly modified (such as "black earths"), and in some volcanically active tropical regions (e.g. Indonesia, Hawai'i).

The key characteristics of the world's tropical rainforest biome are the lack of seasonality and the constant and even distribution of temperature and precipitation throughout the year. There are no killing frosts, few limitations based on external factors (except perhaps for soils), but competition for ecological space (or niches) is fierce at every level of the forest. From the soil surface, to the trunk, up to the canopy, the richness of species means that many species overlap the same space and range of ecosystem use. Tropical rainforests contain the vast bulk of the Earth's biodiversity in all its forms (species and genetic). Diversity can be further enhanced in such locations if elevation (altitude) is a factor, such as the so-called cloud forests in tropical locations.

Although the "tropics" technically extend away from the equator to 23.5° north and south of the equator, tropical rainforest usually has a more confined location closer to the equator. The steady climate is a function of the **Intertropical Convergence Zone**, or **ITCZ**. The ITCZ is located near the equator where the trade winds converge in a low-pressure zone. Solar heating in the region forces air to rise through convection, which results in a very humid climate with high cloud cover and frequent rainstorms.

If the early human–environment concerns about tropical rainforests revolved around biodiversity in the 20th century, now we are just as worried about the role of carbon in these forests. The rainforests are one of the primary components of the **global carbon cycle**, meaning they are both a source and a sink for carbon, thus any net change (up or down) in the coverage of forests has consequences for the terrestrial part of carbon exchanges on Earth.

Savanna and tropical dry forest

Savannas are unique among biomes in that they are the only one defined by the coexistence of two vegetation forms, trees and grasses. In general savannas include scattered trees with a near-continuous herbaceous (grassy) understory. The density of trees varies as a function of a number of key variables, including the precipitation, soil conditions, and disturbance history (e.g., fire regimes, grazing, and tree-cutting). The ratio of trees to grasses can vary dramatically (see Figure 3.12). This has created confusion over the meaning of the term savanna, and some have argued that it be replaced by "woodland" or "wooded grassland." However, the term savanna captures one of the most novel aspects of this biome – tree density varies as a function of the disturbance regime. For example, fires are extremely common in most savannas. During the long dry season grasses desiccate and are frequently burned by fires.

Savannas are found just poleward of the equatorial zone (from 10° to 30° latitude) where rainfall is seasonal. They are characterized by an intense (usually winter) dry season lasting at least three months with one or two wet seasons. Annual precipitation varies from about 400 to 1200 mm per year. The variation in savanna climate is largely a function of the annual migration of the ITCZ. During the wet season, the ITCZ moves over the savanna, bringing rainstorms. After several months of rains, the ITCZ shifts toward the equator leave the savanna dry. During the long dry season, savannas are often desiccated by dry winds blowing from neighboring deserts. The winds leave the vegetation parched, and most

Figure 3.12 Photos of varying savanna types. *Source*: Paul Laris. Used with permission.

grasses die while most trees shed their leaves. During the dry season fire is a common event. Temperatures vary markedly, with daytime highs well over 30° C during the hottest months, with temperatures dropping into single digits at night in the coolest areas.

Savannas with drier climates have shorter grasses and trees that are adapted to extreme drought. These arid savannas, such as the Sahel of West Africa, experience prolonged dry periods of over six months. Moister or **mesic** savannas (where precipitation is above 650 mm) have taller grasses, and trees are adapted to frequent fires.

In the moister regions, savannas may give way to tropical **deciduous** forests, which are usually found in hot lowlands just poleward of the equatorial zone sandwiched between savanna and rainforest (between 10° and 30° latitude). Rainfall is higher than in the savanna, but quite seasonal and also dictated by the migration of the ITCZ. A key characteristic of the climate of a tropical deciduous forest is a pronounced and extensive winter dry season. The term deciduous means that trees shed their leaves during one season, usually in response to stress from cold or dry periods. Thus, during the dry season many trees lose their leaves. Perhaps the easiest way to describe a tropical deciduous forest is to contrast it with a tropical rainforest. Compared to a rainforest, these forests have a lower and more open canopy, allowing more light to reach the ground and supporting more understory vegetation.

In general savannas tend to have very heterogeneous vegetation formations. For example, changes in soil moisture can result in dramatic shifts in tree density and type. **Riparian** forests – areas alongside streams where there may be more and larger trees – are common and often have a structure and species similar to tropical dry forests. This heterogeneity is one reason that savannas have relatively high levels of biodiversity. Savannas are composed of very large areas that support huge human populations as well as some of the Earth's most charismatic mammals such as lions, elephants, giraffes, and rhinos. It is believed that humans evolved in the savanna of Africa. Large portions of the savanna have long been used for agriculture and livestock production. Today, many savannas, especially those in South America, are under threat from rapid increases in agricultural areas, such as in the *cerrado* (savanna) of Brazil.

Desert and semi-desert

Desert plants are characterized as **xerophytes**, organisms able to tolerate long stretches of drought, and are typically able to survive months without any precipitation. Thick leaves, waxy plant surfaces, thorns instead of leaves: these are all ways for plants to cope with the short- and long-term stress of low rainfall amounts. Soils are have high mineral content and negligible organic matter accumulation at the surface.

Most of the world's deserts are found from 15° to 30° latitude, as they are strongly shaped by **Hadley cell circulation**. As air rises at the ITCZ, near the equator, it loses its ability to carry moisture. As air parcels then descend, they create downward drying winds that create the world's largest expanses of arid and semi-arid regions (see Figure 3.06a above).

While generally thought of as "hot" regions, deserts are more appropriately thought of as dry regions, since temperature fluctuations and seasonal changes are possible. Areas with less than 500 mm of rain are thought of as semi-arid, while truly "arid" regions receive less than 250 mm in precipitation on an annual basis. The determining factor is the balance between precipitation and **evapotranspiration**. The other remarkable quality about precipitation in true deserts is the high variability of rainfall (and snow). It is unusual for any desert to have an "average" amount of precipitation fall in any given year. The Atacama Desert in Chile, for example, is located where both Hadley circulation and a cold current offshore conspire to create hyper-arid conditions. While the Atacama does receive small inputs of fog moisture (locally called *garua* in Chile and Peru) from the Pacific, true rainfall events are rare.

Compared to other biomes, deserts are not remarkably diverse. They do, however, exhibit species and ecosystem relationships that are not found elsewhere, and for certain groups of organisms, there is notable variety. Deserts are also not equal in terms of biodiversity when compared to each other. Deserts have often been thought of as "wastelands." In the last hundred years, however, as humans have continued to live in arid regions, usually in the form of expanding cities and villages, the amount of groundwater pumping has increased dramatically. Artificial irrigation can free the mineral-rich soil to provide nutrients for plants and cultivated agriculture, but this also frees mineral salts which can then accumulate at the surface of soils (**salinization**) and form a salt crust. In California issues of desert use versus preservation are heating up on another front as stakeholders debate the fate of the deserts. Some see the need to preserve the fragile desert habitat, while others do not wish to lose access to desert lands for off-road vehicle adventures. Finally, energy companies and some environmentalists see the desert sun as a future source of electrical power, but solar panels take up space and may not be compatible with wildlife or nearby communities which may prefer an open landscape.

Mid-latitudes

The Mediterranean biome

The Mediterranean biome is found in areas surrounding the Mediterranean Sea and in the warm temperate areas on the west coasts of continents (although a small patch can also be found in southeastern Australia). The Mediterranean biome is the smallest we will consider. This biome is located just poleward of the subtropical high pressure zone and is moderated by cool ocean currents in the 30° to 40° latitude range. It occupies less than 2 million square kilometers and is found in narrow strips along west coasts in extreme Southern Africa, Australia, North and South America, and around the Mediterranean Sea in Europe. Despite its small size, this biome has been increasingly settled and domesticated by humanity.

Summers are hot and dry and winters are mild. The climate is dominated by the migration of the subtropical high, which brings hot, dry summers and then moves toward the equator, allowing low-pressure storms to bring precipitation in winter. Precipitation is moderate, ranging from 30 to 100 cm, nearly all of which falls in the winter months. Average temperatures range from 15° to 20° C and frosts are rare.

The typical vegetation of the Mediterranean biome is a mosaic of shrublands and wooded grasslands with occasional valley forests. The woodlands in both the Mediterranean Sea area and in California are typically dominated by oaks. In Australia the dominant tree is eucalyptus while the southern beech and acacia dominate in Chile. Conifers are also found in the higher elevations, including pines and cedars. The South African portion of this biome is dominated by shrubs and reeds ("restios"). Much of the shrub vegetation is evergreen and **sclerophyllous**, meaning plants have small, hard (sometimes waxy) leaves that resist water loss, and tough bark to resist fire. Many of these plants are adapted to fire. A common shrubland in California is known as Chaparral. Chaparral is a type of sclerophyll forest sometimes called an elfin or dwarf forest. Chaparral is dense, closed canopy shrubland found on the slopes of coastal mountains in California between 150 and 1500 m in elevation. Fires are common, but not annual, in chaparral, as they are in most Mediterranean environments. Long, hot, dry summers leave the vegetation parched. Annual grasses typically make up about half of the vegetation and these grasses dry shortly after the winter rains. Many areas of Mediterranean climates also feature high wind speeds, so leaf adaptations to retain moisture are not solely attributable to ambient heat. The shrublands of Mediterranean biomes are fire-dependent, meaning they require fire to persist. However, scientists have recently learned that too frequent fire can result in changes from native shrublands to exotic grasslands.

Temperate grasslands

Most of the biomass in grasslands hides beneath the surface. Even in short-grass prairies, which can look especially unremarkable in appearance and height, some three to five times of the biomass is underneath the soil surface.

Temperate grasslands are found throughout the mid-latitudes (30° to 60° N and S), often located in the interior of large continents on the leeward side of mountain ranges where rainfall levels are too low to support woodland and/or where periodic fire keeps tree growth in check. Annual precipitation averages from 30 to 100 cm annually. Potential evapotranspiration often exceeds precipitation during the growing season, and droughts are common. While grasses and herbs dominate this biome, trees and woodlands are often found in valley bottoms and along streams (riparian vegetation) where soil moisture is higher.

While principally shaped by fire and grazing disturbance, the biodiversity of temperate grassland areas is lower than in their tropical counterparts (savanna). Disturbance factors, especially fire, create temporary and local biodiversity pulses, until the dominant grasses can recolonize the affected areas over time.

Figure 3.13 Prairie dog colonies in a temperate grassland environment. Cheyenne Mountain State Park, Colorado Springs, CO. *Source*: Eric Perramond. Used with permission.

A final factor that affects grasslands and their diversity is the increasingly studied role of **bioturbation**: the animal disturbance and mixing of soils. For the prairies and steppes of many regions, colonies of rodents such as prairie dogs have profoundly influenced the well-known richness of grassland soils. By creating large colonies, these rodents mix organic matter grazed at the surface with the mineral components of the surrounding soil (see Figure 3.13). The frequent establishment, and die-off, of animal colonies in these environments spreads the cumulative effects of bioturbation across large expanses of grassland. Like termite colonies in the tropics, prairie dogs shape the soils that humans later depend on. That role has diminished in historic settlement of grasslands, as humans have typically viewed these animals as pests (at best), or as dangerous to livestock at worst. Many countries have pursued systematic campaigns to eliminate ground rodents despite their important role in creating soil conditions attractive to human cultivation.

Grassland was perhaps the dominant land cover type prior to significant human modification. The perception of grasslands as difficult to use persisted well into the 19th century, but they are now the most transformed of Earth's biomes, and are undoubtedly the most valuable and productive agricultural land globally.

Temperate and deciduous forest

Temperate and deciduous forests are noted more for their range of seasonal leaf-losing vegetation and relatively milder climates. They are found on all continents except for Africa and Antarctica between the tropical latitudes and the harsher cold climates in the north and south. These forests typically receive between 750 and 1500 mm of rain per year, a large range, with a good percentage arriving in the form of snowfall and occasional ice storms. A key characteristic of the mid-latitudes (30° to 60° N, S) is that there are frequent alternating incursions or clashes of tropical and polar air masses and the distinct changes in season they cause.

The rhythm of temperature is more prominent than that of rainfall. Whereas seasons in the tropics are often characterized by alternating wet / dry periods, those in the mid-latitudes are characterized by temperature shifts from winter to summer. These regions are subject both to cold fronts from the Arctic and Antarctic as well as subtropical moisture incursions, creating a true mix of precipitation and temperature throughout the year. And this seasonal mix of temperature and precipitation causes is reflected in the vegetation characteristics.

The variability of temperature and seasonal conditions has created the adaptation of deciduous seasonality: deciduous species save their energy by losing leaves in the fall, shutting down their **photosynthesis**, or the conversion by plants of carbon dioxide (CO2) into plant sugars using sunlight. During the winter, deciduous species re-emerge as buds, and leaves reappear in the spring seasons. This has profound implications for the ability of forests to use, store, and transfer carbon in deciduous regions: since CO2 uptake by forests is only active for six to seven months, the vegetation is unable to constantly photosynthesize, unlike tropical forests. It should be noted that deciduous forests can have several species of conifer within the forest canopy, and thus they are also often called "mixed forests" in many regions. So patches of evergreen are clearly visible in many areas of deciduous forest.

While the total number of species in deciduous and temperate forests pales in comparison to the number in tropical locations, as the name implies, these forests are home to a number of both endemic species as well as generalists from other regions that have naturalized successfully in the forest. As a percentage of global land cover (vegetation), deciduous and temperate forests have seen the brunt of deforestation in recorded histories of many regions around the world. Temperate forest soils are generally supportive of cultivated agriculture and were thus extensively used in the historical development of many temperate regions of the world. For developed countries in western Europe, North America, and parts of Australia, the bulk of deforestation occurred well before 1900. In fact, for many regions, the regrowth of now secondary forest is notable for its density. Unlike tropical forests, which continued to decline as a percentage of cover in Africa, Asia, and Latin America, deciduous forest cover in North America and western Europe grew in the last 20 years of the 20th century (FAO 2001). Much of this has to do with the agricultural shift to former grasslands, where even more fertile soils were available to exploit.

High Latitudes

Boreal forest

Boreal forests are high-latitude (50° to 65° N) coniferous forests. In Russia, these areas are named "taiga." Boreal forest is found principally in the northern latitudes of North America and Eurasia; it forms a wide, nearly continuous belt in these two regions: from Alaska to Newfoundland, and from west of the Ural mountains all the way to the Kamchatka Peninsula in eastern Siberia, respectively.

Boreal forest ecosystems see long, bitter, cold winters with average temperatures hovering below the freezing point. They have remarkably short, cool summers, with little spring or fall transition periods before winter conditions arrive again. Forest cover is possible because of the mild summer temperatures, which can last four to six weeks, when soil moisture is thus easily accessible to plants, but anything but the hardiest pines and spruces has difficulty growing the remainder of the year. Precipitation is meager, varying between 20 and 200 cm, although is often higher in coastal areas. In general the air is too cold to hold enough moisture for precipitation. Much of the precipitation falls as snow, with a maximum in winter.

As is the case with tundra (see below), the lack of available moisture for plants is one of the principal challenges for growth. Water is rarely in liquid form for transpiration and plant use. Snow and ice may be a constant for nine to ten months, but solid water does little for plants until the summer thaw. Because of the continental effect, areas of boreal forest can feature the highest seasonal swings in temperature, ranges of anywhere over 30° C (86° F) between the lowest and highest annual temperatures, because of the sheer size of the continental land mass surrounding them (Canada, Russia). The data below display this wild variation in seasonal temperature swings:

Vevkoyansk, a weather station in Siberia

January temperatures: –50° C mean monthly average (–70° C for the coldest January)

July temperatures: –/+13° C mean monthly average (+ 32° C for the hottest July)

Boreal forest vegetation is quite simple and uniform, composed of about 75% conifer trees and 25% deciduous trees in homogenous stands that are slow-growing because of long winters. Trees are taller near the southern fringe, while there are fewer, stunted, trees near the northern edge. While low in overall diversity, even when compared to temperate forests, boreal forest has long been a source of industrial timber and local fuel wood, and serves multiple ecosystem functions for the colder regions. Since these regions are frigid for much of the year, the density of human population has been historically low. More recent disturbance has been triggered by widespread industrial deforestation in large concessions to private companies. A vegetation type common to the transition area between boreal forest and tundra is sometimes called **krummholz** ("crooked wood" in German) since wind and cold conditions stunt the tree forms to a gnarled, dense, appearance. High-altitude can substitute for high-latitude conditions, once again, and these gnarled, bent plant forms can be found in mountain regions.

Tundra

The tundra areas are composed of treeless plains, and are located north of the boreal forest in latitude and at high elevations in mountainous regions. Areas of tundra are typically found on the fringes of the Arctic Ocean in the north and a ring surrounding Antarctica, respectively north and south of boreal forest (see above). The 10° C isotherm identifies the limit of this zone, which is the same as the poleward limit for trees (the tree line). On the poleward edge, the limit of this climate is the 0° C isotherm for the warmest month. North of this limit is the permanent ice cap. With little available moisture for plants, harsh winds, and poor root access to nutrients, vegetation here is typically sparse, stunted, and small in stature. On mountains, one can find patches of tundra above the tree line, where conditions (slope, wind, soil, and temperature) conspire to make plant life difficult. Grasses, some mosses, and lichens are the dominant and most common form of plant life in both high-altitude and high-latitude tundra biomes.

In tundra, daily weather and longer-term climate are dominated by polar easterlies. Tundra is perhaps best thought of as a "cold desert." Low precipitation rates, a maximum of 250 mm, falling mostly in the high-sun (summer) period, and year-round freezing or frost conditions, combine to create brutal growing conditions for most vegetation forms. Insects and animals must also be able to tolerate these conditions. Much like boreal forest conditions, there are only one to four months of temperatures above freezing. Frost can occur at any time.

Permafrost is permanently frozen subsoil; only 8-30 cm (4-12 in) of the surface soil melts in summer. The presence of this frozen subsoil has much to do with the treeless aspect of tundra, since deep-rooted plants have little chance to establish any foothold. The landscape is water-starved and highly mobile. Soils, since they thaw and freeze seasonally near the surface, "starve" the larger plant forms of water when it is held in ice. The bulging movement of permafrost, when thawed in the summer, also favors shallow-rooting plant types, since they can accommodate the soil creep inherent in permafrost soils. Small trees that can survive are often bent, toppled, and at awkward angles because of this creep in the top active layer of normally frozen soils.

Both tundra and boreal forest are of increasing interest as scientists learn more about their important storage of greenhouse gases. Principal among these is the amount of methane hydrate (technically a "clathrate" combination of natural gas and water) held in ice storage. Since methane (CH_4) is a very effective warming gas, tracking even minor temperature changes in both of these biomes is of vital interest. As warming occurs, that methane trapped in ice air bubbles may release slowly or suddenly, and tundra and boreal forest may be significant new sources of methane.

Disturbance, Succession, and Change

Although critical factors determining the range of a given biome, such as climate, topography, and soil conditions, may be relatively stable over the medium term (say tens of years), this does not mean that vegetation patterns are stable or in equilibrium. Over long periods, climate change results in vegetation shifts. The most dramatic of these changes occur in response to the repeated advance and retreat of ice sheets in the so-called **glacial periods**, or **ice ages** as they are commonly known. The last 2.6 million years of Earth's history are known as the **Pleistocene** epoch, and during these last few million years the Earth has been witness to repeated glacial and interglacial periods. It should be noted that ice ages are not only colder but *drier* than **interglacial** periods such as those experienced today. We are currently in what is known to geologists as the **Holocene** epoch, equivalent to the last 12,000 years of Earth history (Roberts 1998), also an interglacial period. The cooling and drying of the climate during an ice age dramatically alters the kind of vegetation found in a particular place. Change is the norm, not the exception. This long-term perspective is necessary to understand the notion of **disturbance**, perturbations to natural systems that are usually short-lived, such as fires. It is also crucial for understanding longer cycles of climate change (see Chapter 8) and how scientists go about understanding the past through a variety of methods (see Chapter 5).

While biomes help us understand the relationships between climate and vegetation, they represent a rather static view of plant communities. As noted, biomes were based on the dominant vegetation type found in a geographical area which was often, but not always, thought to be the **climax community**. The climax pertains to the end-point vegetation community that theoretically perpetuates itself under prevailing climatic and soil conditions in the absence of disturbance. In the classic view of Clements (1916), a climax community is relatively stable and in steady-state equilibrium with the surrounding physical environment. We know, however, that the Earth's climate is not static. Climate changes, and therefore biomes must change as well. Thus, the concept of climax community is less useful today, but still prevalent in our discussions of environment, notions of wilderness (see Chapter 2), and even in natural resource policy language.

Many plant communities progress through a series of relatively recognizable and predictable changes in structure over time, resulting in a stable plant community. **Succession** is the concept that communities proceed through a predictable, step-by-step process of change over time. Through the process of succession, the original species colonizing a site may be replaced completely or they simply become less numerous as different species emerge.

Succession is generally divided into two types: primary and secondary. Primary succession is a progression that begins on bare surfaces such as rock that are totally lacking in organic material. Such conditions can occur following a volcanic eruption, the passing of a glacier or some other major geologic event. Primary succession gradually transforms the substrate into a soil that can support a living ecological community. Succession is driven by changes brought about to the environment by the organisms themselves making it suitable for different species to take over and dominate. Each step in the succession process is known as a **seral stage** (see Figure 3.14).

Secondary succession is a progression that begins with a disturbance that causes the destruction of, or damage to, an existing ecosystem but leaves much of the soil

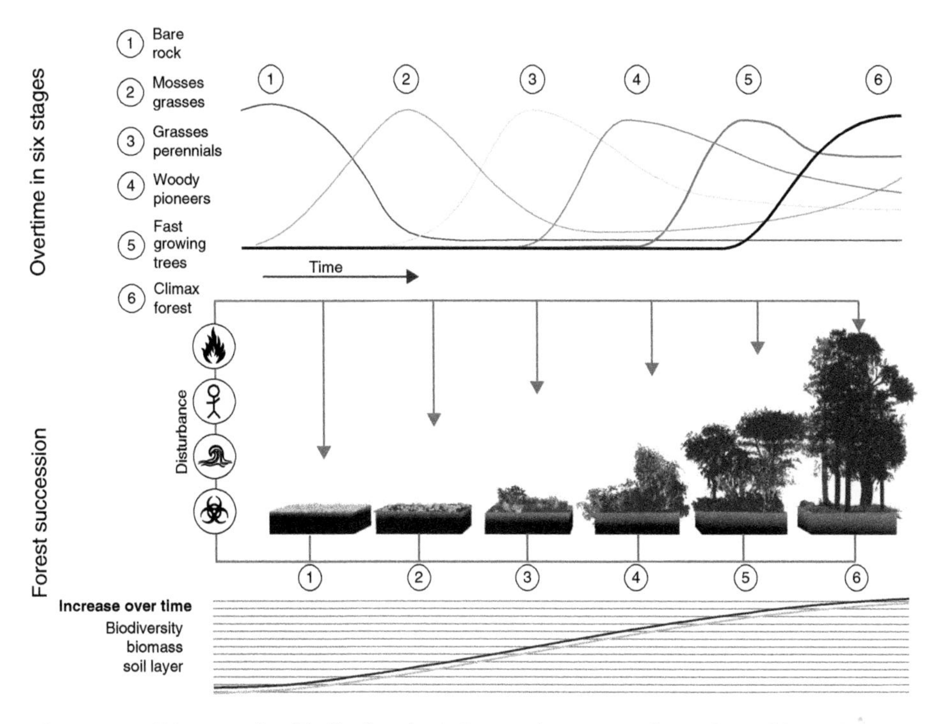

Figure 3.14 Diagram of an idealized ecological succession process. *Source*: Lucas Martin Frey, http://en.wikipedia.org/wiki/File:Forest_succession_depicted_over_time.png (accessed April 18, 2011). Licensed under Creative Commons Attribution 3.0 Unported.

and some organisms intact. Climate shifts such as ice ages can be thought of as major disturbances to ecosystems. Over shorter time spans, disturbances such as fire, flood, windstorm, or human activity can cause dramatic short-term changes to plant communities. Such disturbances shift succession processes back to earlier phases, and the process of succession begins anew.

In the idealized case, secondary succession results in a climax that is identical to the original. Each ecosystem is thought to have a single climax state based on unique climate and soil features of the site. As is the case with many **models**, the succession model is an oversimplification and the real world is more complex (Botkin 1990). In fact, secondary succession rarely produces a community identical to the original. In some cases, disease might wipe out a key species, while in others the type of disturbance dictates which species colonize a site first, influencing the entire process. In other ecosystems there does not seem to be any recognizable pattern to succession. It is also important to note that the only thing that differentiates a climax community from a successional one is the time scale over which change occurs. Climax communities also change, although not as rapidly as other seral stages, so ecologists and biogeographers often talk about multiple stable states instead of a single climax community. Despite these limitations to the ideal model of succession and climax, in many cases there is an identifiable, predictable pattern of change during succession, and the later stages in succession are more stable and longer-lasting than earlier ones. Notably, if succession is rapidly induced by humans (through deforestation), the capacity of the environment to produce a desirable climax community is reduced. So the scale and the pace of disturbance are both keys to understand what will happen to vegetation in any particular place.

Disturbances can be conceptualized in terms of regimes. A disturbance regime is defined in terms of timing, frequency, predictability, intensity, and extent of a regular pattern of disturbance. The disturbance regime of an area can have a dramatic effect on the types of organisms found there. The timing or seasonality of a disturbance influences its impact on vegetation. For example, a fire burning at the end of a long dry period when vegetation is moisture-sapped will burn hotter and more intensely than one following a rainy period. The impact of a single, infrequent but intense, disturbance that impacts a wide swath of a landscape, such as a volcanic eruption, will have very different effects than a series of relatively frequent, low-intensity disturbances that impact smaller portions of the landscape.

Disturbance regimes, especially those with low intensity, often have uneven impacts on the landscape, creating patterns. Even micro-scale disturbances, such as the death of a single tree in the forest, can play a critical role by opening up space and light in a closed canopy. Periodic disturbances may be sufficiently frequent and severe so as to prevent the expansion of some species or communities into areas where they could otherwise survive. In Yosemite Valley of California, for example, Native Americans historically burned the valley floor to prevent the encroachment of trees. Later, under park management, fire was suppressed and trees took over some of the meadows. Most recently, park managers have reintroduced burning in the valley in an effort to recreate the landscape as it would have appeared during the time it was occupied by Native Americans (see Figure 3.15).

Figure 3.15 Prescribed burn in Yosemite Valley, 2003. *Source*: Paul Laris. Used with permission.

Disturbance: Questioning the Human/Natural

As can be seen in the example above from Yosemite, disturbances can be either natural, anthropogenic (human-caused), or both. Fire, for example can be caused naturally by lightning, but throughout large areas of the world people have modified the natural fire regime by either increasing the frequency of fires (such as in the chaparral of California or the African savanna) or decreasing the frequency, such as in the ponderosa pine forests of the American West. Species often adapt to a specific disturbance regime; either increasing or decreasing the frequency and intensity of fire can critically impact certain species. In many areas of the American West, for example, fire suppression has resulted in the buildup of flammable fuels, creating an extreme fire hazard. When these fires occur in these environments, they can burn with high severity, causing damage to species adapted to high-frequency, lower-severity fires.

Although disturbances cause the partial destruction of ecosystems and can result in the temporary elimination of some species, other species are entirely dependent upon disturbance for their existence. In southern California, for example, numerous wildflowers sprout only following a severe fire event. These plants live for only a short time, sometimes a single season, but during this time they produce seeds that fall to the ground to await the next disturbance event. These plants may or may not serve an important successionary function, but they are certainly a sight to be seen when in full bloom.

Key Ecological Concepts

As we focus in and move down in scale from the biome to the landscape and finally to the individual species, we see that different factors become important for explaining patterns at different scales. Geographers and ecologists refer to this as **scale dependency**, meaning that certain variables or factors appear to have a greater effect on vegetation formations depending upon the scale of analysis. As we have already seen, at the global and regional scale, global climate patterns, elevation differences, and distance to water bodies play important roles, but at the landscape scale one begins to see the effects of such factors as topography, hydrology, and soil structure. **Topography** can be thought of as the description of shapes and features on the landscape, especially those related to changes in **relief** (the relative difference in elevation between locations). Two key elements of topography are **slope** and **aspect**. The slope is the gradient of a hill or mountain and describes its steepness. Aspect refers to the direction toward which a hill or mountain slope faces.

In areas outside of the tropics, one finds that the north-facing slopes of hillsides or mountains are often composed of different plant communities than those found on south-facing slopes. In the northern hemisphere, for example, south-facing slopes receive more direct sunlight and are thus warmer and drier than north-facing slopes (see Figure 3.16).

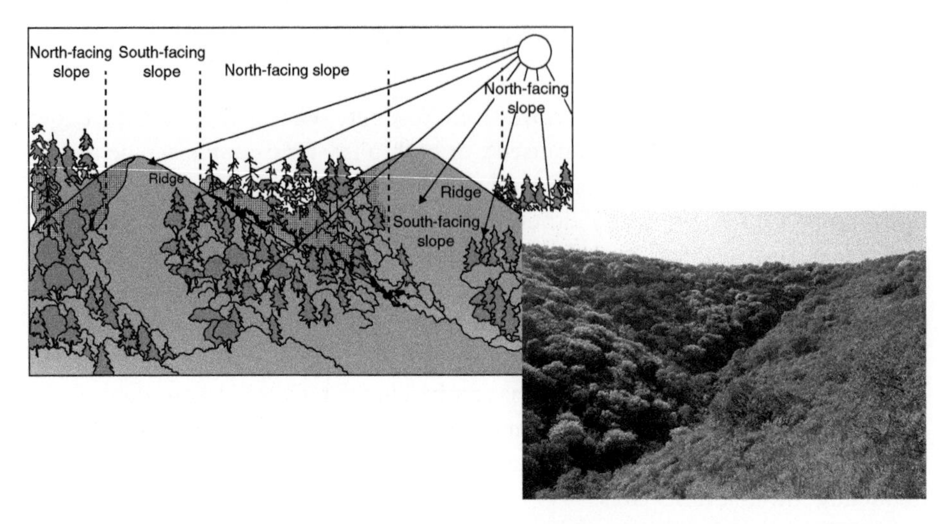

Figure 3.16 The effects of topography on solar incidence and subsequent vegetation cover. Note that the southern faces of slopes have far less vegetation cover due to the drying influence of direct solar incidence. *Source*: Photo by Noah Elhardt/Creative Commons.

Limiting Factors and Niche

Zooming in even closer, down to the level of the individual organism, we find that additional factors come into play, such as limiting factors and species interactions. **Limiting factors** are those that restrict the success of particular individuals in a larger biological population. Long ago, Shelford (1913) recognized that an organism performs best between limits of certain environmental factors called a **range of tolerance**. Above and below this range the performance of an organism is reduced to a point where it cannot survive. The relationship between performance and an environmental factor often takes shape of a bell curve (see Figure 3.17). At certain limits there is a zone of intolerance where individuals can survive, but growth or reproductive performance is impaired. Each species may exhibit a range of tolerance for a number of limiting factors. Although a given organism interacts with its environment in numerous ways, certain factors may be more critical than others to a particular species' success. For example, many plants are limited by the scarcity of water and others cannot survive where temperatures are too low. The fact that organisms perform best under a specific set of environmental conditions should come as no surprise to us. Just think about how few world records are broken at track meets when the weather is very cold or very hot.

Although a single factor, such as freezing temperature, can limit the range of biological individuals (the tree line on a mountain ridge is a good example), the range of most species is governed by a complex mix of factors. If we mapped an "n" dimensional space for a species based on many factors including moisture, temperature (low and high), light, nutrients, etc., we could explain quite a bit

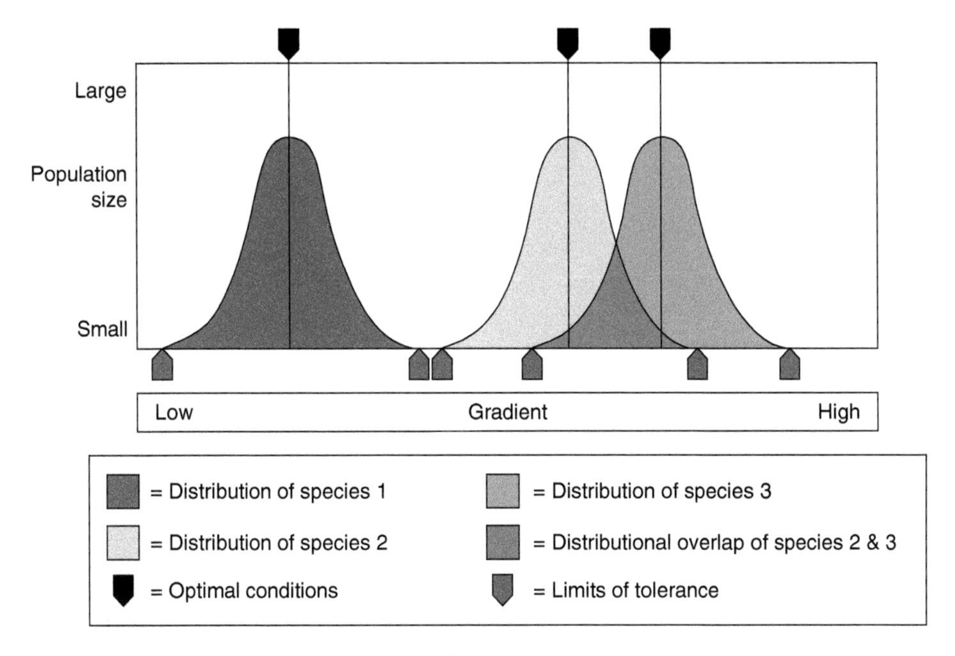

Figure 3.17 Overlapping ranges of tolerance for different species. *Source*: http://www.pc.maricopa.edu/Biology/ppepe/BIO145%20Black%20Board/notes/note08_1.html.

about the distribution of that species. Indeed Figure 3.17 represents the overlapping ranges of several **niches**. The niche is defined as the total physical environmental requirements (or limitations) of a population or species. Hutchinson (1957) developed this concept to help us conceptualize how environmental conditions limit the abundance and distribution of species.

Some species exist under a very narrow range of environmental conditions (imagine squeezing the curves on both sides). These are called **specialists** because they have very particular niche requirements. Those that exist under a wide range of conditions are referred to as **generalists**. The spotted owl, found only in the old-growth forests of the Pacific Northwest, is a good example of a specialist, while the dandelion is a great example of a generalist as it is found on every continent (except perhaps Antarctica) and in many different environments. Whether a species is a generalist or a specialist affects its ability to cope with changing environmental conditions, including climate change.

Interactions: Competition, Predation, Mutualism

While the limitations imposed by physical environmental factors have important impacts on geographic patterns, in many cases these factors alone do not determine a species range. By focusing in at the level of the individual we see how the interactions between species and even between individuals of the same species affect species distribution. There are three principal ways in which species interact: competition, predation, and mutualism.

Competition is the mutually detrimental interaction between individuals in which individuals compete over a limited resource. Competition may be interspecific (between individuals of different species) or intraspecific (between individuals of the same species). Competition affects the range in which a given species is found. We can see the impacts that competition (or the lack of it) have by visiting a botanical garden, where one can find many more species in an area than would naturally occur. A botanical garden must be carefully maintained to prevent competition between species. If abandoned, the biodiversity of the garden would decline because species would interact and some would out-compete others for resources, causing their extinction from the garden. Optimal ranges of species (Figure 3.17) and overlapping competition for resources help to explain patterns found in nature's geography.

Predation is any interspecific interaction in which one species benefits and another suffers. Examples of predator–prey relationships include the relationship between a lion and a zebra, between an herbivore and the grasses it consumes, or between a parasite and a host, such as the case of mosquitoes and humans. Predator–prey interactions can limit the distribution of species. This relationship works both ways. The range of a predator can be limited by that of its prey, especially when a predator relies on a limited variety of prey. Interesting examples of predators limiting the range of prey have resulted following the artificial (human-caused) introduction of prey populations into new areas. For example, following the introduction of the Nile perch into Lake Victoria in East Africa many of the endemic species of cichlid declined.

Mutualism is interspecific interaction between at least two species in which each species benefits the other. A good example of mutualism is the relationship between plants and animal pollinators such as flowers and the honey bee. In many cases of mutualism the relationship is obligatory in the sense that neither species can exist without the other. In others, the species involved can exist alone but they are more successful with the help of mutualism.

Finally, species known as the **keystone species** play a critical role in maintaining entire ecosystems. Keystone species can limit the distribution and community structure of many species in an ecosystem through direct or indirect effects. For example, the American bison has a dramatic influence on the structure of the tall grass prairie. Not only do bison consume prairie grasses, they also trample and modify the grassland, creating a heterogeneous patchwork. Areas that have been eaten and/or trampled cannot support fire the way untrammeled patches can. Thus, a herd of bison can modify the disturbance regime a large landscape. If the buffalo were to be removed, the disturbance regime would be altered and the species content and structure of the ecosystem would change. The concept of the keystone species is useful to resource managers because it illustrates that all species cannot be treated equally: some are more critical to the persistence of an entire ecosystem than others.

Chapter Summary: Linking Back to Human–Environment Relations

We now recognize that our environment is subject to periods of stability amidst constant change. Depending on the period or spatial scale of analysis, the same area may appear to be stable or undergoing change. How we think about stability and change shapes our understanding of the human–environment relationship (Kricher 2009). Both perspectives offer us insights into how the Earth's different biophysical environments are maintained and are modified through time. Early models of the biophysical world were built upon key concepts of stability and equilibrium. This way of viewing the world increased our understanding of how ecosystems work and why we find particular environments where we do on the Earth's surface (the Koeppen map of biomes is a great example). It is important to note, however, that this view aided in developing an ethic for the appropriate role of humans in the environment. Stability was often thought to be natural and desirable, while change was viewed as unnatural and the result of human action and the cause of degradation. The lesson for humanity from this early view of ecology was to strive for balance and equilibrium because that was what was thought to be natural.

More recently, as the environmental sciences have matured, we have been forced to develop more subtle and sophisticated ways of understanding the human role in terms of environmental change as well as stability (see Botkin 1990, for example). It is clear that *the environment is always changing*, and while we consider some changes to be benign or even necessary, such as evolutionary change, other changes are negative, especially if they are too rapid, intense, and/or extensive. So while we know that there are natural rates of climate change and extinction, humans are speeding up these rates to unprecedented levels and the potential for catastrophic damage is increasing (see Chapters 8 and 11 on climate change and biodiversity, respectively). This also means that pure "physical geography" concerns are of importance to policymakers. For example, there is an entirely new set of debates going on right now on the proper role and agency for so-called Earth System Governance (see http://www.earthsystemgovernance.org/) and how we can make appropriate and transparent decisions based on our best understanding of what is happening to the planet.

The shift in thinking has led geographers to focus on change and to put the emphasis on how the natural world is always in flux, unstable, and even chaotic. Equilibrium and stability may be neither possible nor desirable in most situations. The lessons for how we should think about the human–environment relationship from this newer perspective are less clear-cut and more challenging. But as you read the remaining assigned chapters, try to think back to these general principles as a template for thinking about difficult questions and alternative perspectives. Scholars treat the role of the physical environment differently, some sounding deterministic (we can't escape the confines, it makes us who we are) while others are far more relativist in approach (the environment has nothing to do with what we become). Our opening example of the dark earths of Amazonia shows that nothing is preconditioned or preordained, but these soils also illustrate that we understand poorly our own past environmental histories (Chapter 5) and the confines of our human–environment imaginations (Chapter 2). Read carefully, and think about your own response and position within these debates.

Critical Questions

1. Why is it so difficult to separate or distinguish between "natural" and "human" causes of environmental change? Name one example where biophysical change is hard to attribute to either side of the human–environment coin.
2. After reading this chapter, what factors or "drivers" of biophysical change seem to be the most important at the global scale? Think about a local place: is the same factor the dominant aspect of what determines that place?
3. What biome are you currently living in, and what kind of ecological disturbances are common nearby? How does this biophysical setting matter to the local human–environment relationships?
4. Do humans occupy a narrow niche like many other biological species? What is our relationship with other species based on the concepts of competition, cooperation, mutualism, dependence, and predation?
5. Is the "keystone species" a useful concept in your opinion? Can you think of how they keystone species would vary for any given area depending on the scale of analysis?

Key Vocabulary

abiotic and biotic
albedo
angle of incidence
anthropogenic
arid and semi-arid
aspect
atmosphere
biogeography
biome
biosphere
bioturbation (of soils)
climax community
community
competition
condensation
climate and weather
closed (forest) canopy
deciduous (forest)
dew point
disturbance regime
ecology
ecosystem
ecotone
environmental lapse rate
equinoxes
evapotranspiration
generalists
glacial (period)
global carbon cycle
Hadley cell circulation

Holocene (epoch)
hydrosphere
ice ages
interglacial (period)
Intertropical Convergence Zone (ITCZ)
keystone species
krummholz
landforms
leeward and windward
life layer
limiting factors
lithosphere
mesic (forest)
models
mutualism
niche
orographic effect
permafrost
photosynthesis
physical geography
Pleistocene (epoch)
predation
rain shadow
range of tolerance
relief
riparian (forest)
salinization
scale dependency
sclerophyllous
seasons
seral stage
slope
solar radiation
spatial scale
specialists
specific heat
stability and change
succession (primary, secondary)
temperature regimes
terra preta
topography
transpiration
urban heat island effect
xerophytes

References

Balée, W. and Erickson, C.L. (eds.) (2006) *Time and Complexity in Historical Ecology, Studies in the Neotropical Lowlands* (New York: Columbia University Press).

Botkin, D.B. (1990) *Discordant Harmonies: A New Ecology for the 21st Century* (New York: Oxford University Press).

Clements, F.E. (1916) *Plant Succession: An Analysis of the Development of Vegetation* (Washington, D.C.: Carnegie Institution of Washington).

FAO (2001) *Global Forest Resources Assessment 2000: Main Report*. FAO Forestry Paper No. 140 (Rome). Also available at www.fao.org/forestry/fo/fra/main/index.jsp.

Glaser, B. and Woods, W.I. (eds.) (2004) *Amazonian Dark Earths: Explorations in Space and Time* (Berlin: Springer).

Goudie, A. (2005) *The Human Impact on the Natural Environment: Past, Present, and Future*, 6th edn. (New York: Wiley-Blackwell).

Goudie, A. and Viles, H. (2010) *Landscapes and Geomorphology: A Very Short Introduction* (New York: Oxford University Press).

Hutchinson, G.E. (1957) Concluding remarks. Cold Spring Harbor Symposia on Quantitative Biology, 22(2), pp. 415–27.

Kricher, J. (2009) *The Balance of Nature: Ecology's Enduring Myth* (Princeton, N.J.: Princeton University Press).

McDonald, G. 2003. *Biogeography: Introduction to Space, Time and Life* (New York: Wiley & Sons).

Roberts, N. (1998) *The Holocene: An Environmental History* (Oxford: Blackwell).

Shelford, V.E. (1913). *Animal Communities in North America* (Chicago: University of Chicago Press).

Strahler, A. (2011) *Introducing Physical Geography*, 5th edn. (New York: Wiley & Sons).

Part II

Contemporary Perspectives in Human–Environment Geography

4

Cultural and Political Ecology

Local Human–Environment Interactions in a Global Context

Icebreaker: A Farmer in Her Field

Imagine a poor farmer somewhere in the world trying to feed his or her family (see Figure 4.01). He or she works long hours every day tilling a small plot of marginal land just trying to make ends meet. The household cultivates a variety of crops on its plot using simple, handheld tools. The farmer also relies on the neighboring forest for a number of products, from firewood to wild leaves and berries. Coming up short, this farmer may head into the woods from time to time to hunt an animal or harvest a few trees that bring in extra

An Introduction to Human–Environment Geography: Local Dynamics and Global Processes,
First Edition. William G. Moseley, Eric Perramond, Holly M. Hapke and Paul Laris.

Published 2014 by John Wiley & Sons, Ltd.

Figure 4.01 A female farmer working her peanut field in southern Mali. *Source*: W.G. Moseley. Used with permission.

income. This household, with several children, only seems to grow poorer over time. In assessing this situation, resource managers and development experts undertaking conventional analysis often arrive at the following conclusion. First, the household is using primitive farming techniques which lead to limited productivity and are degrading the soil. Second, its poverty is both a product of backward farming practices and of having too many children. Third, this poverty is a problem because it is driving environmental degradation, such as the overuse of soils, wildlife, or forests. Finally, this situation is best addressed by introducing modern farming practices to increase food production, family planning to limit the number of children, and an emphasis on cash crops that will provide the income necessary to reduce poverty and the necessity to overuse natural resources.

The diagnosis of this situation may or may not be accurate. Cultural and political ecologists ask us to consider other factors and ponder further questions. It may be that the household's farming practices are mischaracterized as inefficient. Perhaps, for example, the corn (or maize) yield is low relative to a field of mono-cropped (or single-cropped) plot of hybrid corn. However, the standard analysis often ignores all of the other crops this farmer may be growing in his or her field with the corn, not to mention the additional cost of inputs associated with growing hybrid corn (such as the cost of seed, fertilizer, and pesticides). The farmer may be using a hand-held hoe because heavier mechanized plows are not just expensive to purchase, operate, and maintain, but because they destroy shallow tropical soils over time. The causes of poverty in this context are also unclear and may or may not have any relation to the conventional diagnosis. The assumption that population is a cause of poverty rests on a one-sided assessment which emphasizes the cost of raising an additional child, with little or no attention to the economic benefits of doing so. The cost of raising children varies greatly from context to context, as do the benefits. Given that agriculture is often a very labor-intensive endeavor, many rural households need the labor of children to help run the farm. In fact, surveys of farming households in rural areas of the Global South reveal that larger households tend to be wealthier. As discussed above, the estimates of this household's agricultural productivity may be artificially low as well. A broader question to consider is why this household is on a small, marginal plot to begin with. In many instances, poor farmers have been displaced onto less productive lands by powerful interests. As such, the marginal character of their land is not a product of their own doing,

It is also questionable whether poverty is driving environmental degradation in this instance. In this case, poverty is – at best – only the immediate cause of environmental degradation. This poverty may be a result of longer-term factors such as the displacement of the farmer onto more marginal land (as discussed) or the low prices offered for agricultural crops. The decision to go into the forest and cut down a tree may also be tied to structural factors. There is a global market for tropical hardwoods such as mahogany. Without high external demand for tropical timber, there may not have been a decision to harvest wood as an income-generating strategy. Finally, the policy remedies put forth to address this household's poverty may have limited effectiveness or even be detrimental. Modern farming approaches are often prohibitively expensive for poorer farmers (leading to indebtedness) and may have unintended consequences on the environment. The increased availability of modern family-planning approaches is good if it is desired by the family – and it will help improve child and maternal health. It is not clear, however, whether the introduction of these practices will help reduce poverty. Finally, a greater emphasis on cash crops may not reduce poverty and it could be deleterious for food security. In sum, a story of environmental degradation begins to change when we approach it with a fresh set of questions provided by cultural and political ecology.

Chapter Objectives

The objectives of this chapter are for the student to be able to understand and critically discuss:

1. The conditions which gave rise to cultural ecology.
2. The key elements of the cultural ecological lens for understanding human–environment relations.
3. The major critiques of cultural ecology.
4. Key theories from development studies which contributed to the emerging field of political ecology.
5. Key themes or approaches within political ecology.
6. The major critiques of political ecology.

Introduction

This is the first of four chapters which introduce students to different theoretical traditions within geography's human–environment tradition. In exploring cultural and political ecology, we examine the intellectual and societal conditions which fostered the rise of **cultural ecology** and the contributions of this interdisciplinary subfield. We then look at radical development geography and how it mixed with cultural ecology to produce political ecology. The chapter ends with an exploration of the major themes within political ecology and subsequent critiques of this approach.

Setting the Stage for Cultural Ecology

The 1950s and 1960s represent a period of rapid change for environmental thinking in the Global North and the Global South. In the case of the former, the immediate post-World War II period in North America and Europe was a period of great optimism. This optimism was often characterized by a belief in the ability of modern technology to solve humanity's problems. Many of the innovations developed during the war years were now being applied to civilian uses, including advances in engineering and chemistry. One of the best examples of this was DDT, a pesticide developed for use to control mosquitoes (the major vector for malaria) during fighting in the Pacific. DDT was found to be extraordinarily effective and killing and controlling all types of insects. As such, it began to be widely used in US agriculture for pests or for mosquito control in urban communities.

Americans' trust in technology, and modernist development more generally, began to falter in the 1960s. A seminal publication, which would sow the seeds for the contemporary environmental movement, was Rachel Carson's *Silent Spring* (1962). In this text, Carson began to raise concerns about increasing use of pesticides in US agriculture. In very accessible language, she described the many problems with pesticides, most notably, their ability to bioaccumulate in animals, and biomagnify as one moved up the food chain (see Chapter 10). This problem not only silenced the birds (hence the title of the book), but created a whole host of problems for other animals at the top of the food chain. While Carson was writing about a particular issue, the broader impact of her work was to raise questions about the modernist path to development taken in the Global North and its concomitant technological innovations. Perhaps technology could not solve all of humanity's problems? Maybe there was a dark side to this particular development trajectory which had not been fully explored?

If we now turn to the situation in the Global South at this time, many countries in the tropics had recently become independent (especially in Asia and Africa), ending seven to eight decades of European colonial rule. A major aspect of colonialism was the control and extraction of natural resources for the benefit of the colonial power.

These could be mineral and energy resources, or agricultural commodities. Getting colonies to produce these commodities, and setting up mining operations, was often a long and brutal process which involved the wholesale reorganization of economies away from production for subsistence and local markets towards production for European and global markets. What is often less understood is that this massive reorganization of tropical economies was often accompanied by a whole series of justifications provided by European natural resource managers and administrators. As such, European agricultural and forestry experts critiqued most indigenous natural resources management practices as backward, primitive, inefficient, and wasteful. For example, the traditional practices of mixing crops or leaving trees in fields (polyculture and agroforestry), both of which are very common in the tropics, were negatively portrayed (see Chapter 10). Furthermore, production for subsistence was consistently devalued in comparison to production for the market. In 1925, a white South African economist wrote:

> Three-quarters of the population is in the process of emerging from a primitive culture in which the distribution of labour between different uses is effected not by the pecuniary [or financial] incentive upon which a capitalist economy chiefly relies, but by totally different arrangements. Among these natural obstacles are traditional attitudes of Natives towards land and cattle; conservatism; family ties; ignorance of conditions prevailing or opportunities offering. (quoted in Bundy 1979: 3)

In many instances, these negative reviews of traditional agricultural and forestry practices were used to justify colonial efforts to take land away from local people or to impose new farming regimes. Local people lost land when state forests or game reserves were created. They also were forced to cede land to European settlers. While we often think of colonialism as a form of political domination, it was also a form of ideological control as the aforementioned European ideas about traditional management strategies in the tropics were often internalized by local people. Over time, local people sometimes came to believe that their management strategies were backward and primitive. Decolonization would not only create a space for political change but it would create a space for, as the Kenyan author Ngũgĩ wa Thiong'o (1986) would say, the decolonization of the mind.

Cultural Ecology

It is out of this intellectual environment (a questioning of modernization in an era of decolonization) that the interdisciplinary subfield of cultural ecology emerged in the 1960s and 1970s. These scholars had a preference for studying small-scale subsistence societies in the Global South. Geographers and anthropologists were probably the most active in this area – but other disciplines were contributing as well.

Cultural ecologists did at least two things which were important for advancing the understanding of tropical agriculture and natural resource management practices. First, they set out to study these systems on their own merits, often revealing that these approaches were ecologically and economically rational. Second, these scholars were not afraid to question so-called modern practices (often originating in the temperate zone) to determine if these really represented an advancement over practices developed locally.

There are at least four core themes that appear repeatedly within cultural ecology scholarship. First, society and nature are seen as intimately connected. Particular attention is given to how people manage resources via a range of strategies in regard to diet, technology, reproduction, and settlement. The variability of the biophysical environment (in time and space) is also an integral component of all such discussions, as is the role of environmental constraints.

Second, cultural behavior is often considered in its functional role. As such, there might be a traditional cultural practice, such as a prohibition on eating pork, which historically was grounded in the fact that there was a real risk of contracting trichinosis from consuming undercooked pork. In other words, whether or not contemporary individuals were aware of a biophysical rationale for why they engaged in certain practices, cultural ecologists sought to understand if there was a functional explanation for them. For these scholars, it was important to understand if there were biophysical reasons why people did something, rather than it just being a traditional or cultural practice. Comprehensive studies were undertaken to gather empirical data on this behavior. Time and energy studies were very common, and one particular version of this type of study was known as *cybernetics*. In cybernetics, human cultural systems essentially are treated as closed ecosystems. Energy transfers are carefully measured within the system and there is an assumption that an equilibrium or steady-state is the ideal. The classic example of this type of study is Roy Rappaport's *Pigs for the Ancestors: Ritual in the Ecology of a New Guinea People* (1968) – see Box 4.01.

The third and fourth themes that are quite prominent within the field are an interest in food production and in population dynamics (as well as how these two variables are interrelated). For example, a number of cultural ecologists were keenly interested in how growing population densities influence the approach to food production. These scholars were inspired by a long-running debate in population studies. Since 1798, Thomas Malthus had argued that food production essentially controlled the level of one's population. The level of food production was determined exogenously, and if an area's population exceeded this capacity then there would be a natural check on famine. Then, Esther Boserup published the *Conditions of Agricultural Growth* (1965) arguing the complete opposite, that it was population which determined the level of food production. Using historical evidence, she documented how increasing population densities led to agricultural intensification and greater levels of food production. Turner, Hyden, and Kates (1993) undertook a series of case studies around Africa to test the Boserupian versus Malthusian hypotheses. Mortimore and Tiffen (1994) carried out a famous case study in Machakos, Kenya, which showed how increasing population had led to higher levels of environmental stewardship and greater food production.

Box 4.01 A seminal work in cultural ecology: Roy Rappaport's *Pigs for the Ancestors: Ritual in the Ecology of a New Guinea People* (1968)

Roy Rappaport, an anthropologist at the University of Michigan, was a leading figure in cultural ecology. Rappaport mainly worked in the Southeast Asian nation of Papua New Guinea, a country which only recently gained independence from Australia, in 1975. This nation was a hotbed for cultural ecology research in the 1960s and 1970s, and a place of frequent collaboration between geographers and anthropologists. Other prominent cultural ecologists working in the area at the time included Harold Brookfield (1962) and Larry Grossman (1981). Papua New Guinea was appealing to cultural ecologists who had a preference for studying small-scale subsistence societies. Papua New Guinea is one of the most culturally diverse countries on the planet, with over 850 indigenous languages out of a population of just under 7 million. The country has a rather unique physical landscape (covering half of a large island it shares with Indonesia) which lends itself to the development of diverse cultural systems. Most of the country's inland areas are highly inaccessible and mountainous. At the time Rappaport was doing his research, the majority of the inland human communities lived in a series of valleys which were separated by ridges. This highly segmented, and not easily traversed, landscape lent itself to a patchwork of relatively isolated agricultural communities and a plethora of languages and dialects which had evolved somewhat independently. Rappaport was studying a Maring-speaking people known as the Tsembaga (numbering 204 people), who farmed tubers and raised pigs. He was particularly interested in this group's ritual slaughter of pigs. What Rappaport came to argue, through the careful documentation of energy inputs and outputs (e.g., caloric measurements of human labor expended, crops produced, pigs raised), was that the ritual slaughter of pigs kept the system in equilibrium. The substance of his argument was that pig populations would build up over time and put pressure on land and farming systems in the community. This might eventually lead to warfare with a neighboring community. While this warfare would eventually produce a victor, a key element to putting this fighting to rest was a massive, ritual slaughter of pigs. This slaughter would serve to redistribute calories within the community in the short term and relieve pressure on the system as a whole over the medium term.

In addition to the different themes pursued by cultural ecologists, scholarship in the field may also be broken out into the more contemporary developmental (or synchronic) cultural ecology and the archeology-based historical, or diachronic, cultural ecology. Developmental cultural ecologists studied contemporary situations, seeking to document local or indigenous knowledge as well as processes of adaptation. In *Indigenous Agricultural Revolution* (1985), for example, Paul Richards argued that local farmers are "scientists" who draw on their knowledge of ecology in a particular place and constantly experiment. He suggested that development projects and programs would be much more effective if, in addition to drawing on outside expertise, they worked with local people and incorporated such understanding into their work. Cultural ecologists were keenly interested in studying how livelihood systems adapted to environmental stress. Their findings contradicted an older notion that such systems were static or incapable of dealing with change. Denevan (1983) argued, for example, that lots of potential **adaptations** existed within any given livelihood system (almost like recessive genes) as there were always individuals in the group who would undertake slightly different approaches to farming. These approaches could then be drawn upon by larger numbers of people during times of stress. Historical cultural ecologists have studied past civilizations in order to see how they have dealt with environmental change and managed natural resources (e.g., Butzer 1990). This work has, for example, shown us how Mesoamericans were able to support much denser populations than live in the area today.

Watts (1983) and others began to critique cultural ecology in the early 1980s. The main concern was that the field did not take account of broader-scale political economy. By analyzing situations in a vacuum, Watts argued, cultural ecologists were missing the root cause of many environmental problems.

The Emergence of Political Ecology

Within geography, after a critique of cultural ecology by Watts (1983), a new trajectory developed in the mid-1980s which came to be known as **political ecology**. Within geography, the geomorphologist Piers Blaikie is generally acknowledged to have first described this approach as "the political economy of human–environment interactions" in his seminal work *The Political Economy of Soil Erosion* (1985).[1] Tom Bassett is the first to have used the actual term "political ecology" in a geographical publication (1988).[2] What is important to remember is that early political ecologists were not abandoning cultural ecology, but building on it. Political ecologists would use many of the same approaches and concepts to understand local-level human–environment interactions. The difference, or innovation, was to situate these local-level dynamics in a broader web of provincial, national, regional, and international forces (see Figure 4.02). As Blaikie notes, his analysis would start with the land manager, and then move out from there to consider the potential influence of political economy. These early political ecologists drew heavily on concepts from radical development geography (and related thinking in the social sciences more generally) to broaden their scope of analysis. As such, it is now necessary to take a foray into development theory in order to understand early political ecology.

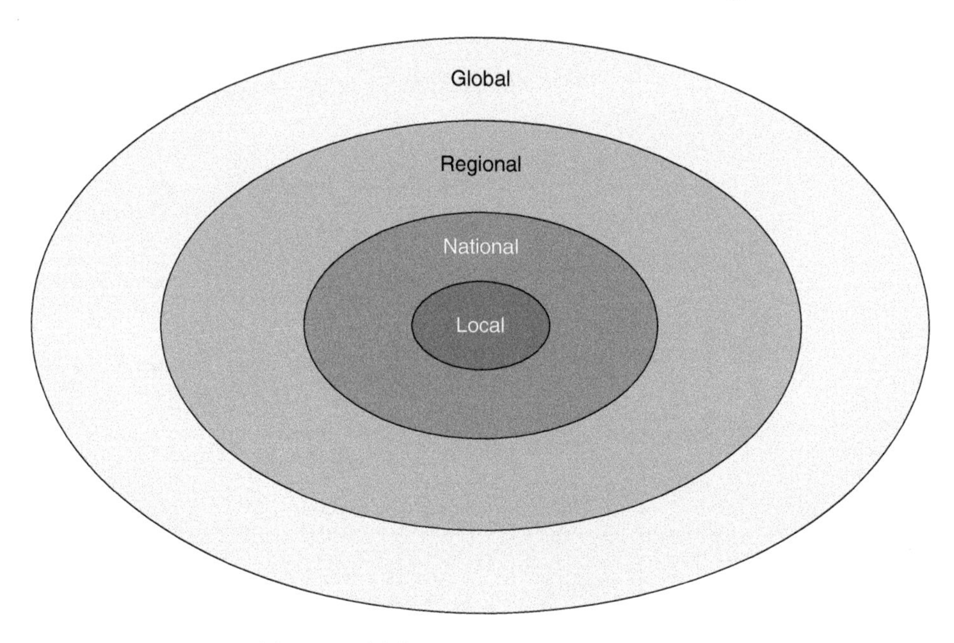

Figure 4.02 Conceptual diagram of different scales of analysis.

Linkages to Development Studies

While it was the more structural approaches to development which would influence the early political ecology of the 1980s, these (structural) approaches were a reaction to a suite of development ideas collectively referred to as **modernization theory** which prevailed in the first decades following World War II. Modernization theorists posited that the European industrial economy was the ideal or pinnacle state of development. These theorists argued that with the right combination of capital, know-how, and attitude, economic growth would proceed. Countries would make the transition from a traditional to a modern state. While these theories were most popular in economics, they did influence thinking in other social sciences. Rostow's stages of economic growth is perhaps the best-known of these modernization theories within economics (1959). Rostow was a macroeconomist writing in the 1950s. He posited that a country went through several stages, starting with a traditional agricultural economy (see Figure 4.03). As agriculture was commercialized, a surplus of funds would accumulate. This surplus was critical investment capital for a take-off phase in which a country begins industrializing. The subsequent stages of increasing industrialization would then follow suit. The metaphor was an airplane lumbering down the runway for take off. The biggest challenge was actually gathering enough speed to get off the ground.

The practical implication of modernization theory in many areas of the tropics was a big push to industrialize agriculture and to build infrastructure such as dams and roads. The idea was that such big infrastructure investments – especially dams – could jump start an economy and put it on the path to industrialization by providing

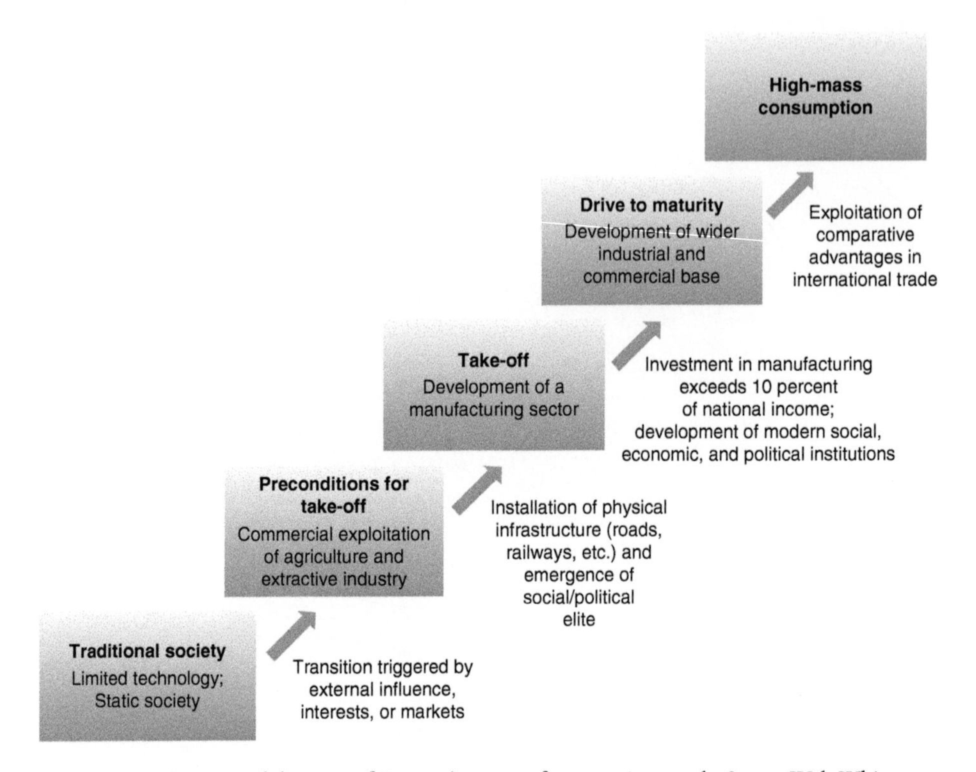

Figure 4.03 Conceptual diagram of Rostow's stages of economic growth. *Source*: Walt Whitman Rostow, *The Stages of Economic Growth* (Cambridge: Cambridge University Press, 1960). Reprinted by permission of Cambridge University Press.

irrigation for commercial agriculture and cheap electricity for manufacturing. Roads could open up new areas of a country and connect them to the national market economy. Modernization theory employed a "pull yourself up by the bootstraps" type of philosophy. If a country adopted the right type of policies, and made the right type of investments, then it would develop. While such autonomy was conceptually appealing because of its tidiness, many argued that it was not the way development occurred in the real world. Multiple critiques of modernization theory emerged in subsequent decades, including structuralist theories of development.

In reaction to modernization theory, a new set of development ideas began to arise in the 1960s which emphasized structure, or the global framework under which countries operated. This emphasis on structure meant that the relationships between countries were as important as, or more important than, internal policies for determining the future development of a country. **Dependency theory**, originally conceived by Frank (1979), is one example of such a theory. This approach suggested that economies in the tropics (Africa, Latin America, Asia) were essentially "underdeveloped" during the colonial era as European countries refashioned these economies (through a combination of taxation policies and forced coercion) for their own benefit. According to Frank, elites in developing countries often colluded with the colonizers to organize primary extraction in exchange for wealth and power. In many instances, farmers in tropical countries were encouraged or forced to produce commodity crops for the European market (often at the expense of subsistence production) and/or pushed to leave their own farms to pursue wage labor on large plantations or in emerging mining sectors. These primary commodities from the tropics were traded for manufactured goods from Europe and other industrialized countries in a global system of unfair exchange that favored Europeans.

Wallerstein (1979) expanded on Frank's ideas though his **world systems theory** which basically gave a spatial face to dependency theory by depicting the world in terms of the core, semi-periphery and periphery (see Figure 4.04). Under this schema, the core countries represent the most developed nations which are dominated by industry, financial services, and an information-based economy. The semi-periphery is an emerging group of nations where high levels of cheap manufacturing increasingly take place. Finally, the periphery represents those nations whose primary role in the global economy is to provide raw materials. All three of these regions operate as a world system, with deep connections existing between each sphere. While Figure 4.04 is schematic in nature, it suggests that most nations in the tropics still play the role of the peripheral producers in the global system. The problem, as world systems theorists see it, is that many tropical countries find it difficult to break out of their role as a producer of primary goods. In other words, Rostow's stages of economic growth will not occur because the least developed countries are locked into a set of relationships with more developed countries. So, for example, several Central American countries have been stuck producing bananas and have found it more difficult to make manufactured goods. Both dependency and world systems theories are sometimes collectively referred to as **structuralism**.

Related to structuralism, and a concept that is useful for understanding differentiation at various spatial scales, is the notion of dualism. **Dualism** refers to situations where two areas are in relationship with one another (though trade for example), and one area is developing at the expense of the other. This concept may be examined at a variety of different scales: international, national, urban, and rural. We already considered dualism at the international scale when we examined dependency and world systems theory above. As such, international-scale dualism refers to different countries which are in relationship with one another via unequal exchange or trade. What is important to understand is that such dualistic relationships also exist at the subnational scale. Within most countries, there are often dualisms which exist between urban and rural areas. Within the United States, for example, many would argue that there are core areas where a disproportionate amount of investment and wealth accumulation occur. These areas, at least historically, are/were in relationship with peripheral zones of resource extraction. In *Nature's Metropolis* (1992), environmental historian and geographer William Cronon writes of the relationship between Chicago and vast interior regions of the Midwest. Timber, iron ore, livestock, and grain historically came to Chicago from the Upper Midwest in exchange for manufactured items. While Chicago

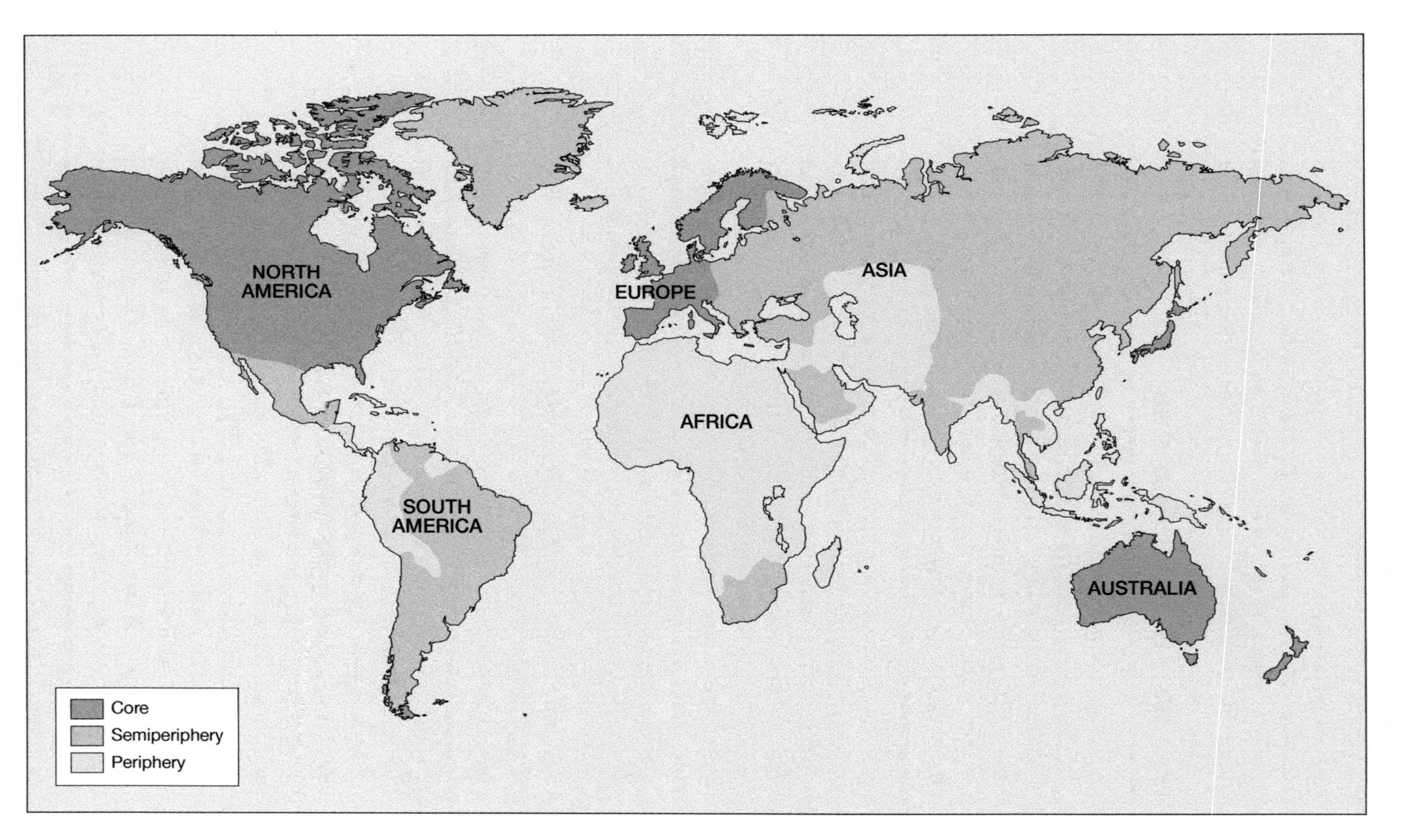

Figure 4.04 Visualizing the world system: map of periphery, semi-periphery, and core. *Source*: The Fuller Projection Map design, from which this rendering derives, is a trademark of the Buckminster Fuller Institute. ©1938, 1967 & 1992. All rights reserved, www.bfi.org.

became fabulously wealthy during the 19th and 20th centuries, the regions of extraction (such as northern Minnesota, southern Illinois, and northern Wisconsin) have little to show for their efforts today.

Dualism also is present in rural areas where one may have an export-oriented commercial agricultural sector which exists alongside smaller-scale, mostly subsistence farming. The classic situation is one wherein small subsistence farmers spend the day laboring on nearby large plantations, returning to work on their own farms in the evenings and during the weekends. The linkages between the two sectors are often pernicious for the small farmer / laborer. Large commercial farms often need access to cheap labor. The advantage of laborers who moonlight as small farmers is that they need not be paid a living wage (as they have their own production to cover a portion of their annual food needs). Many small farmers who end up working on larger farms may initially take on such employment in order to bridge a food production shortfall. In so doing, however, they may embark on a slippery slope of declining production as they spend critical time off the farm which compromises their own ability to produce food. A very similar set of dualistic relationships may exist in urban areas, often between the formal and non-formal employment sectors. Formal sector employers (such as large export-oriented firms) may draw on temporary laborers who are otherwise employed in the informal sector (which includes a wide array of undocumented full- and part-time employment in activities like petty commerce and artisanal production). While it is debatable whether such exchanges actually hurt the informal urban sector, such a system does allow for a supply of cheap labor. In sum, as discussed in relation to dualism between countries and between urban and urban and rural areas, within both rural and urban areas the dualism concept suggests a process of unequal exchange which slowly undermines one group or sector in favor of the other.

Political Ecology

Contemporary political ecology may be characterized by a number of major themes or ways of looking at world which build on insights from cultural ecology and development studies, as well as series of more recent innovations from other areas of geography and the social sciences. These approaches include: (1) thinking across scales, (2) understanding processes of marginalization, (3) attention to social differentiation, and (4) an understanding that power may be expressed in policy documents or texts (i.e., discursively).

Thinking Across Scales

As discussed above, political ecology seeks to understand the potential links between local human–environment interactions and processes operating at broader spatial scales. Structuralism helps us makes some of these links in an increasingly globalized economy. Another way to conceive of these linkages is to acknowledge that most human–environment interactions no longer occur in a closed system, but in an open one. To be more specific, long ago one could imagine a closed, subsistence system wherein the land manager could directly observe the impacts of his or her actions on the environment. If one managed soils or forests in a way that reduced their productivity, then the manager could conceivably observe the cause and effect of such strategies. The land manager might change his or her practices and consumption patterns in response to this feedback. In many instances today, such feedback loops are severed (or made more complicated) in a world where production and consumption are increasingly separated by great distances. Consumers often have a very limited understanding of where their goods are being produced, let alone the environmental impacts of this production. Absent such feedback, consumers are unlikely to adjust their consumption even if it may be leading to the demise of some ecosystems. While such environmental costs are theoretically transmitted to the consumer via prices, this rarely happens as many such costs are not born by the companies that process and distribute products. Similarly, at the level of producers, many farmers, fisherfolk, animal herders, foresters, or hunters increasingly operate within a market system and respond to price cues rather than ecosystem feedback. To be clear, it is not that land managers no longer detect ecological feedback, but that their ability to respond to this feedback may be mitigated by: (1) demand for products that is increasingly exogenous and unresponsive to local conditions (rather than for home consumption which could be varied in terms of types of food); (2) increasing pressures to generate cash income for debt repayment, medical and school fees, consumer items, etc., coupled with low returns for producers of many commodities; and (3) growing inequity in access to resources leading to overuse of a smaller resource base by some producers.

One concept that is useful for thinking across scales is the distinction between **proximate and ultimate causation**. The proximate cause is the immediate and often most apparent cause of a problem or situation. As such, the immediate cause of the person cutting down a tree in the chapter opener appears to be poverty as it is a need for money which leads to the harvesting of the mahogany tree for sale. The ultimate cause is often less obvious and more distant in space and time. As such, our woodcutter's poverty may have been related to a series of historical changes which pushed him or her off prime agricultural land in favor of a larger, more powerful farmer who was producing for export markets. Furthermore, it is not only the poverty resulting from being on a small plot of marginal land which is a product of ultimate causation, but also the options to address this poverty. The opportunity to cut down a mahogany to earn income ultimately lies in the robust global markets for tropical hardwood species such as mahogany, and the purchasers of such wood who rarely understand the environmental toll of such consumption.

One approach to making connections across scales, and connecting proximate to ultimate causes, is by using so-called **chains of explanation**. Such a chain essentially shows how a cause or factor at one scale logically connects to a process at a different scale. One example of this approach is provided in Figure 4.05, on cotton

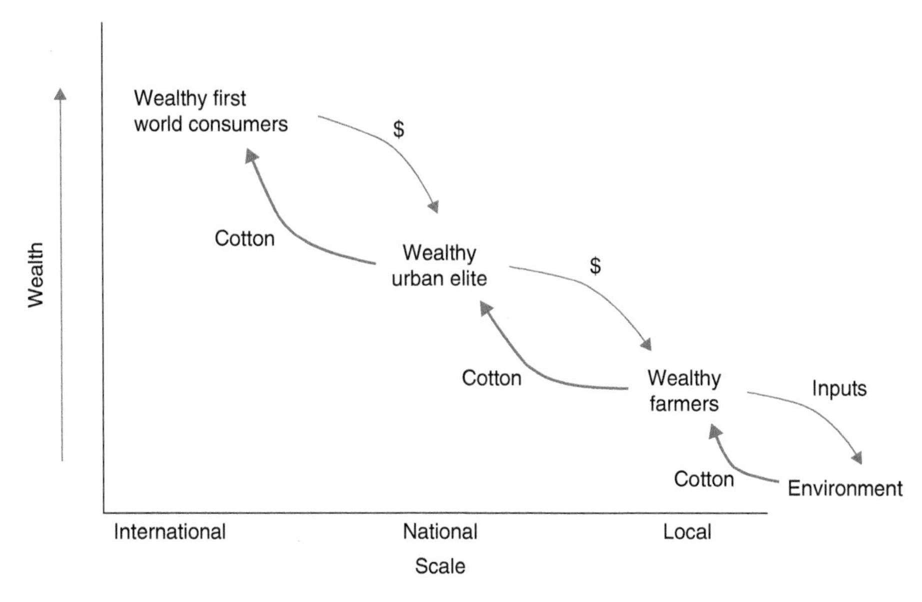

Figure 4.05 Cotton and the three scales wealth-induced environmental degradation. *Source*: W.G. Moseley. Used with permission.

and three scales of wealth-induced environmental degradation. At the local level, we start with soil degradation in southern Mali. This is an agricultural landscape dominated by smallhold farmers (as opposed to large plantations) where cotton is the major cash crop. The major food crops include sorghum, millet, maize, peanuts, and cowpeas. While these are smallhold farmers, there is quite a difference in wealth and farm size between households. Research suggests that it is large (and wealthy) cotton-growing households, using more pesticides and inorganic fertilizers, whose approach to farming is more deleterious for local soils (Moseley 2005, 2008). Chains of explanation help us understand how this approach to soils management does not occur in isolation. The decision to grow cotton in a certain manner is motivated by a series of policies generated at the national level by a relatively wealthy group of urban elite bureaucrats. These policies include an entire infrastructure to support the growing of cotton which does not exist for food crops, including credit (or loans) for agricultural inputs, agricultural extension services, and a guaranteed price and market for the cotton which is produced. The interest of the elite in promoting the growth of this crop relates to fact that it (until recently) funded a large share of government operations in Mali. As such, civil servants understand that they are paid because of cotton. Finally, these decisions and policies generated at the national level in Mali also do not occur in isolation but are linked to decisions and actions at even broader scales. Chief among these are policies of the World Bank and International Monetary Fund (IMF) which encourage Mali to be export-oriented. Consumers in the Global North are also implicated as it is they who consume this cotton, often with little to no understanding of how it is produced.

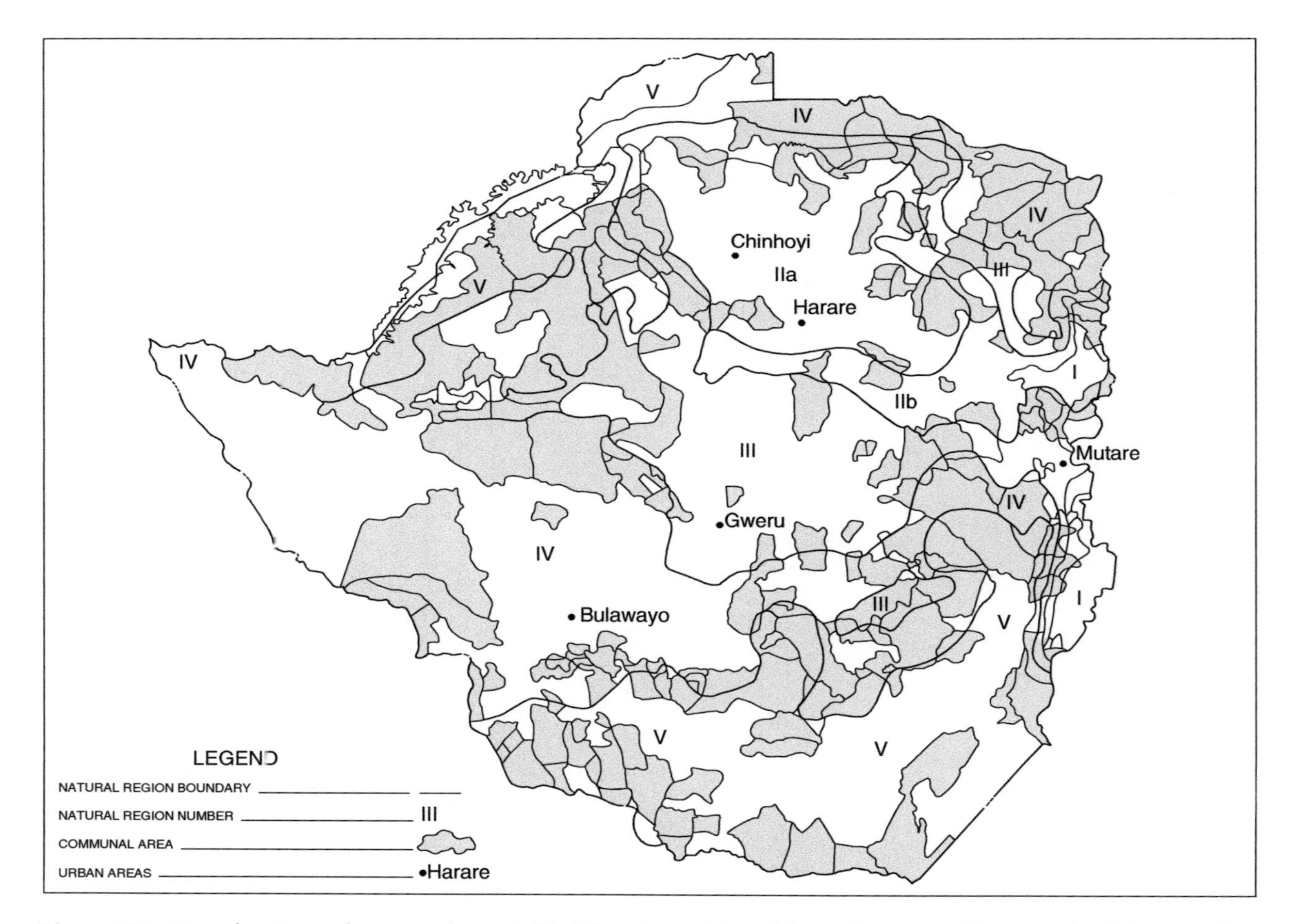

Figure 4.06 Natural regions and communal areas in Zimbabwe *Source*: Adapted from B.I. Logan and W.G. Moseley, "The political ecology of poverty alleviation in Zimbabwe's Communal Areas Management Programme for Indigenous Resources (CAMPFIRE)," *Geoforum*, 33(1) (2002), pp. 1–14.

One specific type of chain of explanation is a **commodity chain**. A commodity chain is the path a product takes from the producer to the consumer. Commodity chains can be quite short or extremely long. An example of the former is the tomato you might grow in your backyard and then consume in your salad. An example of the latter is the Granny Smith apple that an American consumes in the winter, which has traveled thousands and thousands of kilometers from an orchard in New Zealand, passing through several hands along the way. While it is not always the case, shorter commodity chains tend to be more transparent. Such transparency is critical for helping consumers understand the impact of their behavior. Unfortunately, many commodity chains are so opaque that they cannot even be traced after a great deal of effort.

Marginality

In his classic text *The Political Economy of Soil Erosion* (1985), British geographer Piers Blaikie deployed the concept of **marginality** to describe how land degradation occurs in many contexts. He identified multiple forms of marginality – social, environmental, and economic – that often work together. European settler colonies in East and Southern Africa offer good examples of Blaikie's multiple forms of marginality. Zimbabwe, formerly Rhodesia, was a British colony named after the British mining magnate and politician Cecil Rhodes. British surveyors zoned the entire country in terms of agricultural potential, from natural region I (the best land) to V (the least arable land). They then created commercial agricultural areas, mostly in regions I, II, and III, for European farmers. So-called communal areas were created for African farmers, mostly in regions IV and V. These communal areas persist today and can be seen in relation to the natural regions. (see Figure 4.06). The marginality concept is useful for understanding contemporary patterns of degradation and poverty in the communal areas. Essentially what happened is that brute colonial force ensured that Africans (made more socially marginal) were forced onto more environmentally marginal land (in terms of agricultural productivity). As Africans were crowded into these drier areas, they became further impoverished or economically marginalized. Not surprisingly, land degradation has also been high in many of these communal areas – a problem often blamed on population growth or traditional agricultural practices. However, it is the history of marginalization that best explains the degradation in these areas.

While it may be less obvious, the marginalization process can also operate in the built environment of urban areas. Classic examples of this may be observed in many cities in South Africa, where the legacies of apartheid still persist. During the apartheid era, which ended with election of Nelson Mandela to the presidency in 1994, black South Africans were forbidden to live in white areas of South African cities (such as the neighborhood on the left in Figure 4.07). Rather, blacks were forced (or marginalized) into living in townships (such as the neighborhood on the right in Figure 4.07). Townships were environmentally marginal in terms of their limited infrastructure and undesirable location. Most townships in Cape Town are on the Cape flats far

Figure 4.07 Different standards of urban living in Cape Town, South Africa. *Source*: Photos by W.G. Moseley. Used with permission.

from the city center, poorly drained and windy in the winter time. These environmentally marginal areas further compound people's economic marginality.

Attention to Social Differentiation

Political ecology took on board many of the insights provided by the explosion of feminist scholarship which occurred in the social sciences. While some feminist scholars argued that women have a fundamentally different relationship with the environment than men because of differences in biology (which is one strand of ecofeminist thinking), most work in **feminist political ecology** argued that women experience the environment differently because of their different roles in society. Thus, if it is women's responsibility to collect firewood or fetch water (as it is in many societies), they may be more attuned to the decline of forests or a drop in the water table. The attention to difference brought to the forefront by feminist scholarship was not solely limited to difference based on gender, but a whole list of other differences such as age, class, ethnic affiliation, and race. As was the case with gender, these different identities often produced different roles in society and different types of interaction with the environment. Therefore young boys, who often tend cattle, might be more attuned to changing grassland conditions; or older women, who farm rice, may better understand changes in wetland areas. One's race or ethnic identity may also have an impact on where one can live, and these areas may offer different environmental opportunities, challenges, and risks than those experienced by the general population. In the United States, the **Environmental Justice** Movement has argued that many communities of color bear a disproportionate burden of industrial pollution because of where they live.

Power and Discourse: Environmental Narratives

As discussed previously, structuralist approaches within development studies were key for understanding global systems. However, over time, structuralist understandings gradually came to be critiqued for being overly deterministic, insufficiently recognizing the power of local actors to influence outcomes. This power of the local agent is often referred to as **agency**. Furthermore, it was increasingly argued that the power of elites was not just expressed materially (i.e., by doing something to another person or to their environment) but discursively as well. In other words, one could express power by controlling the public **discourse**, narrative, or story about a situation. Collectively, this new understanding of power came to be called **post-structuralism** and it had a tremendous influence on an emerging political ecology.

A key insight inspired by poststructuralism was the notion of an **environmental narrative**, or the dominant environmental story that persisted in a region. Beginning in the mid-1990s, there was a flurry of scholarship in this arena. This scholarship demonstrated the longevity of many regional environmental narratives, many of which dated back to the colonial era. The problem is that these environmental stories, received as environmental fact, were often based on very little empirical evidence. Furthermore, new scholarship suggested that the reality might be quite different from that depicted in these environmental narratives. One of the earliest and best-known examples of this type of scholarship was that produced by Fairhead and Leach (1995). These scholars were investigating claims of long-term deforestation in the humid savannas of West Africa. They specifically were looking for evidence of this in Guinea, where forest islands in the midst of savanna grasslands were thought to be relics of a previously forested landscape. What they found, through a combination of ethnographic research and an analysis of a time series of aerial photographs from the early to late 20th century, was that these forest islands were actually the product of anthropogenic afforestation. In other words, local communities had created these forest islands over time through a combination of protection and tree planting.[3]

Political Ecology Critiques

Political ecology is not without its critics. Peet and Watts (1996) argue for an even more political, political ecology, and suggest that poststructuralist insights are key to realizing the "emancipatory" potential of the field. In other words, political ecology ought to embrace social change as an explicit goal. Furthermore, the role of local agents, and the ability for social movements to produce counter-narratives, are both central to producing such change. In contrast, Vayda and Walters (1999) have argued that that the field has become overly deterministic and insufficiently based on empirical analysis. They point to some authors who engage in very limited environmental analysis, and bemoan the rising influence of poststructuralism within the field. Finally, Walker (2005, 2006) has published a pair of articles which critique the field for having insufficient amounts of ecology and policy.

The reality is that political ecology is now a very large and diverse field. All of the critiques have merit, or could be contested, depending on where you look in the field. We believe this diversity is an asset rather than a problem.

Chapter Summary

The interdisciplinary field of cultural ecology came to prominence in the 1960s and 1970s in an era of political decolonization and a questioning of the inherent superiority of technology and modernization. The field played an important role in the intellectual decolonization process by documenting and explaining local agricultural and resource management practices, particularly in the Global South. In reaction to a series of critiques leveled against the field in the 1980s, it extended its analytical reach and eventually came to view local human–environment interactions within the context of a broader political economy. This new field is now known as political ecology, and it emerged from a synthesis of cultural ecology and radical development geography. Political ecology continues to grow and change, incorporating insights from poststructural and feminist scholarship, and turning its analytical gaze to new regions (the Global North), environments (cities), and topics (suburban lawns, energy use, and transportation).

Critical Questions

1. Before reading this chapter, what did you believe to be the major causes of environmental degradation in the Global South? The Global North? Have the perspectives and insights of cultural and political ecology changed in any way your understanding of the causes of these problems?
2. What were the major intellectual contributions of cultural ecology?
3. What were the main elements of Michael Watts' critique of cultural ecology?
4. Think about what you had for breakfast this morning. How many different regions or countries do you think your food came from? Do you know how your food was produced and the impact of this production on the environment?
5. Can you think of an environmental problem in your neighborhood, community or region to which the insights of cultural and political ecology could be applied?

Key Vocabulary

adaptation
agency
chains of explanation
commodity chain
cultural ecology
dependency theory
discourse
dualism

environmental justice
environmental narrative
feminist political ecology
marginality (social, environmental, economic)
modernization
political ecology
poststructuralism
proximate and ultimate causation
structuralism
world systems theory

Notes

1 A very different tradition (associated with the Malthusian view), also known as political ecology, had developed much earlier in political science.
2 For example, the anthropologist Wolf (1972) had used the term earlier.
3 Unfortunately, since the time of publication, Chinese loggers and refugees from Liberia have harvested some of these forest islands.

References

Bassett, T.J. (1988) The political ecology of peasant–herder conflicts in the northern Ivory Coast. *Annals of the Association of American Geographers*, 78(3), pp. 453–72.

Blaikie, P.M. (1985) *The Political Economy of Soil Erosion in Developing Countries* (London: Longman).

Boserup, E. (1965) *The Conditions of Agricultural Growth: The Economics of Agrarian Change Under Population Pressure* (Chicago: Aldine).

Brookfield, H. (1962) Local study and comparative method: an example from Central New Guinea. *Annals of the Association of American Geographers*, 52(3), pp. 242–54.

Bundy, C. (1979) *The Rise and Fall of the South African Peasantry* (Berkeley: University of California Press).

Butzer, K.W. (1990) The realm of cultural ecology: adaptation and change in historical perspective. In B.L. Turner II et al. (eds.), *The Earth as Transformed by Human Action*, pp. 685–702 (New York: Cambridge University Press).

Carson, R. (1962) *Silent Spring* (Boston: Houghton Mifflin).

Cronon, W. (1992) *Nature's Metropolis: Chicago and the Great West* (New York: W.W. Norton).

Denevan, W.M. (1983) Adaptation, variation, and cultural geography. *The Professional Geographer*, 35(4), pp. 399–406.

Fairhead, J. and Leach, M. (1995) False forest history, complicit social analysis: rethinking some West African environmental narratives. *World Development*, 23(6), pp. 1023–35.

Frank, A.G. (1979) *Dependent Accumulation and Underdevelopment* (New York: Monthly Review Press).

Grossman, L. (1981) The cultural ecology of economic development. *Annals of the Association of American Geographers*, 71(2), pp. 220–36.

Malthus, T. (1987) Essay on the principle of population (1798). In S.W. Menard and E.W. Moen (eds.), *Perspectives on Population* (New York: Oxford University Press).

Mortimore, M. and Tiffen, M. (1994) Population growth and a sustainable environment. *Environment*, 36(8), pp. 10–20, 28–30.

Moseley, W.G. (2005) Global cotton and local environmental management: the political ecology of rich and poor small-hold farmers in southern Mali. *Geographical Journal*, 171(1), pp. 36–55.

Moseley, W.G. (2008) Mali's cotton conundrum: commodity production and development on the periphery. In W.G. Moseley and L.C. Gray (eds.), *Hanging by a Thread: Cotton, Globalization and Poverty in Africa*, pp. 83–102 (Athens: Ohio University Press).

Peet, R. and Watts, M. (1996) Liberation ecology: development, sustainability, and environment in an age of market triumphalism. In R. Peet and M. Watts (eds.), *Liberation Ecologies: Environment, Development, Social Movements*, pp. 1–45 (New York: Routledge).

Rappaport, R.A. (1968) *Pigs for the Ancestors: Ritual in the Ecology of a New Guinea People* (New Haven: Yale University Press).

Richards, P. (1985) *Indigenous Agricultural Revolution:* Food and Ecology in West Africa (London: Hutchinson).

Rostow, W.W. (1959) The stages of economic growth. *Economic History Review*, new series 12(1), pp. 1–16.

Turner, B.L., Hyden, G., and Kates, R. (1993) *Population Growth and Agricultural Change in Africa* (Gainesville: University of Florida Press).

Vayda, A.P. and Walters, B.B. (1999) Against political ecology. *Human Ecology*, 27(1), pp. 167–79.

wa Thiong'o, N. (1986) *Decolonising the Mind: The Politics of Language in African Literature* (London: James Currey).

Walker, P. (2005) Political ecology: where is the ecology? *Progress in Human Geography*, 29(1), pp. 73–82.

Walker, P. (2006) Political ecology: where is the policy? *Progress in Human Geography*, 30(3), pp. 382–95.

Wallerstein, I.M. (1979) *The Capitalist World Economy: Essays* (New York: Cambridge University Press).

Watts, M. (1983) On the poverty of theory: natural hazards research in context. In K. Hewitt (ed.), *Interpretations of Calamity*, pp. 231–62 (Boston: Allen & Unwin).

Wolf, E. (1972) Ownership and political ecology. *Anthropological Quarterly*, 45: 201–5.

5

Environmental History

Icebreaker: The Sonoran Desert Past and Present

In the dry valleys of the Sonoran Desert, stretching from northern Mexico into parts of the US Southwest, many people seek solace in desert landscapes. What visitors perceive, and what they expect to see, play a role in the experience they have in this environment. These arid landscapes, now riddled with thorn scrub plants that tear at clothing and livestock, have changed over time. Early Spanish documents from the 17th century suggest a different landscape of oak grasslands. By the mid-1700s, a Jesuit missionary noted that "on the hills,

An Introduction to Human–Environment Geography: Local Dynamics and Global Processes,
First Edition. William G. Moseley, Eric Perramond, Holly M. Hapke and Paul Laris.

Published 2014 by John Wiley & Sons, Ltd.

as well as on the plains, there are the most excellent pastures, where grow in superabundance the choicest grass and all kinds of healthful herbs" (Pfefferkorn 1989: 43). Even as late as the 1880s, there were common descriptions in the Sonoran Desert hills of lush grasslands broken by clumps of small-standing oak trees.

Today, grass and oak still exist of course, but the patterns have changed over time, as has the species composition of the landscape. Remarkably, elderly ranchers and farmers living in these same Sonoran valleys remark on the decline of small oaks, state that grasslands are in poorer shape today than they were in the past, and often view the past as the proverbial good old days. But the old documents and more recent oral statements are problematic. If we choose to believe the documents and the oral accounts, they refer to lush pastures and common oak tree groves, and we may be adding, to this vision of a fallen landscape, one that has declined in quality. Are the documents accurate? Since no 17th-century Spanish document describes grass cover in terms of a percentage of ground cover, what does "lush" actually mean in this context? Can we put a number to "superabundance" in Pfefferkorn's description? These documents are obviously vital for understanding the region's environmental history, but how accurately do they capture what actually happened?

Using local knowledge or the memory of long-term residents in any particular place is a good starting point for understanding environmental change. But human memory is a fickle thing, so human–environment scholars are interested in using multiple sources of historical information to understand the past. This can involve documents written during the period of interest which can provide direct, textual evidence of some changes, or it can involve more advanced techniques, that rely on indirect evidence, to reconstruct those past environments. If we use the present landscapes as a basis for understanding, to argue that much has changed since the 1600s, are we telling the human–environment truth? Since change is the norm, not the exception, how can environmental history as a perspective help us understand the human–environment condition?

Chapter Objectives

1. To understand the importance of change over time in human–environment relationships.
2. To select the right "tools" for studying specific historical environmental issues.
3. To appreciate how the story-teller in environmental history affects the kind of environmental history narrated.
4. To apply some lessons of environmental history to contemporary human–environment challenges and problems.

Introduction: What Is Environmental History?

The broad objective of this chapter is to introduce students to the interdisciplinary subfield of environmental history. After setting out what constitutes environmental history, the chapter explores the contributions of geographers to this interdisciplinary arena. The chapter then examines different methods and approaches that have been applied to understand environmental history, dwelling in particular on those approaches frequently adopted by geographers. The chapter ends by detailing evolution environmental history as a subfield, the various "conversations" or research areas within environmental history, and the value of this approach to understanding contemporary environmental issues.

Environmental history can be broadly defined as the study of human–environment relationships as they change over time. Environmental history offers two unique perspectives for exploring the human–environmental nexus. First, it explores how humans have modified the natural environment over time. Second, it examines changes in the ideas we hold about the environment, and how those changes in ideas over time may shift our understandings of human–environment relations. Most environmental historians also give nature a clear "role" in historical events, rather than just seeing nature as a backdrop to the past. In this chapter we will see that several flavors of environmental history now exist within geography itself and contribute to the larger literature in different ways. We take a biographical and story-telling approach, since it is impossible to survey all of the important works in environmental history.

How does one write a history of the environment, if it's only humans writing it? Is history useful to understand our changing perceptions and relations to the environment? How can we write environmental histories in and about cultures that were, or largely remain, pre-literate? Here, we explore the variants of environmental history, as being practiced today in a variety of fields, such as anthropology, geography, and history. Environmental history can help us:

1. Understand environments through time: reconciling our understanding of **paleoecology** (long-term ecological changes, often without humans involved) with the more recent human–environmental impacts (environmental history as recorded in documents).
2. Understand the radical and profound changes to human environments caused by industrialization and modernization, at multiple spatial and temporal scales.
3. Provide a useful context for nature's agency in environmental history: What role for nature itself? Can we produce transnational environmental histories?

Environmental history is an expanding field in its own right, even if it is logically associated with the larger field of history, since it draws on almost every discipline with a human–environment facet, from anthropology to zoology concepts. Geographers are actively creating, reconstructing, and telling compelling environmental histories grounded in particular places, regions, or global contexts (see Williams 1994). Here, we will review some of the important authors who have shaped concepts of environmental history and emphasize environmental history as a perspective, a way of thinking about past (and current) human–environment relations.

For more than 20 years now, geographers have self-identified with the interdisciplinary field of environmental history. Arguably, geographers have been doing a form of environmental history for much longer. We can consider the **narratives** of the early 19th-century German natural scientist Alexander von Humboldt

Figure 5.01 Alexander von Humboldt, explorer and encyclopedic author. *Source*: © Archive Pics/ Alamy.

(1769–1859) as a version of proto-environmental history (see Figure 5.01). Von Humboldt studied the physical and cultural characteristics of landscapes as they had changed through time, even if his systematic nature was also reminiscent of early biology and ecology research. Von Humboldt's more holistic vision of the Earth included humans, and he was among the first to examine how particular human environments are shaped by both natural and cultural factors. While reflecting a clear influence from the biological and physical sciences, Humboldt's work also included human impacts. His early mapping of altitudinal gradients in South America, for example, formed much of the basis for early understandings of Andean biogeography (see Chapter 3) and agricultural geography. He wrote encyclopedic and ambitious natural histories of regions with a unified science perspective, and is considered by many to be the first ecologist and biogeographer. Were he alive today, von Humboldt would certainly make use of biophysical methods (see below) and proxy data to reconstruct environments.

The forms of human–environment analysis used by Humboldt, however, would not be repeated to any great extent until the writings of George Perkins Marsh (1801–82), an ambassador and insightful commentator, who traveled from his childhood Vermont environment to make observations on the Mediterranean Basin and elsewhere (see Figure 5.02). Marsh was greatly concerned with the perceived amount of environmental degradation in the Mediterranean. Marsh's writings on human–environment relationships, captured in *Man and Nature* (1864) later republished as *The Earth as Modified by Human Action* (1874), were not widely read or heeded at the time.

Indeed, it would take almost a hundred years before the reissue of *Man and Nature* would have some impact, but the book is considered by some to be the first synthesis of ecology sensitive to human impacts. A widely traveled diplomat and native of humid and temperate New England, Marsh's criticisms of human–environment relationships in the Mediterranean were notable for their clarity, and he decried the overuse of agricultural lands he observed around the region. His critiques, however, also reveal that the notion of "degradation" is in the eye of the beholder. It was easy, in other words, coming from lush and green Vermont to critique a much drier climate and its peoples. Yet agriculture, animal herding, and overall land use in the Mediterranean Basin have endured for millennia (Butzer 2005). Marsh's work, however, remains important because even in the 19th century (much like today), environmental issues were surprises to most readers. Both von Humboldt and Marsh were also influential in shaping late 19th- and early 20th-century authors writing about environments and human environments. Their respective observations and perspectives also clearly inform current work, even if our descriptions of historical human environments have improved.

In the late 1800s, **environmental determinism**, a school of thought that argued that environments directly shaped and pre-defined human behaviors and capacities, emerged in geography and related disciplines. This worldview was informed by certain elements of **social Darwinism**, a racist and unfortunate view of humanity that emphasized the superiority of dominant European groups at the expense of subject populations around the world. For example, one of the proponents of this view of environmental determinism, Ellsworth Huntington (1876–1947) viewed climate and human development as intimately tied together. In 1915, Huntington commented on the situation of white colonists in South Africa: "Thus the two races face each other under conditions which lessen the white man's energy while they stimulate the black man. The whites are still far ahead and will doubtless continue to be so indefinitely" (1915: 27). So even if his concern about whites succeeding in tropical climates was merited because of various tropical diseases which were clearly a concern at the time, Huntington still wrote off black populations, suggesting they would never catch up. What would Huntington think of the country of South Africa today, after the fall of apartheid?

Figure 5.02 George Perkins Marsh, author of *Man and Nature* (1864). *Source*: Library of Congress Prints and Photographic Collection.

Viewed in hindsight, environmental and climatic determinist work coincided with the height of European **colonialism** in parts of Africa and Asia, and human–environment writers of this period reflected the lingering influence of social Darwinism. This deterministic perspective, however, would wane as scholarship challenged the flawed bases and biases of this work. Anthropologists, geographers, and historians in the 1930s and 1940s proposed that **cultural possibilism**, that environments influence but do not determine cultural patterns, shaped human behavior. And there was a growing realization that colonialism produced damaging environmental effects (Sauer 1938).

Clarence Glacken's *Traces on the Rhodian Shore* (1967) added another perspective on Western society's relationship to nature (see Figure 5.03). What distinguished the book was the focus on how perceptions of the environment shaped attitudes and actions in Western societies. So Glacken was not simply discussing human *impacts* on the environment, like Marsh (see above), but how historical *perceptions* of environments shaped various trajectories, different track records, of human–environment interactions. Unfortunately, Glacken (1909–89) did not live to complete his intended full treatment of human–environment relationships into the modern period, although the book's first volume was undoubtedly one of the key texts for environmental history as a field. Glacken's work is also a useful reminder about the distinctions between approaches that emphasize our biophysical impact as a species (Chapter 1), in contrast to the work that concentrates on the power of ideas and culture in shaping our perception of those same environments and impacts (Chapter 2). These distinctions, or contrasts, remain useful for understanding the various kinds of environmental histories.

Figure 5.03 Clarence Glacken, proto-environmental historian and geographer. *Source*: Collection of Department of Geography, University of California, Berkeley.

The Range of Environmental Histories

In history, Alfred Crosby's *The Columbian Exchange* (1972) influenced a new generation of geographers, and historians showed renewed interest in the long march of human–environmental affairs. Crosby triggered a new focus for telling human–environment stories: the role of non-humans themselves in shaping these relationships. Microbes, pigs, and cattle have all shaped particular regional histories. The indigenous population crash in the New World, for example, was largely (but not only) caused by pathogenic bacteria and viruses like smallpox. Within a century of contact with Europeans, it is highly probable that some 80% of indigenous populations in the western hemisphere were wiped out. The arrival of Old World pigs in Florida, to use another example, decimated native animals, birds, and plant life. And the use of cattle in Mexico was an effective way to displace Natives from their land, and control large portions of territory. Crosby used the term **portmanteau biota** to describe the variety of biological baggage that colonizers bring to new environments. But this baggage can have deleterious consequences in new environments. You can probably think of non-native species that arrived in your region, and how this is (or is not) tied to particular arrivals of other peoples in your area or country.

Yet environmental history as it evolved in the 1970s and 1980s did resemble work in geography in certain ways. Contemporary work in environmental history treated either a thematic resource (forests) or a specific region (say, New England) to focus on for their analysis. Let's use some early examples to see the similarities to Crosby (1972) and the emerging differences. Richard White's early work on the social and ecological relations of Native tribes (1983), and his later work on the ecology of colonial encounters between tribes and Europeans, were influential (1991). Likewise, William Cronon's *Changes in the Land* (1983) and Carolyn Merchant's *Death of Nature* (1989) were clear calls that environmental history could be done on a regional scale (Worster, Merchant) or with distinct cultural groups in mind (White). Merchant's work still resonates today, as she was one of the first writers in environmental history to include a strong critique of ignored gender relations in human–environment stories.

Unlike Crosby, however, these later historians did not overly emphasize the role of biology or ecology in shaping their arguments. These were not deterministic histories, in other words, but ones that did treat the environment as some kind of actor in historical affairs. Humans and their environment were in a mutual relationship. Crosby's microbes, in other words, were not the only characters in the later histories. And this trend towards adding the social and political factors in environmental history was reinforced with Don Worster's (1985) *Rivers of Empire*, a work that echoed in fields well beyond geography and history, especially given its radical framing of western (US) water issues. The focus on a single type of resource or ecosystem, however, still remained common. Differences between geographers and historians, however, were clearly visible even then (see Box 5.01).

Box 5.01 Two different environmental histories: Mexico's Mezquital Valley

The Valle del Mezquital, in the state of Hidalgo, Mexico, is located near the center of the country. The state of Hidalgo is semi-arid to arid, with variable topography and a long record of human occupation. When historian Elinor Melville's (1994) *A Plague of Sheep* appeared nearly 20 years ago, it received favorable reviews from her peers in history. Karl Butzer, however, based in geography but long a geoarchaeologist, objected to many of the conceptual and methodological consequences of her approach (Hunter 2009). Melville used the "**irruptive ecology**" model borrowed from an Australian context. The concept of irruptive ecology pertains to new species arriving in new environments with few limitations (or predators); they achieve great population success in a short amount of time. But Melville's analysis was firmly rooted in the use of historical documents, rather than fieldwork in the Mezquital. Instead of **pastoral transhumance**, in which animals are rotated seasonally for pasture and to avoid overgrazing, a common practice in both the Old and New Worlds with grazing livestock, Melville seemingly had sheep swarming like locusts over the Mezquital's valleys and hills, erasing vegetation cover and exposing soils to subsequent erosion.

Butzer and Butzer (1997), in contrast, reconstructed the vegetation of the Mezquital and Bajío regions of the 16th and 17th centuries. Using documents similar to the ones used by Melville, but adding landscape observation and comparison, the Butzers concluded that the modern landscape resembled what is described in 16th-century documents. This example also illustrates the lingering but principal differences between environmental histories written by geographers and those written by historians. Just as individuals within our environmental histories should not be forgotten, we must also be conscious of our own position as we write these environmental stories. The Mezquital case also illustrates how important it is to think of multiple sources of information. How can information from indirect sources, or **proxy methods** as they are called, and direct sources like archival documents be used together for a more holistic understanding of the past? The bottom line is that different use of sources will produce different environmental histories.

Reconstructing Past Environments: Texts and Tools of the Trade

Nature–society geographers engaged with environmental history use a wide variety of sources for reconstructing or interpreting past human–environment relationships. For biophysical reconstruction, we can draw on the field of paleoecology: the study of past physical environments. Geologists and those who study Earth's past environments over millions and billions of years are studying paleoecology. The bulk of the proxy methods discussed below are targeted at the Pleistocene epoch (see Chapter 3), the more recent Holocene, and what Crutzen and Stoermer (2000) have termed the **Anthropocene**. The last term, although largely informal, refers to the more recent past and the foreseeable future, in which humans have played a part in reshaping the planet's biogeochemistry and biodiversity. So paleocology is a bit different since environmental history presumes that humans are not only around, but writing about their experiences in documents.

Ideally, scholars both have access to, and use, written documents that address the very human side of that equation, and also have some facility with at least interpreting biophysical data and other sources from the natural sciences. But as Butzer (2005) has noted, it is rare for anyone to have the toolkit and training in equal portions for the social and the natural. The amount of information available today for environmental historians and geographers of all stripes is nearly overwhelming, and compiling it all is an "almost impossible task" (McNeill 2003). Rarely do researchers have the skills, time, and resources to combine literary and textual analyses with detailed field and laboratory investigations of biophysical proxy data. Because of this, it is increasingly common for environmental historical work to be completed in larger research teams, with different scholars contributing different skills and pieces of the puzzle to answer larger questions together. Historians still tend to focus on the archival, documentary lines of evidence, while geographers and anthropologists working on environmental histories have used other sources of evidence. And these different sources, or techniques, can lead to quite different stories.

Some of these sources are thought of as **direct sources**, such as archival documents written by first-hand witnesses of any event. But **indirect sources** (or **proxy sources**) such as using pollen or soil information, can also be valuable and are now commonly used. Since soils and pollen are not labeled numerically, laboratory and instrumental methods must be used to corroborate their age and what they can tell us about past environments.

Archival documents are some of the most common tools used in environmental history, especially in areas where literacy and record-keeping have preserved information on climate or environmental changes. Archival documents are considered primary sources, in that no other person (in theory) has manipulated the writing or what has been left on the document. To name one example, the longest existing climate records come from central England, but only date to the

17th century (1600s). Both geographers and historians commonly use a variety of archives and source areas for reconstructing local, regional, or even national narratives of past environments. Geographers, however, may be more concerned about policy or immediate relevance (see Endfield 2008 for an excellent example). For historians, these are the starting point of solid environmental histories. For geographers, they may simply be one source of evidence for reconstructing past human–environment relationships.

Government documents and quite recent 20th-century agency policy statements can be used as well to frame how our understanding of a process has changed over time. For example, Mark Carey (2010) analyzed the changing perceptions of glacial melt-water lakes in highland Peru as a way to capture how the "risk" of high-altitude lakes changed in official policy and local stories. Changing perceptions of these lakes as, alternatively, sources of irrigation, hydropower, threats to towns, and now as resources "at risk" because of a warming climate, show the value of environmental history. Settlement maps, forestry records, and historical photo collections are other examples of good potential sources for understanding how environments and human–environment relations have changed over time. This is not an exhaustive inventory on archival methods and documents, but should give you an idea as to the range and creativity of using written sources for good environmental histories. Along with the archives, there are an increasing number of field sources used to understand our past relationships with particular environments. Increasingly, we are also turning to oral history as a complementary method to inform past archival documents as well (see Box 5.01). Now we turn to more indirect (proxy) ways of reconstructing past environmental histories.

Pollen analysis (palynology). Fossil pollen can be used to reconstruct past environments. The outer lining of pollen is made of **sporopellenin** that resists acid degradation. Palynologists study the genus assemblages of preserved pollen for clues regarding past environments and vegetation. Palynology is best suited to moist, tropical environments or anywhere with long-lived lakes where pollen can be trapped and preserved in sediments. Dating these pollen grains left behind in sediments, using carbon dating with charcoal for example, is ideal for dating sequences where pollen is present. Thus, one can associate a cluster of pine pollen for example, with a certain period of time (say, 1280 years ago). Doing so with repeated sequences of plant pollen with dates can lead to a regional picture of vegetation change over time. Pollen can be also be used to understand the history of hurricane impacts (Liu et al. 2008) as well as early cultivation of food crops (Arford and Horn 2004; Sluyter and Dominguez 2006).

Lake varves. These layers of sediment deposited in annual waves on lake bottoms are typically used in stable, temperate settings where seasonal deposition can leave behind noticeable "lenses" of sediment. In the New England region of the United States, bodies of water that have been stable can leave behind a useful signature for the amount of soil erosion and precipitation, and length of growing season, even if local waters can have variable lenses. Essentially, a dark band in the sediment column reflects the growing (read "summer") season, while the light or gray bands reflect

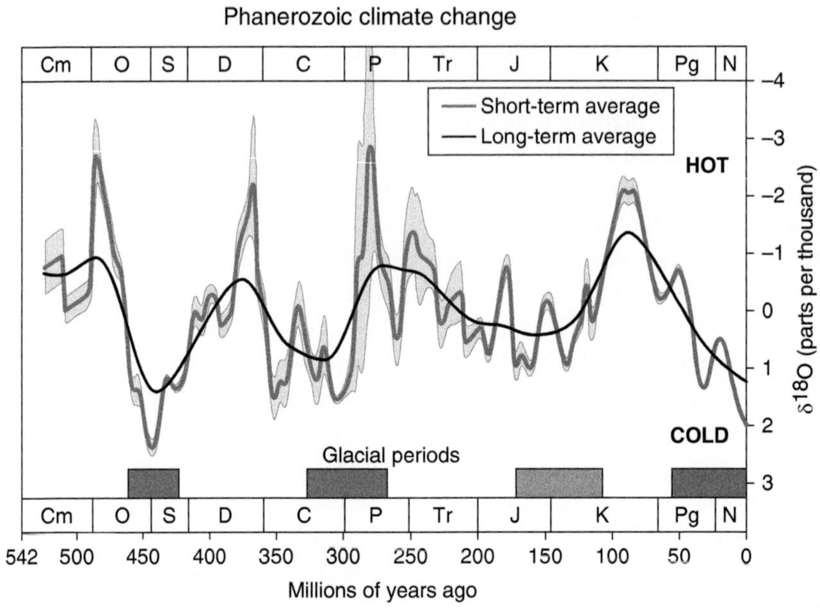

Figure 5.04 Long-term climate variations as recorded by ice cores and the various weights of oxygen isotopes trapped in ice. *Source*: http://en.wikipedia.org/wiki/File:Phanerozoic_Climate_Change.png.

vegetation at rest (winter), when little organic matter is deposited in a lake bed. Thus, a sequence of dark/light in a band represents one "year" in lake varve records.

Packrat middens. Because of the nature of their limited range of plant collection, the detritus collected by packrats (*Neotoma* spp.) serves as evidence of past and localized vegetation around their dens. Unlike pollen, which can be swept along by wind currents for hundreds of miles, these middens of large plant and organic remains stay largely fixed in place, nicely tied to the normal range of a packrat (about 1 square kilometer). Midden contents can be considered fine pixels of vegetation assemblages in larger regional studies where other proxy data may be available, such as tree rings or pollen, to understand climate and plant community changes over time. In deserts or semi-arid regions, packrat middens play a vital role in understanding vegetation and climate dynamics.

Ice cores. These extracted cores from glaciers are useful for long-term climate reconstruction, based on ratios of $^{18}O/^{16}O$ isotopes in seawater. A high ratio (more ^{18}O) means that more water in the past was held in the form of glacial ice, representing a cooler period of Earth's climate history. A lower ratio (less ^{18}O) represents an interglacial period that was relatively warmer (see Figure 5.04). These records are less useful for short-term climate variations because of the vagaries of dating isotopes' signatures, but are crucial in any glacial setting or for understanding the Pleistocene epoch (2.6 million to 12 thousand years ago).

Analysis using **tree rings (dendrochronology)**. Dendrochronologists, and many paleoclimatologists, use tree ring patterns to reconstruct climate variations using a relative scale (warmer, rainier; cooler, drier) in some temperate environments.

Figure 5.05 An example of visible tree-rings in a cut section of an oak tree. *Source*: © Shutterstock.

Like lake varves, it takes a light ring (growing) and dark ring (dormant) to make one "year" of tree growth record (see Figure 5.05). Milder, wetter years, for example, will produce larger (wider) tree rings. Drier years produce smaller rings. Rings can also show evidence of a fire scar, a dark ring that can be useful for building a fire history of a forest patch. Tree rings are not useful for the tropics, where tree rings are lacking, although recent progress has been made in areas where a strong wet-dry season oscillation occurs in many tropical regions. In some regions, tree rings can provide a history extending back 5000 years; these are largely used for reconstructing and understanding Holocene epoch climate and vegetation responses.

Geochronology. This type of dating, based on sequences of deposits, is tied to some other proxy data, such as C14 dates. This includes tephrachronology, the dating of volcanic tephra layers in a deposition sequence. There must be some form of localized carbon in the deposit to allow for dating. Uses of geochronology include landscape and flood history reconstruction.

Carbon-14 (C14) dating. The half-life of carbon provides a useful tool for dating any organic materials found in soils/remains/archaeological sites. Carbon dating is only valid for sites or areas less than approximately 40,000 years old because of the half-life of carbon isotopes. Carbon dating is an important part of any study focusing on the Holocene (last 12,000 years) and the late Pleistocene. It is typically used in combination with other methods, to provide a temporal context, such as for pollen and geochronology research.

Paleosols are buried or "fossil" soils and can provide evidence of sedimentation over a stable landscape; they may be triggered by natural environmental changes or events (volcanic explosion, ash covering a formerly stable landscape), or by human-induced changes (site abandonment, old fields). With high organic matter content (carbon-14), these soils can be easily dated, and are useful for landscape reconstruction or understanding the histories of watersheds.

Remnant shorelines. A more limited tool for reconstructing past landscapes is the use of old shorelines on landscapes, which may be visible along basin edges. Remnant shorelines have been used to determine the extent of pluvial lakes during the Pleistocene and Holocene. An example is the reconstruction of old shorelines of Lake Bonneville in the Great Salt Lake Basin of Utah. The remains of this lake are marked by clearly visible lines, looking like bathtub rings in the geology.

The key aspect to these various proxy methods is that they are usually used in combination. You could use charcoal embedded in soils as a source of information, through C14 dating, found with nearby pollen to provide an estimated age of the pollen. This would give you some idea of the vegetation at that particular time if you find enough pollen grains. Alternatively, those same sources of information could help you reconstruct the fire history of an area.

Box 5.02 Oral environmental histories

Figure 5.06 80-year-old Don Eliseo Martinez, shown here stooping to collect wild chiles in Sonora, Mexico, has much oral-historical knowledge that can inform environmental histories of the region. *Source*: Photo by Eric Perramond. Reproduced with permission.

One of the starting points for geographers working within environmental history is using the available local knowledge. In this case, the local environmental historical lore, **oral traditions** or **oral histories**, it is the people that are the starting point for beginning to understand any story or narrative of how a place or landscape has changed. Oral history uses current memories of lived experience as a method, often the oldest generation available. Oral tradition is best understood as messages transmitted across multiple generations. So using oral history is more common, as it takes special skills to use oral traditions in understanding the past as remembered and transmitted. Understandably, there are risks to counting on the oral stories of others to build a case or larger argument, as it is often challenging to separate rumor from truth, and hearsay from commonly held opinions (Perramond 2001).

Inspired by much of the oral historical lore and traditions kept alive across generations in African villages, some of the best human–environment work has come within these contexts, as best exemplified by Africanist historian Jan Vansina (1984). For example, Kull (2004) used oral history and local knowledge in reconstructing the use and purposes for fire use in Madagascar, even though the national government has criminalized the use of fire. In *Misreading the African Landscape*, authors Fairhead and Leach (1996) emphasized how certain kinds of landscapes were categorically perceived by European ecologists and colonial administrators as "fallen" from some form of ecological grace. Fairhead and Leach argued that the assumption of past and current degradation facilitated certain kinds of (colonial) decision-making for administrators assigned to sub-Saharan countries. In contrast, oral historical accounts painted the picture of a landscape that was now forested around villages *because* of people. It was not that local people had decimated forests in between villages; rather, the local villagers had forested settlements. In other words, the African landscape was not a fallen environment, but a green, humanized one planted with intent. This is but one example of how useful oral history can be in contributing to environmental historical work.

Urban, Industrial, and Bodily Environmental Histories

Environmental history is moving away from the bucolic, agrarian settings that once characterized the interdisciplinary literature towards a larger, more modern context of humanity's relationship with the planet. This move to tackle the modern, industrial, and toxic legacies of our recent past is of course reflected in a great many disciplines. In geography, the move from cultural ecology to increasingly political ecology is one reflection of this push to understand the modern era and globalization. So what are the effects of cities, industry, and pollution in terms of environmental history? Here we will look at a sampling of the kind of works pushing the boundaries in the 21st century. We will then move on, in the next section, to examine recent efforts to understand how "power" is recorded in archival documents, and how such documents can be used in different ways.

Are cities natural? What kind of area do they impact given the heavy demands for food and water in a city? These questions were the context for *Nature's Metropolis*, William Cronon's (1992) masterpiece, on Chicago and its agricultural and environmental hinterlands. And given how urban our global, 21st-century civilization is, it seems vital to understand the urban connection to rural resource bases. Cities do pose special challenges given their far-reaching connections to the global countryside and economies. Even though cities are easily mapped in terms of administrative politics (just think of your own city limits), they are difficult to conceive of as distinct ecosystems that can be interpreted through the lens of environmental history. But industrialization, urbanization, and globalization have distinct environmental histories, set in particular places. They have also had distinct effects on the human body. Here we discuss three examples of how environmental history as a perspective can inform these modern processes, and why that perspective is interesting and useful.

Industrial Histories

As the most recent and influential form of livelihood, industrialization has had consequences for both the planet and the humans living near industrial sites. In many ways, this is about understanding the environmental history of capitalism, and assessing this impact on the Earth (see Hornborg et al. 2007). While the modern context of industrial impacts on human health is often couched in the language of environmental justice (see Chapter 7), environmental history is a perspective that adds depth. It allows us to understand both the radical changes caused by industrialization and modernization, and get a sense of how those long-term pursuits may have unintended consequences for people. Here we use two examples to illustrate how global and nation-state treatments can help illuminate these effects.

One of the most ambitious, and instructive, projects was McNeill's (2000) *Something New Under the Sun*, focused on the environmental impacts of humanity on the planet during the 20th century. Combining both archival documentation and statistics of humanity's impact in that century, McNeill's work reveals a staggering transformation of our planet in a humane way. It also becomes clear that countries in the northern hemisphere created more global changes because of the wholesale pursuit of industrialization, high levels of consumption, and dependency on fossil fuels. One can also narrow down these assessments to nation-state stories.

Brett Walker's *Toxic Archipelago* (2010) recounts the industrial-environmental history of Japan, with a particular focus on pain. Bodily sacrifice, suffering, and pain form a story-telling device for Walker. What the reader gets is a poignant set of stories, mostly focused on mining and extractive pursuits, including one of the first modern signs that humanity's biogeochemical experiment on the planet could go horribly wrong: Minamata Bay and the fishing communities of Kyushu, Japan. The pursuit of industrialization in Minamata Bay led to mercury poisoning of humans and non-humans, as people ate contaminated seafood (see Chapter 7 for more on this case). By using a focus on pain, felt by all sentient beings, Walker is talking to all of us. It creates an intimate, deeply felt, portrayal of how human–environment interactions in history are also sensed by humans.

Natures of Cities

Cities are fascinating places in which to conduct environmental history. Although traditionally considered non-natural and highly human, cities do have profound impacts on the surrounding countryside and on those living inside the city. What is the spatial and ecological footprint of a city? In many ways, they also have their own needs: cities need water, food sources, and an elaborate infrastructure of transportation, sanitation, and water drainage to avoid flooding (Melosi 2008). There are too many examples of urban environmental history to cover here, but chances are one of the cities you live close to has been the subject of study. Two common approaches, however, in urban environmental history have been focusing on particular features common to many cities (such as rivers) and trying to quantify the reach and impact of a city.

Rivers and urban rivers are the subject of multiple works in environmental history, from the Mississippi River and New Orleans, to the Los Angeles River (White 1996; Colten 2000, 2006; Gumprecht 1999). Increasingly, these environmental biographies of cities are taking into account larger watersheds, since clusters of cities can occupy the same river (see Cioc 2002).

Last but not least, the concept of social metabolism, or how much of the Earth's primary resources (e.g. vegetation, water) is consumed by humanity in particular settings (Fischer-Kowalski and Haberl 2007), helps describe and explain the impacts of cities on the environment. This new work is also giving greater agency to non-human nature, such as the role of animals in urban industrialization (see McShane and Tarr 2007 for an example). Maybe the most intriguing part of the metabolic approach is that it's so easily understood, and it is a pragmatic approach to how humans use resources.

Bodies and Health

Boring down to the micro-scale of daily life, environmental history also has something to say about the consequences of environmental changes to the human body. Early works like Rachel Carson's *Silent Spring* (1962) revealed the consequences of chemical biocides on non-human nature and warned of the possible human health consequences. Now environmental historians are adding new and more subtle studies about the consequences of toxins, biocides, and air pollution on our own bodies.

For example, Nancy Langston's *Toxic Bodies* (2010) discusses the history of synthetic estrogen diethylstilbestrol (DES) as a frequently prescribed treatment for pregnant women, to ensure larger, healthier babies. Long-term effects of DES, however, have included post-natal problems for reproduction, and clear links to certain forms of cancer. Two aspects of the book are notable: one is the clear set of unknown consequences of living in this era of chemical prescriptions we give ourselves and that affect other species. Synthetic hormone releases into water systems, for example, can affect the reproductive success of amphibians. The other take-away lesson is that using environmental history to understand a case like DES can illustrate policy lessons useful in the present.

Gregg Mitman (in *Breathing Space*, 2007) has connected concerns over industrialization, and the rise of allergies and asthma, as diseases of modern life, and how early solutions to quell allergy suffering did not work because humans re-created the same conditions (plants, insanitary houses) in new locations. Mitman also argued that allergies are not "things" but relations, and that modern obsessions about pharmaceutical solutions only treat individual suffering, and not the complex causes of allergies that are outside of the human body. Lawns, trees, and ornamental flowers are all members of residential neighborhoods, and are unlikely to be torn out because of allergy sufferers. The cultural conditioning to seed, and maintain, non-edible crops like grass (lawns) is only part of the reason why external solutions to allergies are a challenge (see Robbins 2007). Allergies, then, are just as cultural in origin as they are "natural" to the populations who suffer them. Differences in income, and lack of access to expensive drugs, also exacerbate the unequal ability to decrease the effects of allergies. Combining this historical sensitivity with insights from political ecology, then, may produce a new kind of environmental history.

Power, Economics, and Environmental History

In many ways, environmental history is a combination of political economy, history, and ecology perspectives. Scholars in these fields share a passion for both archival and field research. But writers who are interested in the issues of power in different societies sometimes proceed differently: it is today's field-informed component that may lead to questioning past archival materials, a practice that some traditional historians might find to be heretical (Offen 2004). Yet archives, and official documentation, are narratives themselves, of past relationships of power. Those who are literate, in charge, and part of a larger web of hierarchy are the ones most likely to produce and be reflected in official archival documents (see Box 5.03).

Environmental historians now question some of the very sources that are taken for granted as starting points for environmental histories: the documents, letters, and literary instruments that serve as the basis for writing most good environmental history. A kind of field-informed environmental history is emerging, one that combines the best of archival analysis with a new ethnographic sensitivity for the living people of that place. This welcome change overtly acknowledges that written sources were written with a purpose, they reflect some powerful interest in the past, and they may not give voice to all of the agendas common at the time.

Box 5.03 Profile of Diana Davis' *Granary of Rome**

Figure 5.07 Nomad children in the Maghreb.
Source: Courtesy of Diana Davis.

Davis (2007) argues that the French used classical Greek, Roman, and Arab texts to portray North Africa's 19th-century (1800 s) environments as "fallen" from the grace that once was the fertile "granary of Rome." They later bolstered this environmental story with problematic potential vegetation maps that provided the authority of science. In doing so, the French were seemingly well justified in their arguments to control local and native populations, and assert their own access to and governance over natural resources, farmland, and grazing lands. This extended into labeling patch forest remnants as the sad relicts of a once forested past that may never have existed – climate records do not support these narratives – even as local populations were adept at managing semi-arid environments through pastoral strategies. These colonial narratives of environmental change and related policies led to greater social inequality, a decline in native access to resources, and understandably, growing resentment amongst local populations. While echoing past work on colonial justification for external rule, Davis connects postcolonial studies and concerns with environmental narratives in ways that reveal the unequal power relations inherent in the construction of what became the most widely held, but largely incorrect, notion of environmental history for the Maghreb.

*The 2008 George Perkins Marsh Prize, for best book in environmental history, went to Diana K. Davis for *Resurrecting the Granary of Rome: Environmental History and French Colonial Expansion in North Africa* (Ohio University Press, 2007).

If history is written by the "winners" or colonizers, we need other sources of information to interpret past cultural landscapes. Some are using **landscape archaeology**, the physical imprint of past societies that are traditionally silent in the archives, like African farmers in the Americas (see Carney 2001, and Carney and

Rosomoff 2011). Similarly, the French colonial officials described in Box 5.03 left official letters and documents regarding the "degradation" of the local Maghreb hillsides. It was in many ways in their interest to imagine that locals had destroyed a past forest so that the colonial government could enact new restrictions on forest use, or plan reforestation where no forest may have actually been present (Davis 2007).

These are not just academic exercises in criticism. They can inform our current environmental imaginary, how we think about our surroundings, and how past societies and nations viewed certain kinds of landscapes (with their peoples) as somehow inferior, superior, or changed from previous conditions. They also demonstrate how our current visions of what the environment "should be" are influenced by past arguments that have not disappeared (Davis 2009).

Colonial governments entered other countries and tried to direct locals to behave or submit in specific ways, often undermining already existing livelihoods (Davis 2002). In many developing countries, modern visions of protected areas are shaped by these same arguments (see Chapter 11). We have seen this, in our own work, in the context of United States policies towards indigenous peoples. For example, the US Bureau of Indian Affairs (or BIA) attempted to turn clan-based land-tenure relationships at Zuni Pueblo, in western New Mexico, into individual yeoman farms in the early 20th century. They did so by harnessing the Zuni River behind a dam at Black Rock, to control the flow of the river and create a new and so-called progressive irrigation district. The goal was to break up communal, tribal land-holding practices into individual household farms. While perhaps well intentioned, these policies were also aimed at breaking the power of the tribal council of elders, and the results were nearly disastrous.

Similarly, the BIA and agricultural agents on the Navajo reservation, not far from Zuni, decided to reduce the number of sheep grazing on the reservation. While overgrazing may have certainly been a problem in parts of the reservation, the sheep reduction program was viewed at the time as a form of cultural genocide by the Navajo themselves (Weisiger 2009). Thus, the attitudes of colonial relationships do not require an external colony located far offshore; the US had indigenous reservations, while South Africa had apartheid, to name but two examples (Dover et al. 2003). The stance of superiority over other peoples is present in almost every country, the consequences usually disastrous in the end. Using a combination of history and political economy, **historical political ecology** can illustrate the more complex realities of past human environments.

Transnational and Global Environmental Histories

Another variant of environmental history, and perhaps the most daunting or challenging to do successfully, is global environmental history. In the style of the French Annales school of history and geography, let's use Fernand Braudel's (1958) term "longue durée", or "history over the long haul," in a wider context to simplify. Only a few scholars have taken on the challenge of analyzing environmental history at long-term and global scales. Transnational environmental history is a challenge because, with few exceptions, most authors are specialists in the environmental history of a particular nation (perhaps two, at best). So the nation-state, for better and for worse, frames the view of the particular long-term story of humans in their (national) environment (Guha and Gadgil 1992, for India). And yet we know that environmental processes don't respect boundaries (think water, carbon dioxide, species migrations). Two kinds of transnational environmental historical work are apparent today: those focused on a historical and quantitative accounting of human impacts ("the study of"), and those more interested in how we can explain the differences between human societies ("an explanation for").

Many geographers claim to be doing a form of "historical ecology" that combines their interest in historical geography and ecological history. Much of this work in geography is still focused on studying the range, scale, and timing of human impacts on the Earth (see Chapter 2). This work reflects Marsh's approach about the human impact on the Earth even if it appreciates the holistic vision of von Humboldt's multiple scientific perspectives. The best example of Marsh's enduring perspective is Turner et al.'s magisterial volume, *Earth as Transformed* (1990), an update to the original effort (1956), edited by William Thomas. The 1990 volume was a synthesis of research on human environmental affairs over the last three centuries, a sweeping environmental history of quantified human impacts that informs and complements history's efforts to create a kind of global environmental history of our planet. Other works have selected particular ecosystems to focus on, whether it be wetlands (Williams 1993) or global forests (Williams 2002).

In contrast to this work on landscape and human impacts over time, Jared Diamond has published work that attempts to explain world history with concepts from geography and the literature on sustainability. In his earlier *Guns, Germs, and Steel* (1999) and his more recent volume, *Collapse* (2004), Diamond draws some analogies between past societal challenges and our current, contemporary ones. He also offers some compellingly simple explanations for history as it happened. His conclusion, for example, in *Guns, Germs, and Steel* is that the shape of continents resulted in easier colonization by western Europeans. In *Collapse*, Diamond pushes into past environmental histories to illustrate what aspects led to widespread declines in certain cultures. He argues that five characteristics are shared, in whole or in part, by civilizations that collapsed in the past (see Box 5.04). No historian working within environmental history has reached so wide an audience as Jared Diamond.

Capturing the long span of human presence on Earth, and making sense of what we have done to it, is a complex challenge. Historians tend to proclaim specialization in particular periods of time ("I only do French 19th-century history"), just as geographers may express a regional interest (Russia or Southeast Asia, for example). Overcoming our specialization in time or place is daunting, and it is maybe why so few of us try to write broad, explanatory, synthetic works like the kind by Diamond. Even Hughes (2001), in his attempt at an

Box 5.04 Jared Diamond's five characteristics of failing societies in *Collapse*

Jared Diamond has offered a diagnostic list of five features that many past civilizations exhibited prior to their collapse. They are: climate change, hostile neighbors, collapse of essential trading partners, environmental problems, and failure to adapt to environmental issues.

Can you think of the nearest collapsed civilization? What features from Diamond's list actually "fit" for explaining that collapse? What else might explain the collapse that is not included in this list?

Box 5.05 Reconstructing an environmental history

In a group of four or five, pick a region, people, and time period far from your own home.

Now, what methods discussed above would you use to reconstruct the environmental history of that area, time, and people? Is it a region, time, or people where written documents are available?

Will you suggest biophysical proxy methods to bolster, complement, or fill in the time gaps if few documents are available?

Put together a clear list of methods you would use, and why those methods would be important in your case, and be ready to share with the other groups.

environmental history of the world, prefers vignettes of environmental history episodes as opposed to the kind of sweeping argument and narrative that Diamond offers in his books. Box 5.05 lays out one possible way to make this complex task tangible to you, the reader.

Environmental History as Context or as a Tool?

All notable works in environmental history lean towards either academic context (history as history) or as practical and useful resources for current and future environmental management problems. Early work can be characterized as "**declensionist**," in that these works characterized human–environmental relationships as those marked by environmental degradation or at least those in which humans impacted their local surroundings in rather marked ways. If you are drawn into these works, as you read through articles and books, can you find examples of work that are more ambivalent and less about environmental decline?

Authors choosing to end their period of analysis well before the present may, inadvertently, limit the role or potential utility of their work. This is a legitimate choice for authors who choose to work in the academic tradition of environmental history, as the past continues to inform the present. One successful example of environmental story-telling is John McNeill's (2000) *Something New Under the Sun*, which addressed the impacts and environmental record of the last century. While clearly historical, the book also bridges to the same issues we face today, namely how to achieve forms of sustainable livelihood that don't compromise future societies. So environmental histories don't have to focus on esoteric periods of human history; they can be compelling to us now, here, for what they tell us about our current lives.

As you read through works of environmental history, some central questions could be "Who is the audience?" and "Who might use this information in current context?" What is the relevance of environmental history if it does not inform restoration ecologists working today? This is one of the principal dilemmas for the applied scholars of historical environments – having productive engagements with policymakers and practitioners of applied ecology. For example, park managers aiming for conservation or even preservation of "pre-settlement" landscapes, say, in North America are hard pressed to reconstruct those landscapes as they existed. How close can they come to grasslands circa 1800 or original forests as they existed in 1600? Are these achievable goals? In fact, many of the places that we take for granted, such as wildlife refuges, are anything but wild and have distinct human and political origins to them (Wilson 2010).

The remaining challenges we face, that you face if you're interested in environmental history, have to do with agency, narrative, and critical thinking. First, we cannot forget the actual humans, real individuals, involved in our different versions of environmental history (see Box 5.06 below). The "human–environment" relationship can sometimes be treated so abstractly, so neutrally, that few actual names (of people) will ever appear. For example, to say that "modern society abuses the environment" may capture some degree of truth, but it does not specify who is doing the work, where, and when. Using actual names helps specify and explain who did what, and for what reasons. Second, we can do better in telling actual and

Box 5.06 What is your environmental history?

How has your local community shaped its own environmental history? Take a moment to think of the livelihoods in the surrounding area – are they urban, suburban, or rural? What aspects of pre-settlement landscape history are still evident in the landscape? Who were the local residents prior to, say, colonial-era pioneers, or the first industrial-era farmers? What are the specific aspects about your region and country that distinctively shaped the environmental history of your area? How does your family's history fit into the local environmental history?

compelling stories that others want to read. Much of our geographic perspective can inform a place-based or spatial framework for the stories we tell about past and current humanized environments. Finally, we could make better and more critical use of archival documentation, so as to more thoroughly question the provenance and purpose of original documents. These primary materials, while absolutely essential to the entire enterprise of environmental history, should not be taken as stone-clad facts. They are instruments and reflections of power, of the story-tellers who have control to assert and record certain events on paper. Field-based oral history, or landscape reconstructions, can help balance the understood and well-accepted bias left recorded in archival resources.

Chapter Summary

Environmental history is now widely practiced as a human–environment perspective across a variety of disciplines. Geographers in the past did not think of themselves as historians, even if they were interested in archaic or modern landscape processes. But geographers contribute to environmental history in a number of ways. Some authors focus on the ecology of past environments, some on urban ecologies, but the best human–environmental work in environmental history pays attention to change over time. From primary documents found in archives, to fossil pollen, to buried sediments – a number of tools and sources are now available to make sense of past environments. While it is impossible to summarize in a single chapter the variety and diversity of environmental history, it is the perspective that is important (but see White 2001). Just as Chapter 4 illustrated that writing *apolitical* accounts of human–environment relations is problematic, this chapter argues that *ahistorical* accounts of human–environment relations are dangerous. Environmental histories now exist for almost every major region on the planet, and in many languages. Environmental history was once focused on the ecological, the natural, the rural and agrarian. Environmental history in thoroughly modern settings and time periods is helping to inform our understanding of human–environment relationships. Environmental historians do now specifically address issues of industrialization, modernization, and urbanization, as well as more intimate scales like the human body. The underlying assumption for environmental historians, that change over time is the norm and not the exception, can help us understand sometimes controversial subjects such as climate change (Chapter 8), as Bill deBuys (2011) has done in his recent book *A Great Aridness*.

Critical Questions

1 Many environmental historians seem profoundly ambivalent, at best, about humanity's presence in and experience with nature – what is our role in nature and how has that changed?

2 If the concept of an Anthropocene era is correct, or even helpful, what marks the beginning of the era: the origins of agriculture by humans? Would it be the industrial revolution?
3 Can you think of work that is more about the "environment's" history rather than the history of human–environment relationships? Does this body of work reflect a different discipline or approach to the problem?
4 Much of the work in environmental history is not just anglophone but also Americanist in perspective and origin. What problems does this pose for new kinds of environmental history? What stories are left *out* of the current wave of environmental histories?
5 Many of our best environmental histories are written by physical geographers. What aspects or tools seem to be missing from the more humanist geographers in analyzing past human–environmental relationships? What are the potential limitations of some of the proxy techniques listed in this chapter?

Key Vocabulary

archival (methods)
Anthropocene (era)
carbon-14 (C14) dating
colonialism
cultural possibilism
declensionist stories
direct sources
environmental determinism
environmental history
geochronology
historical political ecology
ice cores
indirect (proxy) sources
irruptive ecology
lake varves
landscape archaeology
narrative
oral history
oral tradition
packrat middens
paleoecology
paleosols
pastoral transhumance
pollen analysis (palynology)
portmanteau biota
proxy methods
remnant shorelines
social Darwinism
sporopellenin
tree-ring analysis (dendrochronology)

References

Arford, M.R. and Horn, S. (2004) Pollen evidence of the earliest maize agriculture in Costa Rica. *Journal of Latin American Geography*, 3(1), pp. 108–15.

Braudel, F. (1958) Histoire et sciences sociales: la longue durée. *Annales. Histoire, Sciences Sociales*, 13(4), pp. 725–53.

Butzer, K. (2005) Environmental history in the Mediterranean world: cross-disciplinary investigation of cause-and-effect for degradation and soil erosion. *Journal of Archaeological Science*, 32, pp. 1773–1800.

Butzer, K. and Butzer, E.K. (1997) The "natural vegetation" of the Mexican Bajío: archival documentation of a 16th century savanna environment. *Quaternary International*, 43/44, pp. 161–72.

Carey, M. (2010) *In the Shadow of Melting Glaciers: Climate Change and Andean Society* (New York: Oxford University Press).

Carney, J. (2001) *Black Rice* (Cambridge, MA: Harvard University Press).

Carney, J. and Rosomoff, R.N. (2011) *In the Shadow of Slavery: Africa's Botanical Legacy in the Atlantic World* (Berkeley: University of California Press).

Carson, R. (1962) *Silent Spring* (New York: Houghton Mifflin).

Cioc, M. (2002) *The Rhine: An Eco-biography, 1815–2000* (Seattle: University of Washington Press).

Colten, C. (ed.) (2000) *Transforming New Orleans and its Environs: Centuries of Change* (Pittsburgh: University of Pittsburgh Press).

Colten, C. (2006) *Unnatural Metropolis: Wresting New Orleans from Nature* (Baton Rouge: Louisiana State University Press).

Cronon, W. (1983) *Changes in the Land: Indians, Colonists and the Ecology of New England* (New York: Hill & Wang).

Cronon, W. (1991) *Nature's Metropolis: Chicago and the Great West* (New York: W.W. Norton).

Crosby, A. (1972) *The Columbian Exchange* (Westport: Greenwood Press).

Crutzen, P.J., and Stoermer, E.F. (2000). The "Anthropocene." *Global Change Newsletter*, 41, pp. 17–18.

Davis, D.K. (2007) *Resurrecting the Granary of Rome: Environmental History and French Colonial Expansion in North Africa* (Athens: Ohio University Press).

Davis, D.K. (2009) Historical political ecology: on the importance of looking back to move forward. *Geoforum*, 40(3), pp. 285–6.

Davis, M. (2002) *Late Victorian Holocausts: El Niño Famines and the Making of the Third World* (London: Verso).

deBuys, W. (2011) *A Great Aridness: Climate Change and the Future of the American Southwest* (Oxford: Oxford University Press).

Diamond, J. (1999) *Guns, Germs and Steel: The Fates of Human Societies* (New York: W.W. Norton).

Diamond, J. (2004) *Collapse: How Societies Choose to Fail or Succeed* (New York: Viking).

Dover, S., Edgecombe, R., and Guest, B. (eds.) (2003) *South Africa's Environmental History: Cases and Comparisons* (Athens: Ohio University Press).

Endfield, G.H. (2008) *Climate and Society in Colonial Mexico: A Study in Vulnerability* (Oxford: Blackwell Publishing).

Fairhead, J. and Leach, M. (1996) *Misreading the African Landscape: Society and Ecology in a Forest-Savanna Mosaic* (Cambridge: Cambridge University Press).

Fischer-Kowalski, M. and Haberl, H. (eds.) (2007) *Socio-ecological Transitions and Global Change: Trajectories of Social Metabolism and Land Use* (London: Elgar).

Glacken, C.J. (1967) *Traces on the Rhodian Shore: Nature and Culture in Western Thought from Ancient Times to the End of the Eighteenth Century* (Berkeley and Los Angeles: University of California Press).

Guha, R. and Gadgil M. (1992) *This Fissured Land: An Ecological History of India* (Berkeley: University of California Press).

Gumprecht, B. (1999) *The Los Angeles River: Its Life, Death, and Possible Rebirth* (Baltimore, MD: Johns Hopkins University Press).

Hornborg, A., McNeill, J.R., and Martinez-Alier, J. (eds.) (2007) *Rethinking Environmental History: World-System History and Global Environmental Change* (Lanham, MD: Altamira Press).

Hughes, J.D. (2001) *An Environmental History of the World: Humankind's Changing Role in the Community of Life* (New York: Routledge).

Hunter, R. (2009) Positionality, perception, and possibility in Mexico's Valle del Mezquital. *Journal of Latin American Geography*, 8(2), pp. 49–69.

Huntington, E. (1915) *Civilization and Climate* (New Haven: Yale University Press).

Kull, C. (2004) *Isle of Fire* (Chicago: University of Chicago Press).

Langston, N. (2010) *Toxic Bodies: Hormone Disruptors and the Legacy of DES* (New Haven: Yale University Press).

Liu, K.B., Lu, H.Y., and Shen, C.M. (2008) A 1,200-year proxy record of hurricanes and fires from the Gulf of Mexico coast: testing the hypothesis of hurricane-fire interactions. *Quaternary Research*, 69, pp. 29–41.

Marsh, G.P. (2003 [1864]). *Man and Nature*. Or, *Physical Geography as Modified by Human Action* (Seattle: University of Washington Press).

McNeill, J.R. (2000) *Something New Under the Sun: An Environmental History of the Twentieth Century* (New York: W.W. Norton).

McNeill, J.R. (2003) Observations on the nature and culture of environmental history. *History and Theory*, 42, pp. 5–43.

McNeill, J.R. (2010) The state of the field of environmental history. *Annual Review of Environment and Resources*, 35, pp. 345–74.

McShane, C. and Tarr, J. (2007) *The Horse in the City: Living Machines in the Nineteenth Century* (Baltimore, MD: Johns Hopkins University Press).

Melosi, M. (2008 [2000]) *The Sanitary City: Environmental Services in Urban America from Colonial Times to the Present* (Pittsburgh, PA: University of Pittsburgh Press).

Melville, E. (1994) *A Plague of Sheep* (Cambridge: Cambridge University Press).

Merchant, C. (1989) *Ecological Revolutions: Nature, Gender, and Science in New England* (Chapel Hill: University of North Carolina Press).

Mitman, G. (2007) *Breathing Space: How Allergies Shape our Lives and Landscapes* (New Haven: Yale University Press).

Offen, K. (2004) Historical political ecology: an introduction. *Historical Geography*, 32: 19–42.

Perramond, E.P. (2001) Oral histories and partial truths in Mexico. *Geographical Review*, 91(1–2), pp. 151–7.

Pfefferkorn, I. (1989) *Sonora: A Description of the Province* (Tucson: University of Arizona Press).

Robbins, P. (2007) *Lawn People: How Grasses, Weeds, and Chemicals Make Us Who We Are* (Philadelphia: Temple University Press).

Sauer, C. (1938) destructive exploitation in modern colonial expansion. *Comptes Rendus du Congrès International de Géographie* (Amsterdam), 2, sect. 3c, pp. 494–9.

Sluyter, A. and Dominguez, G. (2006) Early maize (*Zea mays L.*) cultivation in Mexico: dating sedimentary pollen records and its implications. *Proceedings of the National Academy of Sciences*, 103(4), pp. 1147–51.

Thomas, W.L. (ed.) (1956) *Man's Role in Changing the Face of the Earth* (Chicago: University of Chicago Press).

Turner II, B.L., Clark, W.C., Kates, R.W., Richards, J.F., Mathews, J.T., and Meyer, W.B. (eds.) (1990) *The Earth as Transformed by Human Action: Global and Regional Changes in the Biosphere over the Past 300 Years* (Cambridge: Cambridge University Press).

Vansina, J. (1984) *Oral Tradition as History* (Madison: University of Wisconsin Press).

Walker, B. (2010) *Toxic Archipelago: A History of Industrial Disease in Japan* (Seattle: University of Washington Press).

Weisiger, M. (2009) *Dreaming of Sheep in Navajo Country* (Seattle: University of Washington Press).

White, R. (1983) *The Roots of Dependency: Subsistence, Environment, and Social Change among the Choctaws, Pawnees, and Navajos* (Lincoln: University of Nebraska Press).

White, R. (1991) *The Middle Ground: Indians, Empires, and Republics in the Great Lakes Region, 1650–1815* (Cambridge: Cambridge University Press).

White, R. (1996) *The Organic Machine: The Remaking of the Columbia River* (New York: Hill & Wang).

White, R. (2001) Afterword. Environmental history: watching a historical field mature. *Pacific Historical Review*, 70(1), pp. 103–11.

Williams, M. (1993) *Wetlands: A Threatened Landscape* (London: Wiley-Blackwell).

Williams, M. (1994) The relations of environmental history and historical geography. *Journal of Historical Geography*, 20(1), pp. 3–21.

Williams, M. (2002) *Deforesting the Earth: From Prehistory to Global Crisis* (Chicago: University of Chicago Press).

Wilson, R.M. (2010) *Seeking Refuge: Birds and Landscapes of the Pacific Flyway* (Seattle: University of Washington Press).

Worster, D. (1985) *Rivers of Empire: Water, Aridity, and the Growth of the American West* (New York: Pantheon Books).

6

Hazards Geography and Human Vulnerability

Icebreaker: Cyclones Hitting Land in Bangladesh and Myanmar

The Bay of Bengal, which lies in the North Indian Ocean (see Figure 6.01), is a region of the world frequently hit by tropical storms and cyclones that bring floods and devastation to local communities. On the evening of November 15, 2007, Cyclone Sidr (a strong Category 5 storm), made landfall in the densely populated coastal zone of Bangladesh. Its peak 160 m.p.h. winds, heavy rain, and 16-foot storm surge flattened shacks, blew away schools and houses, and destroyed numerous trees. Damage to homes and agriculture

An Introduction to Human–Environment Geography: Local Dynamics and Global Processes,
First Edition. William G. Moseley, Eric Perramond, Holly M. Hapke and Paul Laris.

Published 2014 by John Wiley & Sons, Ltd.

was extensive – estimated at US$1.7 billion. Over 3400 people lost their lives, and 55,000 others were injured, but this death toll was much lower than in previous storms. For example, 138,000 people lost their lives in 1991 and 500,000 in 1970 because of Bangladeshi cyclones.

In contrast, just six months later Cyclone Nargis, another Category 5 storm, wreaked havoc on the neighboring country of Myanmar (Burma) when it hit land on May 2, 2008. Over 146,000 people perished, and damage exceeded US$10 billion – the highest level on record in this region. In fact, Nargis was the deadliest named cyclone in the North Indian Ocean and the eighth deadliest cyclone of all time. Why did two cyclones of equal strength in the same region of the world have such different impacts? Bangladesh is much more densely populated than Myanmar (1045/sq. km compared to 70/sq. km) and is among the lowest-lying countries in the world. Both countries are relatively poor, and yet over 40 times as many people died in Myanmar as in Bangladesh. What factors might explain this? Furthermore,

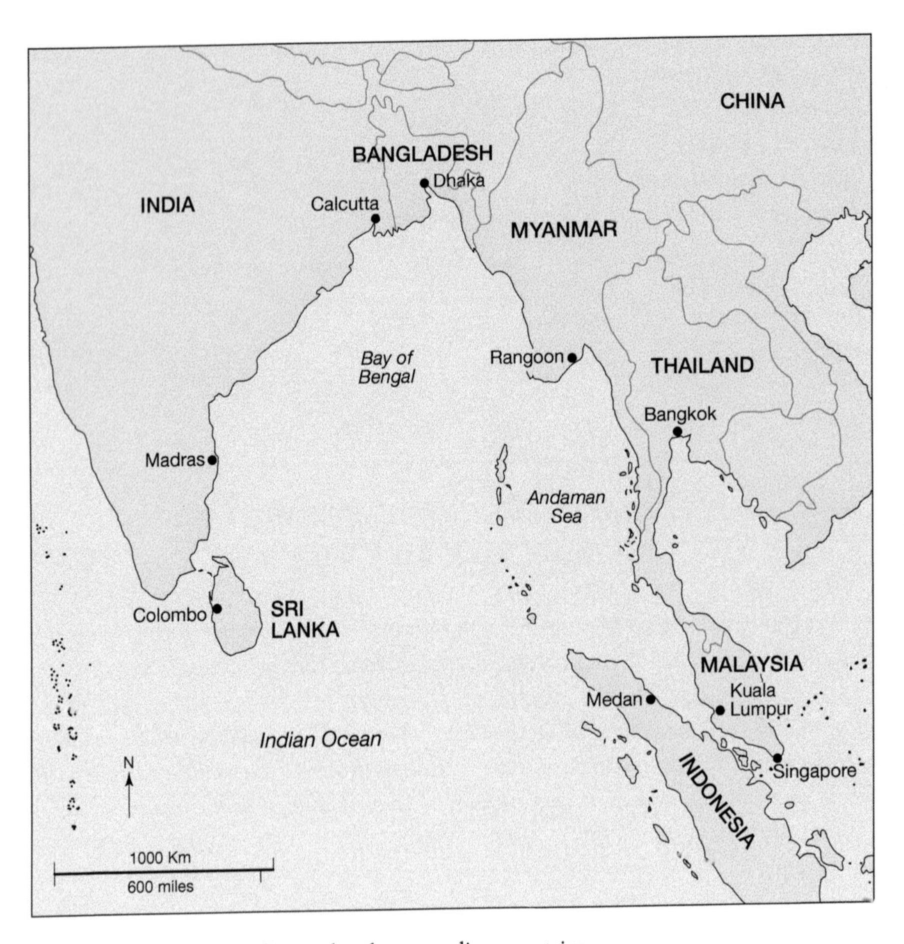

Figure 6.01a The Bay of Bengal and surrounding countries.

what factors explain the tremendous reduction in death toll in Bangladesh over previous cyclones in its history? These are some of the questions we will consider in this chapter.

Figure 6.01b Cyclone Sidr near landfall.
Source: NASA, November 14, 2007, http://commons.wikimedia.org/wiki/File:Sidr_14_nov_2007_0445Z.jpg.

Chapter Objectives

1. To learn about different types of hazards.
2. To critically evaluate the degree to which natural hazards are really natural.
3. To understand how approaches to hazards geography have changed over time.
4. To learn about human vulnerability to hazards and related terms such as risk, resilience, and adaptation.

Introduction

This chapter explores the long-standing and dynamic subfield of hazards geography. We begin by defining what constitutes a hazard and investigating different types of hazards. We then examine the all-important social dimension of hazards which, as the opening example suggests, has as much or more to do with the "hazard" character of the situation as it does with the event that precipitated it. Next we explore the origins of the subfield of hazards geography and how it has evolved over time. The last portion of the chapter is devoted to the related concepts of risk, resilience, vulnerability, and adaptation.

What Is a Hazard?

What is a hazard? **Hazards** are often broadly defined as situations that can cause injury, disease, economic loss, or environmental damage. While geographers have traditionally focused on natural hazards, the term broadly includes a whole range of hazards, such as:

- *Cultural hazards*: unsafe working conditions, smoking, poor diet, drugs, drinking, etc.
- *Chemical hazards*: harmful chemicals in the air, water, soil, and food.
- *Physical hazards*: drought, fires, tornadoes, hurricanes, earthquakes, volcanic eruptions, and floods.
- *Biological hazards*: hazards from pathogens (bacteria, viruses, pollen and other allergens, and animals (e.g., snakes).

Are Natural Hazards "Acts of God"?

We often think of hazards as exogenous events, or "Acts of God." A key insight of geographers is that their destructive or harmful nature frequently has as much or more to do with the structure of society as with the natural environment. How could this be? If lightning strikes and kills or seriously injures a person, isn't this clearly a random event over which we have little or no control? Not necessarily. For example, scholars examining lightning deaths in the United States over the past century have found an exponential decrease in the number of deaths per million people related to such a cause (e.g., López and Holle 1998). One explanation might be that the number of lightning strikes is decreasing. This, however, is not the case. During this time, what the country did experience was a dramatic decrease in the size of its rural population. This is important because lightning tends to strike those who spend a lot of time outside (farmers, ranchers). As the nation became less rural, and more likely to labor inside, we saw a precipitous decline in deaths related to lightning strikes. Therefore, it is the changing structure of American society that has much to do with trends in lightning-related deaths.

Similarly, drought is often considered to be a natural hazard because of the resulting crop or animal loss it can inflict. While people rarely face food shortages as a result of drought in the United States, Europe, or Japan, droughts are often blamed for famine and hunger in other parts of the world. But are famines really caused by droughts? The West African Sahel (see Figure 6.02) is a semi-arid region known for its highly variable rainfall. Farmers traditionally dealt with frequent low-rainfall years via several strategies. First, they would plant different types of crops in multiple dispersed fields. As crops have different water requirements, some plants might do well in low-rainfall years (such as finger millet) whereas others would thrive when rain was plentiful (such as corn or maize). The approach of planting in multiple dispersed fields was also meant to counteract the highly localized nature of convective rainstorms in the region, wherein one side of a village might receive rain while the other did not; and take advantage of the variation in soil types across the landscape which have different water-retention capacities (e.g., sandy soils tend to drain quickly whereas loamier soils tend to retain moisture). While such strategies might not maximize production in any given year, they went a long way towards ensuring a minimum amount of production. Farmers also historically stored grain from year to year (replenishing and expanding stocks in good years, drawing on these in lean times), with household granaries often containing a two-year supply of food (Moseley 1996). These traditional strategies, which evolved in the context of highly variable rainfall, created a situation where hunger was rare.

Many of these practices changed during the colonial period in West Africa. African strategies for mitigating the effects of variable rainfall were disparaged as risk-averse, conservative, primitive, or worse. By implementing – and strictly enforcing – a head tax (tax assessed based on the number of individuals in one's

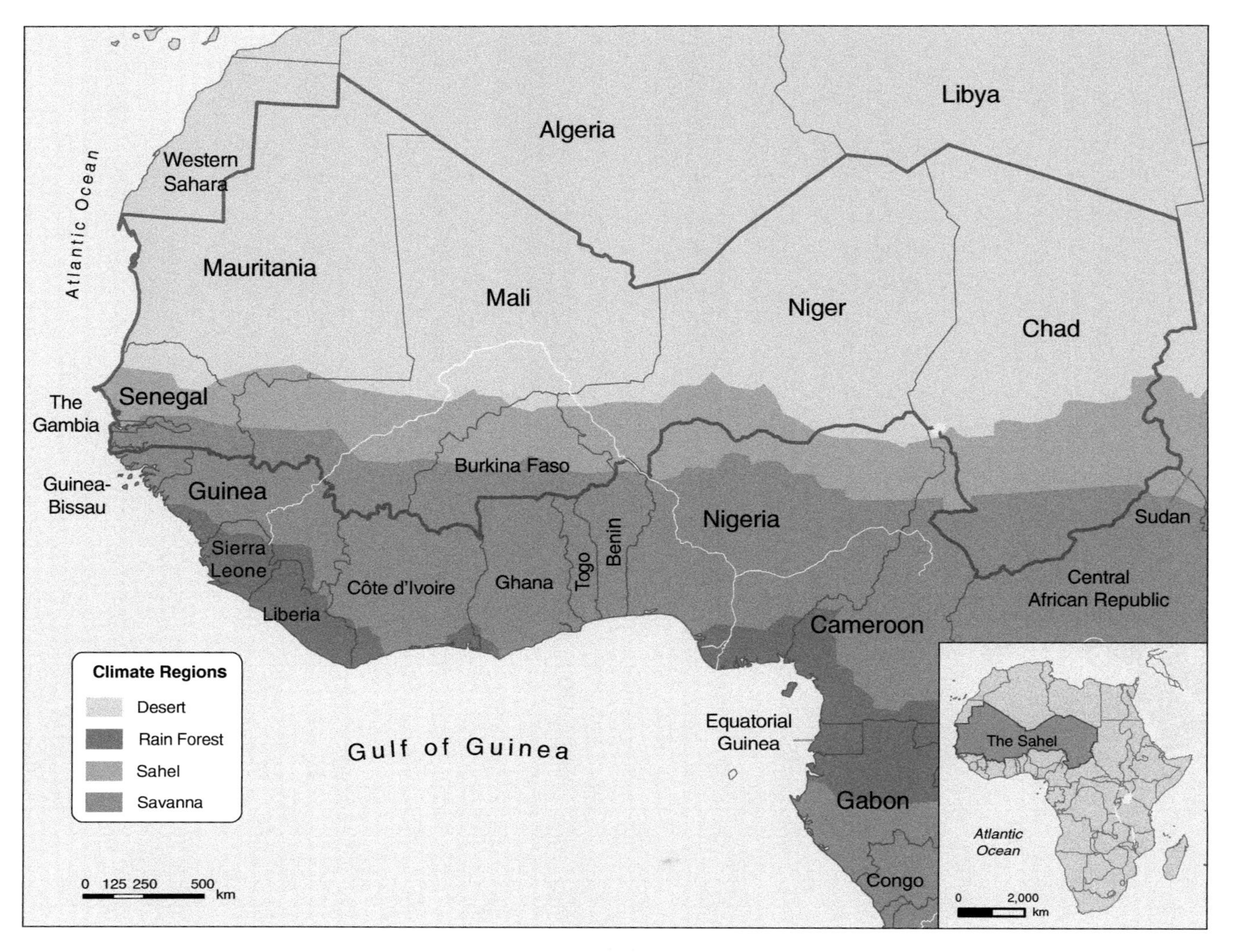

Figure 6.02 The West African Sahel. *Source*: Drawn by Macalester College cartographer Birgit Mühlenhaus.

household), the French and British colonial authorities created a need for cash that could only be acquired via growing cash crops or migrating in search of paid work. The result was a decline in attention to food crops, reduced stocks in household granaries, and increasingly large fields of one type of crop (often cotton or peanuts in the Sahel). The colonial and post-colonial periods have been riddled by droughts, hunger, and occasional large-scale famine. While many are quick to condemn the weather, over-population, or declining agricultural yields for these problems, it is the reorganization of these rural production systems during the colonial period that offers a different explanation (Franke and Chasin 1980; Watts 1983a).

Early Hazards Geography

Natural hazards have long been of interest to geographers and even longer amongst proto-geographers. For example, in his 4th-century BC text, "On Airs, Waters and Places," the Greek scholar Hippocrates suggested that human well-being and danger were strongly related to the natural environment in which they live. Later, in the first century AD, Strabo's "Geography," which describes the then known world around the Mediterranean, had a detailed discussion of earthquake disasters in this region. Geologists are perhaps best known for their attention to the biophysical dimensions of natural hazards, focusing on earthquake detection for example. In contrast to geologists, geographers have tended to focus on both the human and the physical sides of natural hazards. A more formal or organized subdiscipline of approaches to hazards geography began to emerge in the United States in the 1940s and is often most closely associated with the work of **Gilbert F. White** (see Box 6.01). White, whose work focused on flooding in the USA, argued that structural approaches to managing floods (such as dams and levies) could exacerbate these events because they often led to increased human settlement in flood zones and, as a result, greater human loss during flood events. White trained several students, most notably Robert Kates and Ian Burton, who went on to study other types of natural hazards around the world, including droughts, hurricanes, and tsunamis. Even the contemporary work of some geographers like Craig Colten (2006) on recovering from hurricane damage in New Orleans echoes much of the same advice Gilbert White was giving to the US government over 60 years ago.

Evolving Conceptions of Hazards Geography

Hazards geography has continued to mature and expand since it emerged as a subfield in the 1940s. In the early 1980s, a critique began to appear (most closely associated with Watts 1983b; but later with Blaikie et al. 1994) which suggested that hazards geography was ignoring the importance of political economic factors in its analysis. Early hazards geography had typically focused on a natural event (such as a flood) as a "trigger" for a disaster, and then an analysis of the human side of the problem that focused on individual behavior and perception, as well as the

Box 6.01 Gilbert F. White: pioneer of hazards geography

Gilbert F. White is often considered to be the father of hazards geography. This designation dates to his influential dissertation, published by the University of Chicago in 1945, entitled "Human Adjustments to Floods." In this thesis, which focused on the Mississippi River Basin, White argued that an over-reliance on dams, levees, and floodwalls (broadly known as structural works) had increased damage caused by flooding rather than decreasing it. According to White, the restriction of normal flooding through structural works encourages people to settle in floodplains. When such constraints are breached in a high flood year (which become increasingly frequent as floodplains are restricted upstream through structural works), the resulting damage is typically much worse than it might otherwise have been. He famously suggested that "Floods are an act of God, but flood losses are largely an act of man." White often advocated flexibility or accommodation of flood hazards rather than structural solutions. His key insights have endured, and seem especially poignant after disasters such as Hurricane Katrina, which struck New Orleans in the summer of 2005.

In addition to his scholarship, other aspects of Gilbert F. White's career are equally remarkable. From 1934 to 1942 he served in Franklin Roosevelt's New Deal Administration working on various aspects of floodplain and natural resource policy. As a Quaker he was a conscientious objector in World War II, and served with the American Friends Service Committee in France helping refugees. He was interned by the Germans until 1944. After completing his PhD, he became the youngest ever college president in the United States at that time, serving as president of Haverford College from 1946 to 1955. He then served as a professor and chair of the geography department at the University of Chicago from 1955 to 1969 (establishing the so-called "Chicago School" approach to natural hazards), and then the rest of his career at the University of Colorado until his death in 2006. In addition to his work on hazards geography and water resources, White is often remembered for his interest in public policy, serving on numerous US government and international commissions.

rules and regulations of the state. As such, it was argued that people chose to live in potentially dangerous floodplains because governmental regulations permitted them to do so and because they did not fully appreciate the risks involved with such a location decision. The "political economy" critique of this understanding had two components.

First, it suggested that individual "choice" was often circumscribed by a person's position in society and, ultimately, the structure of the economy. As

Box 6.02 The story of Princeville, North Carolina

Figure 6.03 High water in Princeville. A boat carrying a group of Princeville, N.C., residents travels down Main Street after floodwaters from the Tar River completely flooded the town in this on September 17, 1999. Hurricane Floyd dropped more than 20 inches of rain in eastern North Carolina after making landfall. Much like in New Orleans, it was the flooding that caused the damage, forcing more than 100,000 people into shelters, destroying 7000 homes while damaging 56,000. © Alan Marler/AP/Press Association Images.

In 1999, the small town of Princeville, NC (population approx. 2000), made national headlines in the wake of flooding that followed Hurricane Floyd. Because of its low elevation and proximity to the Tar River, the entire town was completely inundated by 20 feet of water for 10 days, and all of its residents were forced to evacuate. Every building in the town was destroyed, and in the weeks that followed, the US government attempted to buy out the town's residents in exchange for their relocation to other communities. But the town's residents resisted. Princeville holds a special place in the history and landscape of eastern North Carolina, and in fact, the United States as a whole. It is the oldest town incorporated (i.e., a town with an official charter and elected officials) by African Americans and the first independently governed African American community in the US. Princeville was established by freed slaves after the Civil War and incorporated in 1885. The town's historical significance, its importance to African Americans in the region, and its strong sense of community were important factors in its residents' refusal to relocate. But the town's location speaks to the role played by race and discrimination in the impact natural hazards can have on different groups within a society. Princeville is located on land that is particularly hazardous – 1999 was the *fifth* time the town had flooded in the past century. One might wonder why anyone would want to live there. But, it was the only land available to African Americans at the time of its founding, and so the legacies of racism and deprivation continue over 100 years later.

such, a poor person might settle in a floodplain because it was the sole place she or e could afford to live or because it was the only place someone of his or her racial category was permitted to settle (see Box 6.02). That person's poverty, or racial stigmatizing, might then be linked back to a certain set of economic arrangements.

The second, and more abstract, aspect of the political economy critique had to do with the conception of a natural event as an autonomous "trigger" to a disaster. Here it was suggested that the human experience of the natural environment (including so-called trigger events) is conditioned by the society in which humans live. As such, a normal natural occurrence – such as the variable annual flood – may be considered a hazard, a blessing, or a neutral event to different human societies. It is a hazard to modern industrial societies that build cities in floodplains or along rivers that have been confined by levees. It is a blessing to societies that depend on floodplain agriculture reinvigorated by periodic nutrient infusions from river silts; or to fishing communities who depend on fish stocks that reproduce in annual flooded zones. Finally, variable river flows may be a neutral event for those who live far from the river. If it is the structure of society that dictates how humans experience a natural event, then we cannot really conceive of natural events as exogenous or independent triggers. In other words, nature and society should not be considered as separate entities, as the shape of the latter influences the experience of the former.

The other major shift in hazards geography over the past 50 years has been an increasing interest in **technological hazards** (as opposed to natural ones). The former include a whole range of human-created dangers, such as toxic waste sites and dirty industries. Some question the distinction between technological and natural hazards, but others find this division useful. While hazards geographers have examined technological hazards (e.g., Cutter 1993), an increasing number of scholars have examined these risks through the lens of environmental justice (e.g., Pulido 2000), an approach which will be examined further in Chapter 7.

Hazards and Society

A hazard is a potential threat to humans and their welfare. **Risk** is the likelihood a hazard will occur. Risk encompasses threat to human life, threat to health and safety, property damage, and economic loss, and is usually expressed in terms of probability, a mathematical statement about how likely an event is to occur and what its probable effect will be. In quantitative terms, risk is proportional to the probability of a particular hazard event occurring and the expected losses the event may cause. Theoretically, if all other factors are equal, the risk associated with a Category 5 cyclone, for example, should be fairly consistent. And yet, as the opening story of this chapter illustrates, the actual threat to human life resulting from two comparable storms differed tremendously. The fact is, "all other factors" are rarely equal. Regional differences in population densities, for example, create differences in risk exposure. Poor construction of buildings can exacerbate the effect of earthquakes, thereby intensifying risk. Also important are other social, economic, and political factors that influence the actual impact of a particular hazard event. Measures taken by the Bangladeshi government and the international community to mitigate the risk posed by cyclones greatly reduced the fatalities from Cyclone Sidr. These included cyclone walls planted with trees, disaster shelters on stilts, early warning systems and timely evacuations (US Department of State 2007). In contrast, the Myanmar government's response during Cyclone Nargis was by most accounts slow to materialize and inadequate. Relief agencies were frustrated by the lack of cooperation from the government, and many of the casualties in Myanmar were related to failures to provide adequate relief in an efficient and timely manner.

An important aspect of risk is the fact that different people perceive risk in different ways. For example, research conducted in the early 1990s revealed that the general public in the United States ranked hazardous waste sites at the top of the list of most serious problems while scientists placed this risk at a much lower level. In contrast, experts were very concerned about exposure to indoor air pollution, but the public expressed less concern for this hazard (see Cutter 1993). What accounts for such divergence in views? Occasionally the language used to name certain events may either exaggerate or understate a particular risk. Sometimes people misunderstand and misinterpret official information about an impending hazard (see Box 6.03). Even when people do understand such information, other factors lead them to behave in ways that fail to reduce risk. Some research indicates that the possession of pets and the inability of shelters to accommodate them is an important reason why people do not evacuate their homes during hurricanes or tornadoes. The presence of a family member who is elderly or handicapped with limited mobility is another factor underlying people's perceptions and decisions. The role played by the media is also a factor. Keith Smith (1992) points out that in the United Kingdom, accidental death accounts for less than 3% of the deaths in a given year, and most of these are due to everyday events like car accidents. Yet the attention given to rapid-onset environmental hazards by the media is disproportionate to the actual death toll of these events. This, of course, is largely due to the fact that such events, although relatively infrequent, can potentially cause catastrophic loss and so loom large in the public's imagination.

More important, though, is that different actors "see" hazards as different types of events based on a whole range of socially and culturally rooted factors, and prepare for them in different ways (Bankoff and Hilhorst 2009). Scientists and disaster management experts tend to define risk in terms of probability and populations, that is, the odds of harm occurring to a certain percent of the population. Their perspective is based on the idea that something negative is likely to happen, and that risk can be both predicted and *managed*. Their priority is to protect the greatest number of people the greatest amount of time. Thus, expert assessments are

Box 6.03 Hurricane maps and evacuation behavior

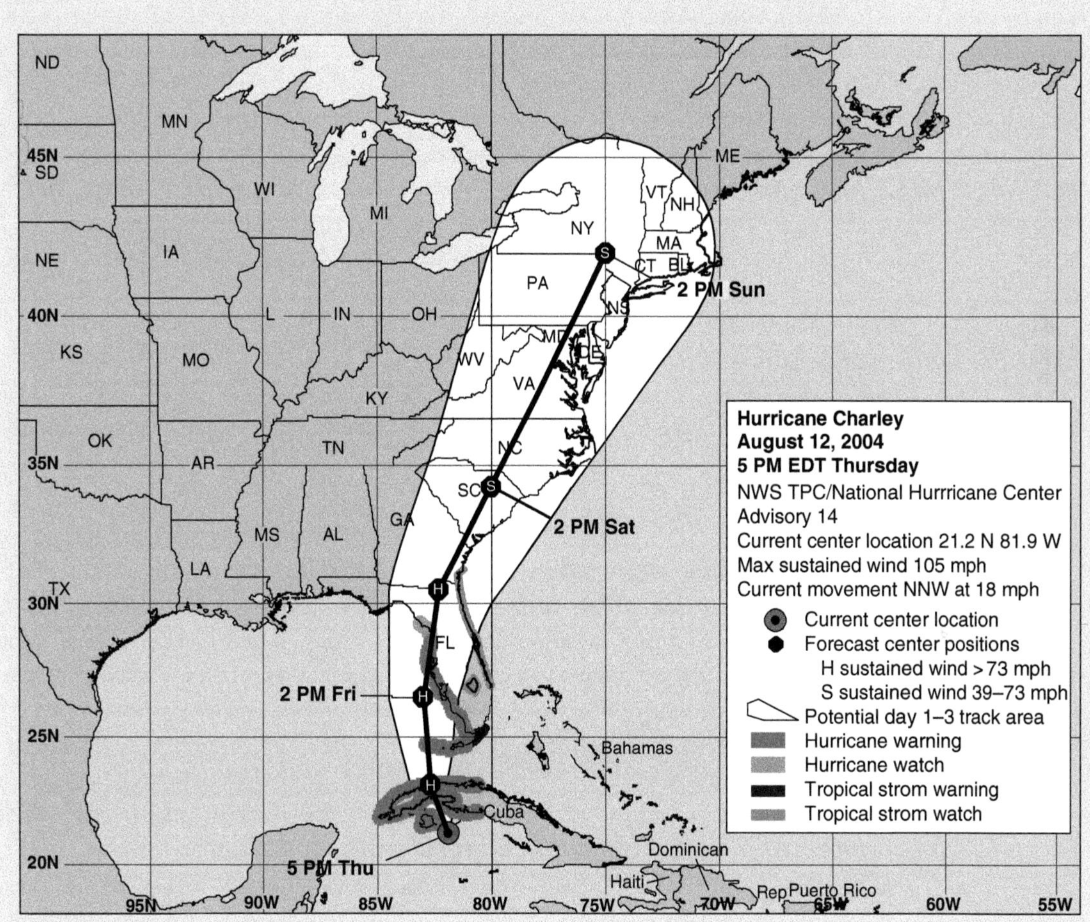

Figure 6.04 Hurricane Charley, August 12, 2004. *Source*: NWS TPC/National Hurricane Center Advisory 14. Reprinted by permission of the National Hurricane Center.

In the US, hurricane watches and warnings are issued when hurricane conditions are expected to affect a region within 36 and 24 hours, respectively. Early warnings are critical for evacuation and other emergency management activities, but the earlier the watch or warning is issued, the more uncertainty there is with respect to the track of the storm. As a result, a graphic showing a cone of uncertainty is usually part of any advisory. An example from about 24 hours before Hurricane Charley made landfall in the US in 2004 is shown in Figure 6.04. This graphic contains a great deal of information about the current location, areas at different levels of risk, and the intensity forecast at different locations as the storm progressed. What most people focus on, however, is the black line that shows the forecast track of the hurricane at the time the warning is issued. The white area represents the forecast error, or the range in which the hurricane could make landfall, as it was understood when the warning was issued. Because people must believe that they are at risk before they are going to take action to protect themselves, it is important to understand how they interpret the cone. Hurricane Charley provides a good example of this. The storm had been forecast to make landfall near Tampa, but Charley suddenly turned toward the Fort Myers–Port Charlotte area and made landfall there on August 13. While this location was well within the cone of uncertainty and had been for several days, landfall occurred south of the black line. Many evacuated from the Tampa region but far fewer did from Charlotte County, where it actually struck. Too many people focused on the black line to evaluate their personal risk rather than the larger area at risk. Given this and many similar stories, there is much discussion about the most effective way to convey uncertainty in a warning so that those who need to can understand the risk to which they may be exposed and take appropriate action.

Source: Contributed by Burrell Montz, PhD, Department of Geography, East Carolina University.

expressed quantitatively as numerical estimates of fatality rates or monetary costs of damage. Average citizens, however, tend to view risk from a personal vantage point, projecting themselves into the situation of people affected by a hazard rather than considering the probability of the event occurring. Their assessments may be much more subjective in nature and include a wider range of factors than just the physical reality of a risk. Public understandings of risk derive from the larger context of daily life in which people face a wide range of issues and concerns that they then prioritize.

Geographers were among the first to study **risk perception**, the individual judgments people make about the degree of danger (risk), and how this then influences their behavior. What we have learned is that a number of factors influence how risks are perceived. These include individual personality, previous experience with an event, the ability to exercise control over a particular risk, the degree to which scientific information becomes part of public knowledge, and how one's friends, family, and neighbors view and respond to risk. People tend to ignore low-probability risks, even if potential damage is high, and they tend to downplay high-probability risks in familiar environments. Daily life presents so many challenges and concerns that people selectively focus on those that seem most immediate and *within their control*.

Risk, then, is defined, or constructed, in ways that reflect moral, political, socio-economic, and cultural situations that are value- and power-laden. One difference between Northern "developed" and Southern "less developed" contexts is that many Southern countries lack a history of strong states and systems for social welfare and protection. The overall environment in such countries is precarious, unpredictable, and insecure, and this generates different risk perception than that in Northern countries (Nathan 2008). For example, poor people living in high landslide risk areas in La Paz, Bolivia, viewed violence, the ability to earn a livelihood, and health problems as more significant risks than the potentiality of a landslide (Nathan 2008). In fact, publicly acknowledging that their homes were at risk of landslide could actually undermine their interests. It could lead to a depreciation of land value and/or forced relocation. In a context in which the poor lack political power and distrust public institutions, denying risk sometimes makes sense. Likewise, political leaders may neglect to acknowledge a risk because to do so would then imply a responsibility to act. The expert scientist perspective on risk – that the negative impacts of a probable event should be avoided through technical measures ("**risk management**") – is itself a cultural product. It evolved in the North over several hundred years, with much initial resistance from the Church, and only became dominant with the rise of Western science and modernity during the Enlightenment when ideas about human ability to control nature took shape.

Vulnerability. Closely related to risk is the notion of vulnerability, or the likely short- and long-term impacts of a particular hazard and society's ability to cope with these in the long term. The United Nations defines vulnerability as a "set of conditions determined by physical, social, economic and environmental factors or processes that increase the susceptibility of a community to the impact of hazards" (World Conference on Disaster Reduction 2005). Thus, vulnerability is not simply a result of environmental extremes or events. It is rooted in the social space of everyday life. It is a *context* in which people live, not an outcome of a particular event. As such, vulnerability varies across time and space, and different groups of people in society usually have different levels of vulnerability based on a range of factors. Typically, those with access to fewer resources from the outset – the poor and socially disadvantaged – face higher levels of risk and suffer more during disasters than those with more resources. Other groups who are particularly vulnerable to natural hazards include women, the elderly, children, members of ethnic minorities, and the physically and mentally disabled.

Factors that create a person's or group's vulnerability encompass physical, economic, political, social, and attitudinal factors that interact in often complex and sometimes hidden ways. Obvious factors include the physical location of human settlements, the construction and type of buildings, and environmental degradation. Some regions in the world are more susceptible to certain types of hazards, and some sites within regions, such as steep mountain slopes and low-lying areas, are more vulnerable to particular hazards than other areas. Less obvious are what we may term "social fabric of place factors," which include things like land-tenure (ownership) patterns, access to resources, social customs, and cultural ideas and attitudes. The 2004 Asian tsunami, for example, revealed that women were more vulnerable to drowning than were men because, as a result of social customs and cultural norms, fewer women than men could swim. Also, inequalities in gender roles and cultural contexts in the region give women less access than men to productive and social resources (income, credit, education, information) and decision-making processes that then placed them at a social, economic, and political disadvantage and made it more difficult for them as a group to weather and recover from the disaster than men (see Mehta 2007).

Sarah Ahmed and Elizabeth Fajber (2009) outline three dimensions of vulnerability. The first, **material vulnerability**, consists of factors related to a person's or household's material situation and well-being such as income source (local/non-local, land- or non-land-based), educational attainment, assets, and exposure to risk (distance from river, coast, or landslide zone). The extent to which one has access to monetary or other productive resources, how diverse these sources are, and on what they are based directly impacts one's vulnerability to a hazard. **Institutional vulnerability** encompasses social networks and kinship ties; infrastructure (access to roads, water/sanitation, electricity, health services, communication); household composition (number of dependants); reliability of early warning systems; and membership in a disadvantaged group (ethnic or religious minority, gender, etc.). Individuals connected to strong social or familial networks of support recover from disasters more easily than those less connected. Access to information, means of escape, and relief assistance also shape vulnerability. Finally, **attitudinal vulnerability** refers to one's sense of empowerment, which comes from access to leadership and knowledge about potential hazards. Those who are politically disempowered or lack access to information about hazards and relief assistance are more vulnerable than those who are not.

Adaptation. The way that peoples and communities recover from an external shock – such as a drought, earthquake, or hurricane – is a key component of vulnerability. If a household has a limited capacity to recover, it generally is considered to be more vulnerable. In areas where people are accustomed to recurring shocks, such as drought-prone regions, households typically have a set of alternative food-procurement methods they fall back on during hard times. These methods, also referred to as **coping strategies**, might include collecting wild fruits and berries which are not normally consumed, skipping meals, temporary migration, selling livestock to purchase food, or sending an older child to another region to work for wages. One distinctive aspect of coping strategies is that they are typically shorter-term, stop-gap measures. Households engage in these strategies to survive, but the goal is to return to the original or "normal" livelihood pattern. In fact, some research suggests that households undergo extreme hardships in the short term so that they may preserve the capital (such a plows and oxen) needed to return to their normal livelihood once a drought has passed (Moseley 2001).

However, sometimes shorter-term coping strategies undertaken to deal with natural disasters evolve into longer-term permanent changes, a process often referred to as adaptation. This is most likely to occur when infrequent occurrences (such as a droughts, hurricanes, or volcanic eruptions) become more common. Longer-term climatic shifts may mean that hurricanes become more regular along certain coastlines or that once infrequent droughts become increasingly common or longer in duration. Underlying geologic changes may also mean that earthquakes become less rare or volcanoes more active. In such cases, households often are forced to adapt, i.e., to make permanent changes to the way they acquire food and income. The most extreme form of adaptation might be to permanently move from an area plagued by natural hazards and take on an entirely new livelihood (e.g., rural farmers move to the city to become street vendors). Some scholars have taken to referring to such households as environmental refugees. However, most households are extremely reluctant to abandon an area and a livelihood altogether, only doing this as a last resort. This is because they have a familial attachment to an area, they are reluctant to give up on a livelihood for which they have knowledge and skills, and/or they may not have the resources (including money) or expertise needed to settle in a new area or take up a new livelihood. As such, many households attempt to adapt existing livelihoods to the new realities of an area. For example, farmers faced with increasingly drier weather conditions may choose to grow more drought-resistant crops, plant on different soil types, invest in irrigation, increasingly pursue non-farming activities, or engage in seasonal migration to other areas during the non-farming season. Coastal fisherfolk challenged by hurricanes may choose to move inland, change the architecture of their homes, or seasonally migrate during more hazardous times of year. Pastoralists (aka herders) threatened by droughts may switch from large stock (cattle) to smaller stock (goats) which are more versatile, change migratory patterns, sink new wells for watering livestock, or engage in non-livestock-related livelihood activities. Urban dwellers hampered by flooding or hurricanes may rebuild their homes on safer ground, reinforce their dwellings, or work collectively with area residents or government to address the hazard.

In addition to this aforementioned dynamic sense of adaptation (i.e., a permanent livelihood change in response to changing environmental conditions), the term adaptation is also used in a more static way in the geography and anthropology literatures to describe how well a society is adapted to its particular environmental setting (Denevan 1983). A well-adapted society is one that employs a range of natural resource management strategies that make sense (in an ecological sense) for a particular area or region. In other words, it is a sustainable system. For example, Nietschmann (1972) described a traditional rural livelihood system amongst the Miskito Indians of Nicaragua that involved a combination of hunting, farming, and fishing (including small-scale sea turtle harvests) in the 1960s. The introduction of improved fishing technologies, along with high demand for turtle meat from a nearby factory, led to the overfishing of turtles and a "maladapted" system. This left the Miskito more vulnerable to malnutrition. In other cases, a farming system might be brought into an area that is inappropriate, or maladapted, for a given set of ecological conditions. For example, Dutch farmers who settled north of Cape Town (in what is now South Africa) in the 17th and 18th centuries brought with them a European farming system that was maladapted to this new environment in the sense that it led to massive soil erosion (Meadows et al. 1996).

The maladapted situations described above hint at the fact that households are not simply adapting to changing environmental conditions, but to external economic conditions as well. Global economic change may bolster markets for some products and dampen demand for others. In adapting to these economic changes, households may engage in practices that are counterproductive or unsustainable in the local environmental context. In some cases, livelihoods may change so much in the face of external economic pressure, that entirely new hazards are created. Such was the case for the drought-induced famine in the African Sahel described at the beginning of this chapter. More challenging yet are situations where both environmental and economic conditions are in flux, forcing households to adapt on two fronts. This situation has been described by O'Brien and Leichenko (2003) as "**double exposure.**"

The Political Economy of Natural Hazards

What becomes clear from the foregoing discussion and the examples discussed above is the fact that hazards are not purely "natural" events. Rather, they are constructed in particular socio-ecological contexts in which a wide range of political, economic, cultural, and environmental processes interact at multiple scales to create different types and levels of vulnerabilities to extreme events. The recognition and analysis of the interplay of these factors in a systematic way underlies what we may call a "political economy approach" to understanding natural hazards. Such an approach begins at the broadest scale with an understanding of the global economy and the ways the demands of its particular production system interact with national, regional, and local factors to generate ecological vulnerability and risk.

The global economy emerged during the colonial era as the industrialization of Europe and North America created demand for increasing amounts of raw materials and foodstuffs from the tropics. As European colonial powers extended their political control over Asia, Africa, and the Americas, locally based, self-sufficient production systems were dismantled and transformed into producers of key commodities such as peanuts, sugar cane, rubber, tea, and cotton, demanded by the colonial market economy. Among the effects of this transformation were the emergence of cash-based commercial economies throughout the world and the disruption of traditional production systems that had evolved in an adaptive manner within local environments over centuries if not millennia. As economic institutions were dismantled and transformed so too were related political, cultural, and social institutions that ensured community sustainability and ability to withstand natural hazards. Traditional social exchange and land-tenure systems that ensured the survival of community members were disrupted, creating economically vulnerable populations, and changing cultural values associated with increased consumerism replaced those more in keeping with a conservationist objective. In some cases the introduction of new crops, land-use systems, and methods of production degraded soil systems (e.g., peanuts in West Africa), forests, and other resources, which accelerated natural processes of climatic change. Hence now the need to adapt to both economic and environmental change we discuss above.

Many of these developments continued through the post-colonial era to the present as newly independent states actively promoted the economic development of their countries along capitalist lines. An added pressure on fragile environments has come from rapid population growth and urbanization. Most recently economic globalization and the rise of **neoliberalism** (economic principles advocating free-market exchange, cutbacks in government spending on social services, privatization, and elimination of the concept of "community") along with unfettered economic growth have further created socio-economic disparities and pressures on resource systems. The huge death toll in the Kashmir earthquake in Pakistan in October 2005, in which almost 80,000 people perished, is attributable in part to poorly constructed, "modern" buildings developers put up in their haste to turn a profit.

Geographers are interested in how such broad-scale economic and environmental processes then interact with national and local economic, political, and cultural processes to shape particular vulnerabilities. While the global economy creates broad contexts for environmental hazards, national, regional, and local institutions give particular form to the influence and impacts of broad-scale processes. How national governments understand hazards, prepare for risk, and deliver relief assistance are crucial influences on how hazards are experienced at the local level. As several examples above illustrate, local political, economic, and cultural factors further inform particular impacts of and recovery from hazard events. If we are to truly understand natural hazards and devise ways of mitigating their effects on human populations, an appreciation for and understanding of the political economy of hazards is an absolute necessity.

Chapter Summary

This chapter has examined the evolution of geography's hazards tradition over time and explored how this has been applied to understand human exposure to events such as drought, tsunamis, earthquakes, floods, and hurricanes. Particular attention was given to understanding the related concept of vulnerability.

Critical Questions

1. What is double exposure? How might farmers be vulnerable to volatile environmental and market shifts?
2. Who is Gilbert White and what were his contributions to the sub-field of hazards geography?
3. How did farmers traditionally deal with variable rainfall in the West African Sahel?
4. How could a very similar natural event, such as similar-sized hurricanes hitting two different areas, produce very different outcomes in terms of loss and life and property damage?

Key Vocabulary

adaptation
attitudinal vulnerability
coping strategies
double exposure
Gilbert White
hazard
institutional vulnerability
material vulnerability
neoliberalism
risk
risk management
risk perception
technological hazard
vulnerability

References

Ahmed, S. and Fajber, E. (2009) Engendering adaptation to climate variability in Gujarat, India. *Gender & Development*, 17(1), pp. 33–50.

Bankoff, G. and Hilhorst, D. (2009) The politics of risk in the Philippines: comparing state and NGO perceptions of disaster management. *Disasters*, 33(2, April), pp. 686–704.

Blaikie, P., Cannon, T., Davis, I., and Wisner, B. (1994) *At Risk: Natural Hazards, People's Vulnerability and Disasters* (New York: Routledge); 2nd edn. published as: B. Wisner, P. Blaikie, T. Cannon, and I. Davis, *At Risk: Natural Hazards, People's Vulnerability and Disasters* (New York: Routledge, 2004).

Colten, C. (2006) *An Unnatural Metropolis: Wresting New Orleans from Nature* (Baton Rouge: Louisiana State University Press).

Cutter, S.L. (1993) *Living with Risk: The Geography of Technological Hazards* (London: Edward Arnold).

Denevan, W.M. (1983) Adaptation, variation, and cultural geography. *The Professional Geographer*, 35(4), pp. 399–406.

Franke, R. and Chasin, B. (1980) *Seeds of Famine: Ecological Destruction and the Development Dilemma in the West African Sahel* (Montclair, NJ: Rowman & Littlefield).

López, R.E. and Holle, R.L. (1998) Changes in the number of lightning deaths in the United States during the twentieth century. *Journal of Climate*, 11(8), pp. 2070–7.

Meadows, M.E., Baxter, A.J., and Parkington, J. (1996) Late Holocene environments at Verlorenvlei, Western Cape Province, South Africa. *Quaternary International*, 33, pp. 81–95.

Mehta, M. (2007) *Gender Matters: Lessons for Disaster Risk Reduction in South Asia* (Kathmandu, Nepal: International Centre for Integrated Mountain Development).

Moseley, W.G. (1996) A foundation for coping with environmental change: indigenous agroecological knowledge among the Bambara of Djitoumou, Mali. In W.M. Adams and J. Slikkerveer (eds.), *Indigenous Knowledge and Change in African Agriculture*, pp. 11–130. Studies in Technology and Social Change 26 (Ames: Iowa State University).

Moseley, W.G. (2001) African evidence on the relation of poverty, time preference and the environment. *Ecological Economics*, 38(3), pp. 317–26.

Nathan, F. (2008) Risk perception, risk management and vulnerability to landslides in the hill slopes in the city of La Paz, Bolivia: a preliminary statement. *Disasters*, 32(3, September), pp. 337–57.

Nietschmann, B. (1972) Hunting and fishing focus among the Miskito Indians, Eastern Nicaragua. *Human Ecology*, 1, pp. 41–67.

O'Brien, K.L. and Leichenko, R. (2003) Winners and losers in the context of global change. *Annals of the Association of American Geographers*, 93(1), pp. 89–103.

Pulido, L. (2000) Rethinking environmental racism: white privilege and urban development in southern California. *Annals of the Association of American Geographers*, 90(1), pp. 12–40.

Smith, K. (1992) *Environmental Hazards: Assessing Risk and Reducing Disaster* (London: Routledge).

US Department of State (2007) Bangladesh: improved flood response (Humanitarian Information Unit).

http://hiu.state.gov/index.cfm?fuseaction=public.display&id=2cc5a2c6-cc56-4ee2-a9d7-279b2d941233. Retrieved August 5, 2009.

Watts, M. (1983a) *Silent Violence: Food, Famine and Peasantry in Northern Nigeria* (Berkeley: University of California Press).

Watts, M. (1983b) On the poverty of theory: natural hazards research in context. In K. Hewitt (ed.), *Interpretations of Calamity*, pp. 231–62 (Boston: Allen & Unwin.)

World Conference on Disaster Reduction (2005) *Proceedings of the Conference Building the Resilience of Nations and Communities to Disasters*, Kobe, Hyogo, Japan, January 18–22 (Geneva: United Nations).

7

Environmental Justice

The Uneven Distribution of People, Pollution, and Environmental Opportunity

An Introduction to Human–Environment Geography: Local Dynamics and Global Processes, First Edition. William G. Moseley, Eric Perramond, Holly M. Hapke and Paul Laris.

Published 2014 by John Wiley & Sons, Ltd.

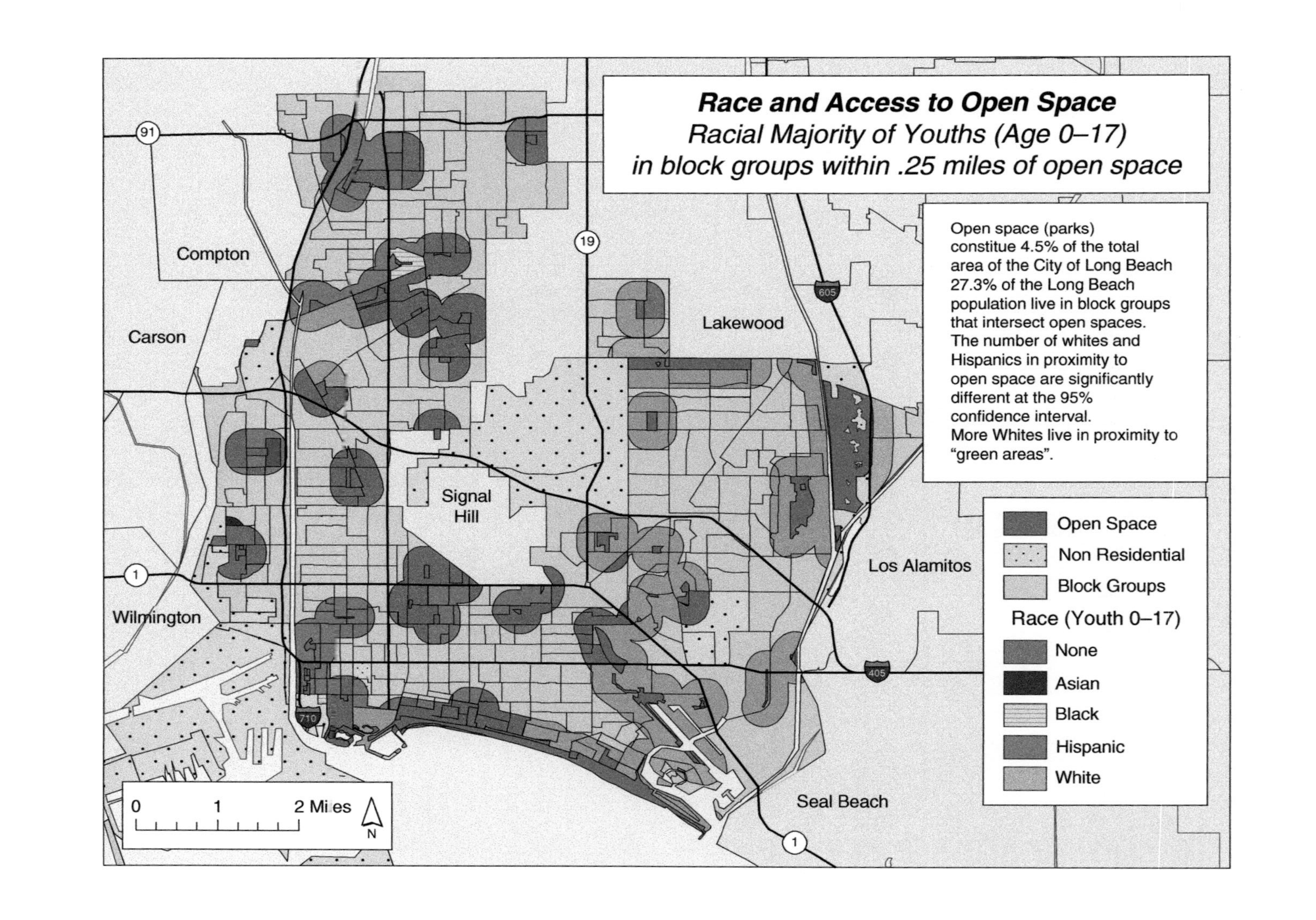

Race and Access to Open Space
Racial Majority of Youths (Age 0–17)
in block groups within .25 miles of open space
Open space (parks)
constitue 4.5% of the total
area of the City of Long Beach
27.3% of the Long Beach
population live in block groups
that intersect open spaces.
The number of whites and
Hispanics in proximity to
open space are significantly
different at the 95%
confidence interval.
More Whites live in proximity to
"green areas".
Open Space
Non Residential
Block Groups
Race (Youth 0–17)
None
Asian
Black
Hispanic
White
Compton
Carson
Wilmington
Signal
Hill
Lakewood
Los Alamitos
Seal Beach
91
19
1
605
405
710
0
1
2 Miles
N

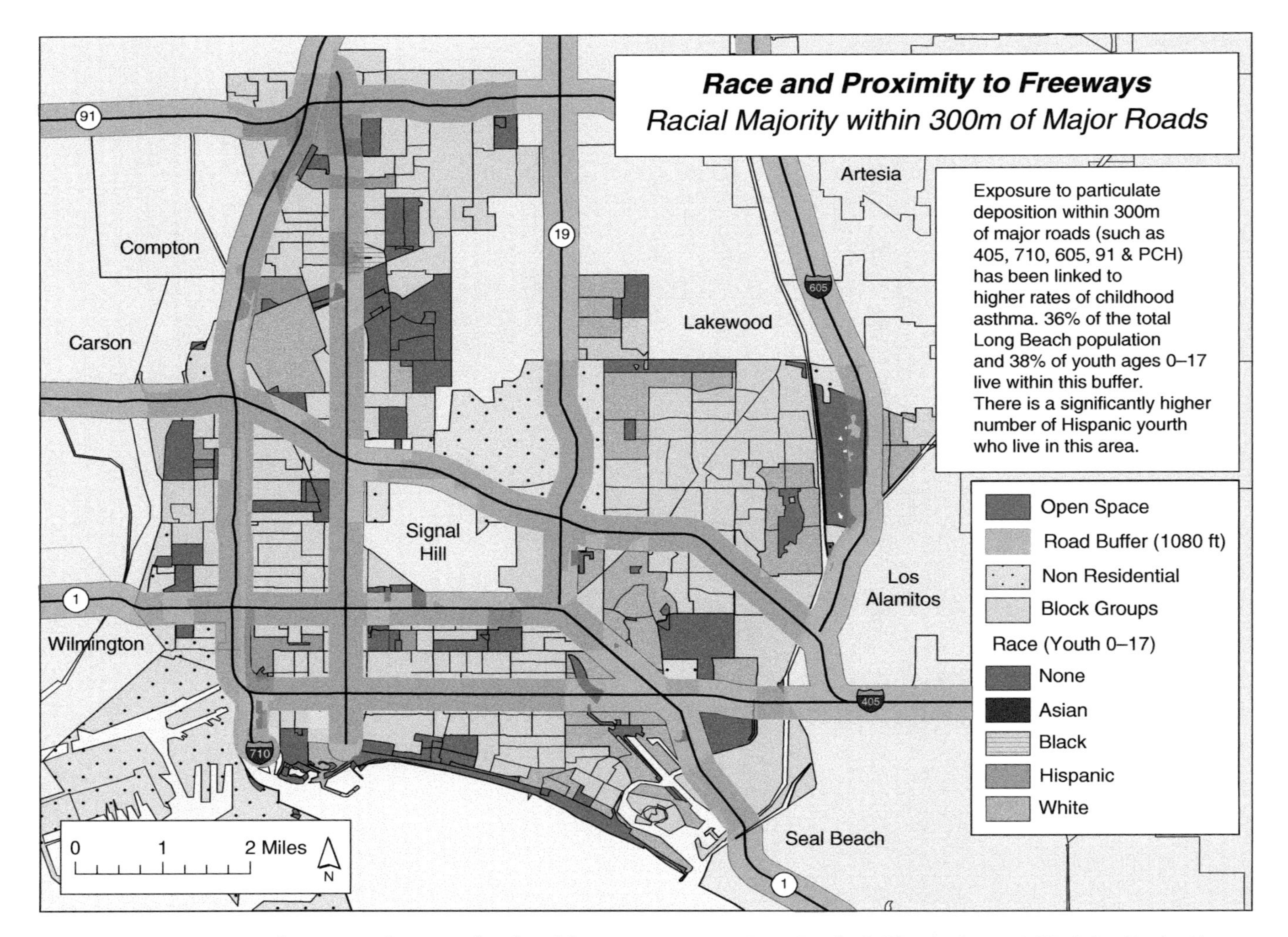

Figure 7.01 Environmental justice as a function of park and freeway proximity in Long Beach, California. *Source*: S. Wechsler. Used with permission.

Icebreaker: Environmental Quality in Long Beach, California

City leaders have made the claim that Long Beach, California, is the world's most diverse city, and indeed they have a good case. In terms of race, there is no single majority group in Long Beach. Culturally, one finds a wealth of different ethnic groups resulting in a wonderful variety of restaurants, art, shops, and food stores. In fact, great diversity can be found along just about any divide – economic, gender, cultural, ethnicity, sexual orientation, and age. Indeed, diversity is one aspect of living in Long Beach that many enjoy. Yet despite the diversity and the relatively small size of the city (50 sq. miles, or 129.5 sq. km for about 500,000 people), Long Beach is surprisingly segregated along two critical fault lines of environmental quality – ***air quality*** *and* ***open space****. Wealthy neighborhoods are located closer to the beaches and parks and farther from the port and major freeways, providing them with healthier air. Poorer neighborhoods are generally located closer to the port and freeways and farther from parks and other green spaces. Data show that this division is not just along economic lines, but along ethnic and racial lines as well (see Figure 7.01).*

Do you ever wonder why a disproportionate number of poor and minority populations seem to live in the most polluted neighborhoods? Did these people move into polluted neighborhoods because they were the only places they could afford to live? Or did the toxic facilities and freeways get built in the areas where people have the least amount of power to resist? Is this question even relevant given that it is not "just" for anyone to live in an unhealthy environment?

Chapter Objectives

1. To define environmental justice and related terms.
2. To describe the history of the development of the field.
3. To understand complex issues of "cause and effect" concerning the link between racism and environmental issues.
4. To understand basic theories of environmental justice.
5. To examine climate change, healthy environments, and globalization from an environmental justice perspective.

Introduction

How and why can a city as a diverse as Long Beach be segregated in such a way? How does your city or town compare to this example? The **environmental justice (EJ)** perspective provides a useful lens into this human–environment relationship. In the broadest sense, environmental justice focuses on the inequitable distribution of environmental qualities and risk exposure among various human groups. In particular, EJ scholars have documented how disadvantaged groups often bear the least responsibility for causing environmental degradation while they bear the brunt of its negative consequences. Similar to the hazards perspective (see Chapter 6), EJ examines the exposure of social groups to human–environmental hazards. Three themes make the EJ perspective unique: (1) it is concerned with documenting unequal patterns of exposure of different groups to hazards (especially hazardous waste and other industrial pollutants); (2) it aims to explain the causes of the patterns, often taking the stance that some groups are relatively disempowered compared to others (usually in terms of race, ethnicity, and class differences); and (3) it takes an explicit stance of **activism** aimed at mitigating or removing the hazards. EJ is thus a politically charged term, one that connotes some form of **remediation** to correct injustices imposed on specific groups such as people of color. As such, we often speak of environmental justice as a **social movement**, a group that cares as much about economic and class justice as it does about environmental issues. This human–environmental perspective thus merges research and activism, creating a dynamic mix. EJ is a body of research, an activist agenda, a social movement, and a human–environment perspective all in one. Although many of the cases here draw on our greater familiarity with US examples, these issues are visible and real for many places and states in different regions of the world.

Definitions and History of Environmental Justice

As illustrated in the previous chapter (Chapter 6) on natural hazards, environmental risks are unequally distributed, and these risks affect human populations differently. Whereas the hazards tradition in geography tends to emphasize "natural hazards," EJ focuses on human-created hazards such as industrial plants, landfills, and other sources of **toxins**, chemicals, or materials hazardous to human health and well-being. These kinds of hazards are a function of wealth and development and as such the field of environmental justice evolved in the industrialized countries. Numerous studies suggest that people of color have **environmental burdens** imposed upon them unfairly due to over-siting of industrial plants and landfills in their communities and from exposures to pesticides and other toxic chemicals at home and on the job. As the last chapter also indicated, some people or communities are better able to cope with the risks of living in an environment with hazards while others are extremely vulnerable. In fact, concerns over unfair exposure for vulnerable populations to natural hazards (Chapter 6), toxins (this chapter), and climate change (Chapter 8) can all be considered dimensions of environmental justice (Leichenko and O'Brien 2008).

EJ is a relatively new field of inquiry and, as is often the case with new fields, the term environmental justice itself is somewhat contested, that is, different groups assign it unique meanings to support their specific causes. For some EJ begins with a question: "Are poor people and people of color subjected to a disproportionate burden of exposure to industrial environmental hazards?" For others it begins with the proposition that people of color and socioeconomically disadvantaged individuals bear a disproportionate burden of environmental pollution and its health consequences. Moreover, they feel that there exists a pattern of environmental inequity, injustice, and racism and that policies are needed to create a more just and equitable human environment. For still others EJ is a social movement that takes an explicitly activist stance, adopting the premise that all people have a right to live in a clean environment free from hazardous pollution or contamination with access to the natural resources necessary to sustain health and livelihood.

What Key Questions Does an EJ Perspective Help Us Answer?

EJ can be defined by the questions it seeks to answer. Four primary ones are:

1. Are industrial hazards unequally distributed?
2. Do impoverished communities, particularly those of color, bear disproportionately high environmental costs and receive disproportionately low environmental benefits?
3. What causes environmental injustice?
4. What policy actions can be used to reverse the trend?

Research Agenda

EJ researchers seek to explain the reasons behind the unequal patterns of hazard distribution. EJ questions why, and through what social, political and economic processes some people are denied basic rights to clean, healthy environments. A notable aspect of EJ research is how recent much of the literature is. For example, it was only in 2008 that an academic journal entitled *Environmental Justice* was started. It is one of the few areas where activist concerns outpaced high-quality research on the following issues (but see Faber 1998): How is it that certain groups of people do not have access to basic resources, or are burdened with pollution or environmental hazards to a greater extent than other groups? What are the social relations of production and power that contribute to these outcomes? Finally, EJ seeks to rectify the situation by asking: What can be done?

Activist Agenda

According to Bullard (1994), one of the notable scholars of the environmental justice movement in the United States, there are numerous problems with the status quo. Low-income and minority communities tend to bear greater health and environmental burdens while more affluent and white communities receive the bulk of the benefits. A big problem in Bullard's view is that the "burden of proof" is placed on the "victims" rather than the polluters because human exposure to harmful substances is legitimated, risky technologies such as incinerators are promoted, the vulnerability of economically and politically disenfranchised people is exploited, and ecological destruction is subsidized. According to Bullard, there has also been a general failure to develop pollution prevention as a dominant strategy, and clean-up activities have often been delayed. These issues will be illustrated in the cases below.

Some Key Terms in Environmental Justice

Part of the complication of defining EJ involves distinguishing it from other closely related terms. Two terms in particular, **environmental equity** and **environmental racism**, have overlapping meanings that create some confusion.

Environmental equity usually implies an equal sharing of risk burdens (although this does not necessarily imply an overall reduction in the burdens themselves) or equal protection under environmental laws. According to Bullard, unequal environmental protection undermines four basic types of equity: geographic, social, procedural, and generational. **Geographic equity** refers to the location and spatial configuration of communities, including their proximity to environmental hazards and sources of pollution such as landfills, incinerators, sewage treatment plants, refineries, major freeway interchanges, and other noxious facilities. As we shall see, research shows that these facilities are not randomly scattered across the landscape; rather, they tend to be located near communities with large minority populations that have low incomes and low property values. The juxtaposition of economic activities, which are largely a function of property values and transportation access, and the social geography of places, creates **landscapes of risk**. Research shows that poor people and people of color often work in the most dangerous jobs and live in the most polluted neighborhoods. As such, their children are exposed to all kinds of environmental toxins. Moreover, they often lack access to healthy green spaces such as parks, or beaches.

Social equity refers to the role of social and economic factors such as race, ethnicity, gender, class, culture, and political power in decision-making about environmental issues. It is recognized that certain groups have obtained more power in decision-making than others and environmental decisions are no exception. Here, we have used, for example, African American (Black) frequently, to reflect the cases we are familiar with. However, Latino populations, indigenous populations and tribes, and other minority concerns vary markedly by region and country. You may have to mentally substitute your own relevant examples for the category of social equity.

Procedural equity refers to fairness in the extent that government rules, regulations, evaluation criteria, and enforcement are applied in a nondiscriminatory way. Unequal protection can result from nonscientific practices and undemocratic decision-making, such as exclusionary practices that prevent certain groups from participating. Examples include holding public hearings in remote locations, at inconvenient times, or in languages that are not understood by those attending.

Certain classes or groups of people may also be made to feel unwelcome in this category of equity issues.

Generational equity is a framework of legal norms designed to bring justice to future generations. Public policy decisions are governed by the concept of fairness to future generations so that our children and grandchildren will have access to the same quality of life as we do – or a better one. Generational equity insures that society does not mortgage the environmental future for present short-term gains. Climate change is a good example; the next generation will bear the costs of our current actions of consumption and burning of fossil fuels as they confront sea-level rises and changing climates that put them at risk.

Environmental Racism or Environmental Justice?

The term environmental racism is a politically charged term that connotes the need for remedial action to correct for past injustice imposed on a specific group of people. It asserts that overt or subtle (often institutional) racism underlies the variation in the distributions of environmental burdens. The term environmental racism overtly creates a linkage between the arenas of environmentalism and civil rights.

Prior to the 1980s, environmental issues and racial justice issues were commonplace in public debate, but they were not often addressed as an interrelated problem. In 1987, Benjamin F. Chavis coined the term "environmental racism," referring to **toxic waste** landfill siting in communities of people of color. These wastes could be biological, chemical, or nuclear in origin but were placed in sometimes unlined pits in many locations. In a later expansion on this concept, Chavis said:

> Environmental racism is racial discrimination in environmental policy-making and enforcement of regulations and laws, the deliberate targeting of communities of color for toxic waste facilities, the official sanctioning of the presence of life threatening poisons and pollutants in communities of color, and the history of excluding people of color from leadership of the environmental movement. (Chavis 1994: xii)

A careful reading of Chavis' definition finds that it includes all four types of environmental equity. His definition of environmental racism is explicitly adversarial. It is also noteworthy that Chavis includes the environmental movement itself as a culprit. According to one view, the pattern of environmental racism is partly the result of a historical failure on the part of interest groups, particularly the mainstream environmental movement, to provide a vision and strategy to address environmental racism and injustice (see Box 7.01 below). But this use of environmental racism as a concept comes from an industrialized, multi-cultural context. Can it be used more widely, or is the larger environmental justice still useful for a wider variety of national and cultural settings?

If environmental racism is an explicitly adversarial term, then environmental justice is quintessentially aspirational. According to Bryant (1995: 6), "Whereas the term environmental racism focuses on the disproportionate impact of environmental hazards on communities of color, environmental justice is focused on ameliorating potentially life threatening conditions or on improving the overall quality of life for the poor and/or people of color. Environmental racism is based on the problem of identification; environmental justice is based on problem solving."

Over time environmental justice has become the preferred term; it is now not only more inclusive than environmental racism (it goes beyond racism and ethnicity to include others who are deprived of their environmental rights), it is also much more palpable politically. And while originally North American in terms of its origins, the concept of EJ is now widespread and widely cited in all kinds of resource and pollution struggles internationally.

The term environmental justice was coined in the early 1980s by Bullard. As is often the case with a new field of research, during its first decades of use this term generated new meanings. In general the definition of EJ has been broadened over time, although some feel it has been watered down in the process. Even the experts have difficulty defining EJ as can be understood from the comments by Laura Pulido, a geographer and prominent figure in the environmental justice field (see Box 7.01).

Box 7.01 Interview with Laura Pulido

Figure 7.02 Laura Pulido. *Source*: Courtesy of Laura Pulido/University of California.

INTERVIEWER: What are the problems with the environmental justice concept?

PULIDO: First of all, it [environmental justice] is an incredibly broad term. When I first began doing this kind of work, environmental justice was about non-white people and poor people organizing around environmental issues. From there, tensions have developed within the movement. Some people would take environmental justice and say, "No, it is just about people of color organizing against environmental racism." In order to make the race and class link, other people would say, "No, it also includes poor white people." Lately, some people are saying, "No, it is also about justice for the fish, justice for the trees." They are really pushing [the term] out and expanding it in another way. On one hand, I am in favor of rights for the fish and trees, but how is that different from the original environmental movement? I think the environmental justice movement started with something really different in terms of interjecting the question of social justice into environmental issues ... I still think that. And, I am comfortable talking about it that way. I can talk about the work of the environmental justice movement, but I have enormous difficulty talking about it as a concept.

INTERVIEWER: Do issues like parks, safety, and street design fall under definitions of environmental justice?

PULIDO: That can definitely be part of environmental justice, but it depends on who is doing it and for what purposes. If it is something from a marginalized community itself, then it has to be respected. If they want to call it environmental justice, then call it environmental justice. I think that's great. This disparity in green spaces and neighborhood resources is just another form of inequality. It is important, however, for another reason too. I think it has been unfortunate that so much of the attention has focused on negative environmental problems. There has been the need to do that, but there is also the question about how to create a more positive environment. And, that gets left out.

Source: Lee and Ehrenfeucht 2001: 17.

History of a Movement

The environmental justice movement combines the civil rights and environmental movements. Some argue that EJ was an outcome of the mainstream environmental movement's failure to address issues of environmental racism, but others point out that civil rights organizers themselves did not see the environment as a critical issue. For example, Whitney Young Jr., head of the National Urban League, made the point that "The war on pollution is one that should be waged after the war on poverty is won" (Foreman 1998: 15). Similarly African American activists fighting the siting of a hazardous waste landfill as an environmental problem for their county did not align this form of environmentalism with their perception of traditional, mainstream environmentalism. Or, as one activist explained, while trying to distance herself from mainstream environmentalism, "African Americans are not concerned with endangered species because we are an endangered species" (McGurty 1997: 314). At the time, the early 1970s, the issue was one of a matter of priorities, and poverty was deemed the most pressing problem. Other minority leaders feared that the increased focus on the environment at this time was distracting Americans from the human problems associated with racism. As such, there was some hostility between the civil rights and environmental movements. This same argument persists in the debate over economic development and environmental justice in the developing world today, as we will see below.

Nevertheless, EJ activists were aggravated by the mainstream environmentalist groups' lack of effort to address important urban environmental issues. In March 1990 Richard Moore, the director of the South West Organizing Project, a grassroots advocacy group in Albuquerque, sent out what came to be known as the "SWOP letter." The letter was addressed to the Big 10 conservation groups and signed by 100 cultural, arts, community, and religious leaders – all people of color. The letter charged the mainstream environmental organizations with a history of "racist and exclusionary practices," a lack of in-house diversity, and an all-around failure to support environmental justice efforts (Durlin 2010).

Many of the environmental groups were caught off guard, and some did not know how to respond. Although perhaps slow to adjust, some mainstream environmental groups have clearly caught up, and groups such as the Sierra Club (once entirely focused on pristine environments) have gradually embraced environmental justice and urban environmental issues. But this issue lingers within other groups. For example, Jake Kosek (2006) has documented how the green rhetoric of environmentalism in the state of New Mexico has affected local Hispanic livelihoods and access to forest resources. A local non-governmental organization, then named Forest Guardians, was instrumental in slowing the pace of regional deforestation in New Mexico. It also produced some unintended consequences, crippling more localized forestry industries in which Hispanic residents worked. Obviously, this example or process of local exclusion from resource use and access is not atypical. Plantation forestry and widespread industrial logging in Southeast Asia, for example, have also displaced long-established customary access and use to local resources.

A key area of hostility derives from the "Not In My Backyard" or **NIMBY** movement. As the name suggests, nobody wants a hazardous site located in their neighborhood but those with power have been more successful at organizing and lobbying to ensure that hazardous waste sites or other hazards were kept out of certain neighborhoods. Initially this movement was concerned with the location or siting of hazards and not necessarily with the reduction of hazards in general. As such, early efforts to mobilize resistance to locating a hazardous industry or waste facility in a particular area meant that the hazard would be pushed into another area. The new area was often one where there was less resistance either because people were less organized or had less power or because the community was willing to accept a certain degree of hazard or risk in return for other economic benefits such as jobs. An example of the latter might be the expansion of a port which could provide a neighboring community with jobs, but which would also subject that community to an increase in harmful air pollutants from trucks, ships, and port equipment.

NIMBYism is one likely reason for the early hostility or disjuncture between the civil rights and environmental movements. The concept of NIMBY has since been eclipsed by BANANA (build absolutely nothing anywhere, near anybody). The idea here is that all people and all communities have the right to a healthy environment and therefore the goal is to reduce the amount of hazards produced. Unfortunately, one negative outcome of such a stance is that polluting and hazardous industries have simply moved offshore to countries where environmental laws are more lax or where people are less organized and powerful, or where people are impoverished and thus willing to accept higher risks and exposure to hazards. This "globalization" of industrial hazards is covered below.

Box 7.02 The roots of EJ in the United States: the cases of Warren County, North Carolina, and Love Canal, New York

Figure 7.03 Warren County warning sign. *Source*: © Karen Tam/AP/Press Association Images.

What originally brought the civil rights and the environmental movements together in the USA? By most accounts the landmark moment in the environmental racism movement occurred in 1982 in Warren County when African Americans protested the siting of a hazardous waste landfill.

As the story goes, Robert Burns and his two sons drove liquid tanker trucks along rural roads in North Carolina and through remote sections of the Fort Bragg Military Reservation. Driving at night to avoid detection, they opened the bottom valve of the tanker and discharged liquid contaminated with polychlorinated biphenyls (PCBs) onto the soil along the road shoulders. This violation of the Toxic Substance Control Act (TSCA) continued for almost two weeks until 240 miles (386 km) of road shoulders were contaminated. Robert Ward of the Ward Transformer Company had paid for illegally disposing the contaminated liquid in an attempt to avoid the escalating costs that were due, in part, to increasing regulation of hazardous waste. Since the contamination occurred on state-owned property, North Carolina was responsible for remediation. Within a few months of detecting the contamination, the state had devised a plan calling for the construction of a landfill to dump the 30,000 cubic yards of contaminated soil in Warren County, a rural area in northeastern North Carolina. The nearby residents, mostly African American and poor, were joined by civil rights groups, environmental groups, clergy, and members of the Black Congressional Caucus in demonstration. The protests ultimately led to a key study by the United Church of Christ which suggested that race rather than income was the single most significant determinant of toxic waste siting.

Figure 7.04 Lois Gibbs, who became an activist in the 1970s when her community, Love Canal, New York, became a national symbol of toxic pollution. Much of the town is still off-limits. *Source*: © Bettmann/Corbis.

Although Warren County is often cited as the first environmental justice case in the US, the seeds of the new movement were likely planted a few years earlier with the hysteria created by the discovery of toxic waste at Love Canal, New York, in 1978. Love Canal was a watershed episode in the environmental movement that created widespread anxiety in the population at large over the issue of locating hazardous chemicals. Love Canal was a suburban, predominantly white working-class neighborhood built on a former toxic waste dump. In Love Canal, a community just southeast of Niagara Falls, a housing subdivision and public school were built adjacent to an area used by the Hooker Chemical Company to dispose of over 40 million pounds (18,143,694 kg) of industrial waste. By the 1970s, this toxic material had seeped into homes and a school, creating significant health problems for residents, including asthma, lethargy, cancer, miscarriages, and birth defects. New York officials decided to evacuate 240 families from the area. Television news coverage of these events showed how toxic contamination had destroyed normal suburban life by financially destroying families and creating significant social disorder. Love Canal became synonymous with toxic waste and remains America's most famous "superfund" site. The Love Canal episode created public awareness and galvanized a movement to clean up toxic sites across the country (see Foreman 1998: 16). While the Warren Country protest was, in part, racially charged, the Love Canal episode was rooted more in terms of class. There remains debate over whether the EJ movement emerged along class or racial lines.

Following the events in Warren County and at Love Canal (see Box 7.02) there were a number of protests over the siting of toxic waste facilities. The United Church of Christ Commission on Racial Justice (UCC) published the findings of the first study on hazardous waste siting in a report in 1978. The study found that more than 60% of African American and Hispanic people lived in neighborhoods near a hazardous waste site. Attention to this form of racial inequality began as an "environmental justice" movement, which emerged in the 1980s and 1990s to protest the placement of waste sites and polluting industrial facilities in predominantly low-income and minority communities.

Two studies, one by UCC and the other by the US government, provided empirical support for the claims of environmental racism (Cutter 1995). Shortly thereafter Bullard published his landmark and aptly titled book, *Dumping in Dixie* (1990), which helped galvanize a movement and provided further empirical support that environmental injustices existed. Since the 1987 publication of *Toxic Wastes and Race in the United States* by the UCC and the 1994 release of Bullard's *Unequal Protection: Environmental Justice and Communities of Color*, the voices of those on the front line of the struggle for environmental justice have been reaching a broader audience within the United States and beyond.

Towards a Global Movement

Some EJ scholars and activists, such as Laura Pulido, link the emergence of environmental justice to a global phenomenon, the rise of the anti-toxics movement, which began in the more industrialized countries but has since spread around the world. The accidental chemical gas poisoning of thousands in 1985, in **Bhopal**, India, produced a new generation of active citizens concerned about environmental justice and location concerns in that country. Many communities have realized that neither the nation-state nor the environmental movement was going to protect them from various forms of hazards and pollution. Perhaps the best-known example of globalization of environmental justice pursuits was the **Minamata Bay** case from the south island of Kyushu, in Japan. Both Bhopal and Minamata are now widely known as place-names linked to particular kinds of chemical and industrial hazards.

During the 1930s, the long-established Chisso Corporation was using inorganic mercury in its plastics production plant in the small town of Minamata, a coastal fishing town on the Shiranui Sea in southern Japan (see Figure 7.05). As a result, high levels of methyl mercury were being piped out directly into the bay that so many people depended on for their fish and shellfish. By the 1950s at least, noticeable effects were common in the town of Minamata and the nearby fishing villages. The link was unclear at first, but mercury is an effective toxin and nerve disruptor even in small doses, and was first noticed for its effects on cats in the area. As cats ate rejected seafood from small fishing expeditions, villagers noticed that the cats were prone to instability, twitching, and collapsing. This "dancing cat" disease was an ominous but unclear warning about mercury's effects on brain tissue. By the mid-1950s, clear and suggestive research concluded that mercury was accumulating in the tissues of fish and shellfish, and then being transferred to humans in the local diet. As a shorthand it became known as Minamata disease since so little was known about mercury poisoning at the time.

Figure 7.05 Minamata Bay on the south island of Kyushu, Japan, where thousands suffered from mercury poisoning. The Japanese government has now declared the bay safe for fishing and for its seafood products. *Source*: Eric Perramond.

Chisso and the Japanese government were skeptical in their response to these claims. Early research was discouraged, results suppressed, and even the national government sided with the private company for well over a decade. Finally, activists from the region were successful in getting the company and the government to take their claims seriously. By 1968, the flow of methylmercury into the bay was stopped, but not before decades of direct point-source pollution had occurred. In this case, then, the "environmental justice" of unequal exposure had to do with location, diet, and class. Fishing villages were not politically powerful, and were often socially constrained to "not complain" about pollution since the industrialization of the region benefited local incomes and the national pursuit of full-scale modernization.

The Minamata case illustrates the complexity and diversity of environmental justice struggles. In this case, it was not a multi-cultural context in which ethnicity or race played a strong role. Rather, class difference and the social mores of certain occupations in Japan were implicated. The lack of political power at the local level led to difficulties in facing the company's clout and getting the national Japanese government to take the issue seriously. The social culture in Japan led to discouragement of activist claims and social marginalization of people afflicted with Minamata disease. A similar mercury poisoning case in Niigata prefecture far to the north (on the main Japanese island of Honshu) did give more evidence and weight to local Minamata Bay residents. The cost of industrialization was clearly felt, however, in Minamata. Local industrialization led to direct, physical, consequences for villagers by way of mercury poisoning. It also produced more subtle, social forms of discrimination against activists and victims of the disease. Neighbors were scared of being "infected" by sufferers of Minamata disease, and there was little public understanding about what caused the disease, and thus social marginalization accompanied the physical poisoning by mercury (Walker 2010). We can only wonder, in light of the recent events at the Fukushima Daiichi nuclear power plant, what implications the recent meltdown will have for local and regional residents for years to come. Is industrial or nuclear poisoning an environmental justice issue if everyone is affected?

But environmental livelihood struggles have an even longer history. For example, it has been shown that that hydraulic mining and commercial oyster production transformed San Francisco Bay from a rich source of local foodstuffs managed communally to an industrial dump. Irrigation canals in Los Angeles were converted into covered sewers, undermining the city's Hispanic and Asian neighborhoods. Such shifts did not occur without protests, but never galvanized into a larger social movement. You can surely think of local examples where you live that are similar. Which ones seem related to minority or indigenous populations? Are there others where class and economic status matter more?

Documenting Disparities: Cause and Effect and the Evolution of a Field

EJ has often been posed as a "Which came first?" or "chicken and egg" problem. In other words, was the hazard sited in a community dominated by a poor and minority group, or did the poor move into the neighborhood of a hazardous site because real estate was cheaper? This question gets to causality – what was the cause of the environmental inequality?

One approach to studying a spatially geographic phenomenon such as environmental injustice is to begin an inquiry by mapping the phenomena and then attempt to explain the observed patterns over time. Mapping is only a first step in explaining why patterns exist, however. By hypothesizing causal relationships between observed patterns and then testing these hypotheses, some possible causes (or explanatory factors) can be accepted and others ruled out. Determining the causes behind the patterns can often be difficult and even controversial. EJ is no exception; tracing the evolution of the field of EJ research provides an illustrative example of how researchers use data analysis in an attempt to prove a point or verify a hypothesis. It also highlights the importance of "constructive critique" in the research process.

As noted, in 1994 Bullard concluded that the geographic distribution of both minorities and the poor is highly correlated to the distribution of major hazards, including air pollution, municipal landfills and incinerators, abandoned toxic waste dumps, lead poisoning in children, and contaminated fish consumption. He argued that virtually all studies of exposure to outdoor air pollution have found significant differences in exposure by income and race. Finally, he found that race more strongly correlated with these hazards than class.

In contrast to Bullard's conclusions, Heiman (1996), for example, in subsequent research found no universal explanation for the siting of hazardous sites across time and space. He and others have argued that the empirical evidence for environmental discrimination remains mixed. The ambiguity largely results from five factors or key issues: (1) the proper scale of analysis or areal unit chosen (for example, county vs. census tract); (2) the characterization of the risk; (3) the affected population (or subgroup); (4) the uncertainty over the intentionality of the siting decisions; and (5) the time frame of the study.

The most common problem or point of contention concerns the issue of scale. Many studies have found that the results of studies tend to be a function of the scale of analysis (or unit of analysis, see below). To give but one example, a study by Anderton and colleagues (1994) using census tracts in their investigation of the siting of hazardous waste facilities found no significant differences in the racial composition of tracts that contained hazardous facilities and those that did not. But when the authors changed the scale of analysis and aggregated to a larger area of a 3 mile (4.8 km) radius, they found a strong relationship between race and location of facilities.

Characterizing risks has also proven difficult, especially where airborne pollutants are concerned. For example, simple relationships of proximity do not account for prevailing wind patterns that disperse pollutants. Moreover, for the case of pollutants such as particulates, long-term studies of the health effects were required before a direct link could be made between a hazard and human health. Only as researchers have closed the links between an emissions source, wind patterns, and health impacts have studies of the impacts of particulates proved conclusive.

The selection of subpopulations for study is also important. Although most EJ studies focus on race and or income, a study by Greenberg (1993) found the factor explaining the inequity was not race or wealth, but age. Accordingly a disproportionate number of the facilities were located in areas with a high percentage of elderly residents. So EJ examples are not only about race or ethnicity, and these examples, together with the Minamata example (above), clearly complicate any straight and simple stories about discrimination based on race alone. All industrialized societies have created hazardous conditions for their own citizens, and it should be clear that even in a relatively homogeneous society like Japan, class and cultural behaviors created constraints for the voice of activists.

Determining the intentionality of siting decisions is perhaps the most difficult issue to untangle. Doing this requires knowledge of how disadvantaged communities are systematically barred from the decision-making process. The issue is also closely tied to the selection of the time frame of analysis. Clearly the time frame chosen for a study of the relationship between the location of a hazardous facility and a minority community is important because urban areas are highly dynamic and communities change over time. Long-term studies of hazardous sites help to uncover the dynamics of change.

Since Bullard's landmark publication in 1994, numerous studies have attempted to test his hypothesis of systematic racial prejudice in terms of hazard siting in the US context. There have also been a number of critical reviews of the body of EJ research. The language in these reviews is sometimes overly harsh, perhaps due to the politically charged aspects of EJ (see Box 7.03 for example). However, the critical reviews served the important purpose of influencing researchers by pushing them and/or directing them to address key gaps and weaknesses in the research. As a result, the more recent EJ research is better and stronger. Most importantly, the evidence produced tends to support the general argument that EJ scholars and activists have made for decades.

Box 7.03 An example of how constructive critique improves research

An analytical review of William Bowen, *Environmental Justice Research: What Do We Really Know?*

William Bowen provided a very thorough and critical review of the EJ literature. Although his review was perhaps overly polarizing it did serve to inspire researchers to focus on key methodological issues and to improve the strength of EJ studies.

Bowen classified EJ studies in terms of the quality of the science: low, medium and high. He ranked Bullard's landmark study in 1983 as scientifically low-quality but noted that it was politically influential. He argued that the 1990s brought a more scientifically focused cadre of researchers to the issue (one could argue this was a result of Bullard's politically charged original work). Although Bowen argues that some of the research was still focused primarily on achieving preconceived political ideals and only secondarily, if at all, on improving the issue's scientific knowledge base, as a rule, the research from both scientists and policy advocates became more sophisticated and convincing. By the late 1990s, Bowen found that the research was reaching "medium quality." For an example, he cites the work by Been and Gupta (1997) who conducted a national-level long-term study covering from 1970 to 1994, based upon census tracts, that looked at the same types of hazardous facilities as were examined in United Church of Christ original study.

In general, however, Bowen had harsh words for EJ researchers as seen in the following statement:

> "Simply stated, the evidence regarding disproportionate distributions is mixed and inconclusive. Moreover, even assuming that findings were strong, clear and distinct ... little to none of the research is meaningfully linked to actual exposure and associated public health effects. As a consequence, little or nothing can be said with scientific certitude regarding the existence of geographical patterns of disproportionate distributions and their health effects on minority, low-income, and other disadvantaged communities." (Bowen 2002:10)

Although perhaps overly critical, Bowen's article garnered several strong responses. One of the more convincing responses was by Mohai and Saha (2007). According to these authors (one of whom was criticized by Bowen for *weak* science), researchers have focused on improving two types of studies over the past decade: (1) pollution dispersion assessments, and (2) site proximity assessments. Pollution dispersion assessment studies involve collecting data about the volumes and toxicities of various air and water emissions, the timing of emission releases, and other key factors such as wind. From these data, estimates are made about the geographic dispersion and deposition of the toxic emissions. Census data are then employed to determine the demographic characteristics of those most likely to live where pollution and toxicity levels are concentrated. A few pollution dispersion studies have gone as far as attempting to conduct risk assessments in which human exposure and expected lifetime cancer risks are estimated. Obtaining complete and accurate information for modeling pollution dispersions and toxicity levels has been difficult, but new technologies such as geographic information systems (GIS) and advanced atmospheric models are improving results. Nonetheless, the authors concluded that, at present, relatively few environmental inequality studies employing pollution dispersion or risk assessment methods have been conducted.

By far the most frequently employed approach for conducting quantitative environmental inequality analyses has been to assess the proximity of hazardous sites to nearby populations, perhaps because this does not require sophisticated modeling of dispersion. As such, nearly all national-level environmental inequality studies have involved proximity assessments. These studies have been very influential in spurring policy development and further research in the area of EJ. Although most have found these disparities to be statistically significant, there has been considerable variation in the magnitude of racial and socioeconomic disparities found. For example, a few studies have even found that there were no race and income disparities associated with the presence of environmentally hazardous sites and locally unwanted land uses.

Mohai and Saha (2007) hypothesized that a likely source of the variation in findings has been wide reliance in environmental inequality research on what has been termed **unit-hazard coincidence** methodology. This approach involves selecting a predefined geographic unit (such as zip code areas or census tracts), to determine which subset of the units is coincident with the hazard and which subsets are not, and then comparing the demographic characteristics of the two sets. The authors point out that two assumptions are implicit in this approach: (1) adverse impacts tend to be concentrated within close proximity of the hazards, and (2) populations living within the host units are located closer to the hazard under investigation than populations living in the non-host units. The authors themselves have demonstrated that this latter assumption is not always the case and that, in fact, the unit-hazard coincidence method fails to control for proximity in several respects. The authors further demonstrated how alternative methods, termed "distance-based" methods incorporating GIS, better control for proximity and produce more accurate results.

In an excellent example illustrating how constructive critique can improve research, Mohai and Saha (2006) responded to the critics by re-examining the original data from the research of others using different methods. In their re-study, the authors demonstrated that when racial disparities around the nation's hazardous sites are analyzed applying distance-based methods as opposed to unit-hazard coincidence methods (see Box 7.04), disparities are found to be much greater. They found that although minorities made up only a quarter of the nation's population in 1990, over 40% of the population living within 1 mile (1.6 km) of hazardous waste sites were persons of color. The nearly 20% difference (43.2% compared to 24%) in the minority percentages between host and non-host neighborhoods within 1 mile of a hazardous site is clearly much greater than the 1% to 3% differences that are found when the unit-hazard coincidence method is applied. The authors concluded that even at a distance of 3 miles (4.8 km), the difference in the proportion of nonwhites in host and non-host neighborhoods is found to be greater. In the Southwestern United States, one can see this remarkable spatial overlap between certain populations and **nuclear waste** and production facilities, like the Los Alamos National Laboratories (in the state of New Mexico). Four nearby Indian Pueblos struggle with the legacy, and unknown consequences, of this facility (see Masco 2006 for more). The atomic era produced risk, of course, for a variety of peoples across the planet. The **Chernobyl** (Ukraine) meltdown in 1986 remains the most infamous example of the nuclear risk. But the recent meltdown at the **Fukushima** I nuclear power plant north of Tokyo (Japan) is another reminder that EJ issues with nuclear dimensions touch all populations.

Box 7.04 The unit-hazard coincidence vs. distance-based approaches

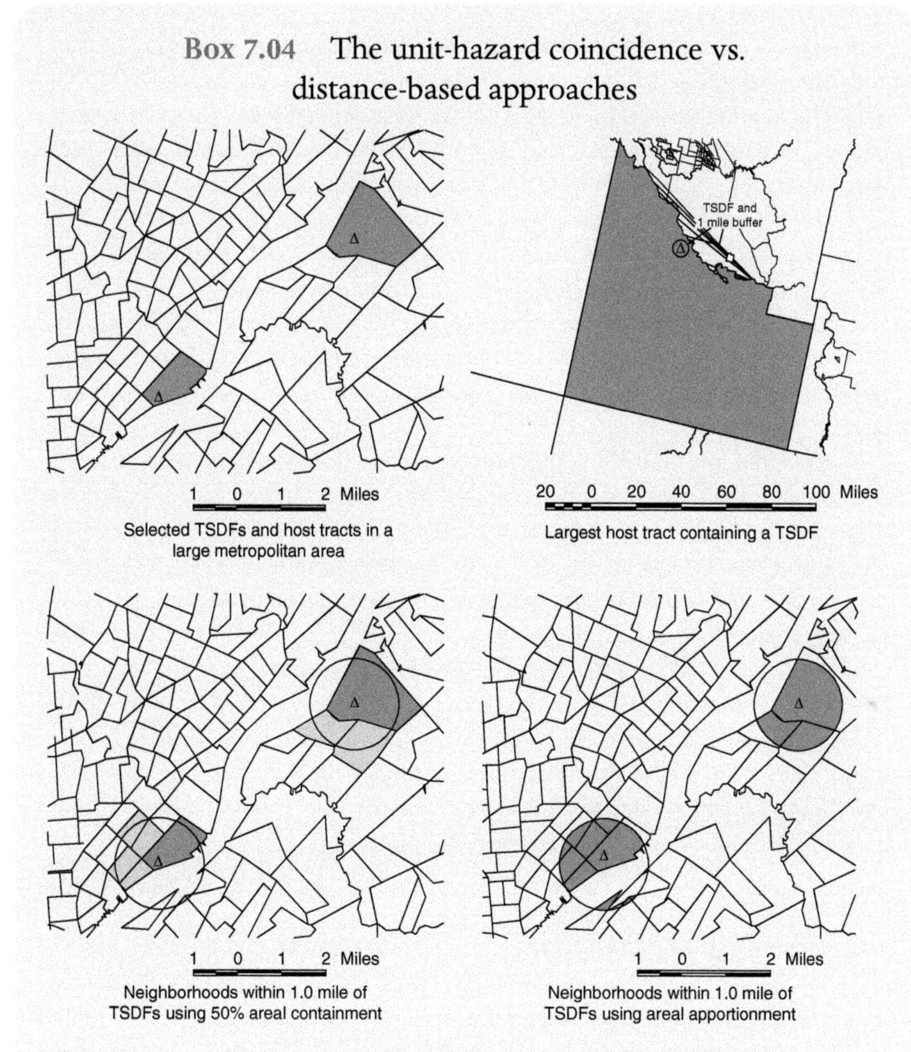

Figure 7.06 The aspects of scale of analysis and proximity to hazardous sites are illustrated in this image. *Source*: Paul Mohai, and R. Saha,"Reassessing racial and socioeconomic disparities in environmental justice research,"*Demography*, 43(2) (2006), p.347. Reprinted by permission of the Population Association of America.

As noted, the unit-hazard coincidence approach has been the most commonly used in conducting environmental inequality analyses, including by the most influential national studies. And as the critics have pointed out, the results of studies using this approach have been contradictory and inconclusive. The problem is that this approach does not take into account the precise geographic location of the hazardous sites, such as the proximity of the site to its host unit's boundary or its proximity to adjacent and other nearby units. When the precise geographic locations of hazardous sites are taken into account, it is often found that they are located near the host units' boundaries and hence very close to adjacent and other nearby units. For example, Mohai and Saha (2007) found 49% of the nation's hazardous waste sites to be within 0.25 miles (0.4 km) of the boundary of their host census tracts, while 71% were within 0.50 miles (0.8 km). Instead of recognizing the proximity of some of the non-host tracts to the hazardous sites, the unit-hazard coincidence method places nearby tracts in the comparison group of tracts, treating them no differently than non-host tracts much farther away. Thus if there is a relationship between the presence of hazardous sites and the demographic characteristics of nearby populations, the characteristics of nearby non-host tracts may be more similar to the host ones proper than to those farther away.

Second, the unit-hazard coincidence method does not take into account the considerable variation in the size of the units such as census tracts. It implicitly assumes that all tracts are of similar size and small enough to ensure that the hazardous sites and residential populations within the tracts are in reasonably close proximity. However, examination of the census tracts hosting the nation's hazardous waste sites reveals that this in fact is not the case. Mohai and Saha (2007) also found, for example, that the smallest tract containing a hazardous waste site is less than 0.1 sq. miles (0.259 sq. km), while the largest is over 7500 sq. miles (19,425 sq. km), with all sizes in between. When a host tract is small, it can reasonably be assumed that everyone living in it is close to the site. However, when the tract is large, it is uncertain how many people in the unit live close by.

Theoretical Explanations of Environmental Inequality

A variety of explanations have been offered as to why environmental inequalities exist. The factors hypothesized to account for the racial and socioeconomic disparities in the distribution of environmental hazards tend to fall into three categories: economic, sociopolitical, and racial.

Economic factors include industry's desire to minimize production costs by locating new facilities in places where land values and operation costs are low. These places may coincidentally be where low-income people and minorities live. Alternatively, the facilities, once sited, may cause a decline in property values and quality of life, motivating affluent people to move away and the poor and people of color to move in because of increased affordability of housing.

Sociopolitical factors involve imbalances in social capital and political power among communities. Disproportionate siting of hazards may occur because poor, minority communities have fewer resources to mobilize and less access to decision makers than do affluent, white communities that would enable them to effectively lobby to keep out unwanted land uses. Even without intent by government and industry to do so, NIMBYism by better organized and more powerful predominantly wealthier communities may lead by default to disproportionate placement of unwanted land uses in minority neighborhoods. In addition, decades of systematic disinvestment in many inner city areas, combined with white flight to the suburbs, have created racial and economic segregation and environmental inequality.

Racial factors are involved when locating hazards in minority neighborhoods is intentional. Even though it may be difficult to find a "smoking gun" of prejudicial attitudes behind siting decisions, deliberate targeting of new facilities may occur because minority communities over time have come to be recognized as the "paths of least resistance" by government and industry. Even if minority communities are not intentionally targeted for society's unwanted land uses, race may still play a role in environmental inequality because housing segregation may limit the ability of people of color to move away from such sites.

There has long been special interest within the EJ movement for understanding whether racial disparities are largely a function of socioeconomic disparities or whether other factors associated with race are also related to the distribution of environmental hazards, but all three factors intermix and overlap. For example, racial inequality in education, employment, health care, land-use planning, and other societal domains can limit the social and political capital of communities of people of color to prevent the siting of polluting facilities and subsequent undesirable neighborhood change. As Pulido argues (see Box 7.01), the institutional and systemic nature of racial discrimination and environmental inequality is inextricably linked to other forms of racial inequality.

Not everyone agrees that hazardous waste disposal companies and public officials in the United States deliberately targeted African Americans, indigenous tribes, or other minority communities for problematic development. While some observers argue that a hazard such as a toxic waste dump would bring jobs to an impoverished community, others argue either that class has influenced hazardous waste-siting decisions more than race, or that poor and minority communities moved in to areas around pre-existing industries.

Some argue that accepting higher rates of pollution is a necessary and reasonable trade-off for economic development. Hollander (2004) controversially makes the case that communities should be able to accept polluting industries if they offer opportunities for employment and development. Take the case of the current economic boom in China, for example. Many argue that this economic growth would not have been possible without a simultaneous increase in pollution and hazardous waste, yet few would argue that the economic boom has not brought major benefits to millions of people. Hollander (2004) makes his case using the **Kuznets Curve** to argue that communities will accept higher rates of pollution in order to advance economically, and that only after a certain level of economic growth has been achieved do people push for pollution controls. As can be seen in Figure 7.07, some environmental degradation seems arguably necessary for economic growth, according to this standard economic model. The model also

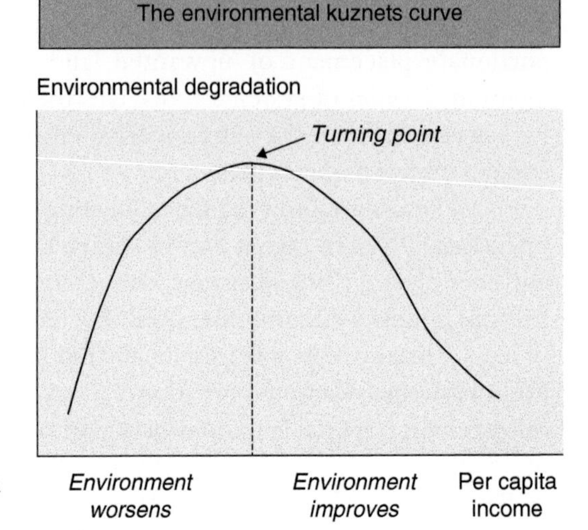

Figure 7.07 A hypothetical Kuznets Curve of environmental quality based on per capita income.

suggests that, after some turning point, environmental conditions begin to improve along with per capita income. Although the model seems to hold for some cases of national development, there is no indication that it holds true at more local scales such as the areas immediately surrounding hazardous sites. Indeed, as much EJ research has demonstrated, it tends to be the poor and disenfranchised that pay the environmental price for the economic developments that are achieved by the more wealthy and powerful groups.

Future Directions for Environmental Justice

> Climate change also threatens the basic life support systems on which humans depend – our water, food, shelter and security. Among the segments of the population that are at greatest risk include the elderly, infants, individuals suffering from chronic heart or lung disease, persons with mental disabilities, the socially and/or economically disadvantaged, and those who work outdoors. (CNRA 2009: 30)

As is the case with all new fields, as EJ matures, new areas of research and interest emerge. There are three key areas of expansion that are of interest to us. The first concerns the globalization of EJ. As domestic policies tighten restrictions on where toxic wastes and hazardous industries may be sited and the cost of safely removing waste rises, industries seek to "offshore" dirty industries and waste. The second new direction concerns the broadening of the movement to include elements of so-called "healthy living." This branch makes the case that the EJ social movement is not simply about toxics and hazards but about access to green spaces and healthy food as well. While EJ advocates have traditionally argued that minority groups are disproportionately subjected to hazardous environments, some now argue that they disproportionately lack other amenities that make up healthy environments such as access to parks, beaches, and other green spaces and even healthy foods. The third and final new direction involves the impacts of perhaps the greatest environmental challenge – climate change. Climate change can potentially have negative impacts on disadvantaged groups in two ways: first, disadvantaged people are likely to suffer more from the consequences of climate change such as extreme weather events (think of how Hurricane Katrina caused greater damage and negative impacts in the poorer African American communities of New Orleans, for example). And second, policies designed to mitigate climate change and produce a more sustainable society may adversely impact disadvantaged groups. The so called **climate justice** movement seeks to address these concerns. Examples of this latest movement focused on climate change can be found world-wide, from Pacific islands clamoring for direct international aid to the more recent lawsuits filed by Alaskan coastal indigenous peoples against the large gas and oil companies. What this movement will produce by way out of outcomes is unclear, but it is a fascinating expansion of the notion of justice to the atmosphere.

Globalization

Only the affluent can worry about effluent…

As globalization has resulted in an increase in the production, distribution, and dumping of environmental hazards, the EJ movement has taken on an increasingly international dimension.

As noted above, Hollander (2004) argues that poorer societies often choose to accept pollution as a trade-off for industrial and economic development. His main point is that only once a society has achieved a certain level of prosperity can it turn to environmental concerns. There are at least two critiques of Hollander's view, however. The first is that there is no guarantee that a country will achieve economic development by accepting more polluting industries. And the second is that while environmental policies in developed countries tend to produce cleaner environments in those countries, they also tend to shift the most toxic industries offshore. As such it is not at all clear that increasing wealth produces an *overall* improvement in the environment; rather, economic development tends to shift dirty industries somewhere else.

This latter issue is once again a matter of scale. Although the US and Europe, for example, have enacted many environmental regulations and "cleaned up" various environments through the implementation of clean air and clean water acts, to what extent have these changes resulted in *global* environmental improvements? Indeed, one could argue that environmental legislation in the US and Europe has resulted in pushing so called *dirty* industries abroad or offshore. Perhaps no example is more glaring than the dumping of toxic waste in West Africa. In August 2006 tons of poisonous chemical sludge were dumped at various sites around the port city of Abidjan in the Ivory Coast, leading to thousands of reports of people suffering from vomiting, diarrhea, and nosebleeds. Investigations traced the dumping to a tanker registered in Panama that had been chartered by a Dutch-based oil trader, as a variety of newspapers reported on the complicated story.

The incident resulted in several environmental groups decrying the lack of proper controls over the international waste-recycling market. As waste management has become globalized, countries with civilian unrest, no environmental law enforcement, or weak legislative frameworks have become prime targets for illegitimate hazardous waste dealers.

According to work by Kwong (2004), empirical data suggest that a developing nation begins to enact environmental legislation once a per capita income reaches a certain level (approximately $5,000 for the case of air pollution, for example). It is not clear, however, whether such a society is responding to increased wealth, or severely degraded environments and human suffering, or both. In addition, it is quite clear that the effects of the pollutants associated with industrialization are not equally distributed among the community members. EJ researchers have shown this to be true for the US, and there is no reason to doubt that similar inequalities are at work in developing countries. There is also growing recognition that EJ issues cannot be separated from a global political and economic context in which they occur (Peet et al. 2011). The poorest always seem to suffer more, and they also tend to be the most powerless to enact environmental legislation. A good example is the case of the Ogoni people of Nigeria described in Box 7.05.

Box 7.05 Ken Saro-Wiwa and the injustices and violence of oil

One would think that the discovery of oil in one of the poorest places on Earth would be a blessing. Unfortunately however, the discovery of oil has sometimes proved more of a curse. Take the case of Nigeria in Africa. Nigeria is Africa's most populous nation, with nearly 150 million people in a country about the size of California. Nigeria is a major producer and exporter of oil and yet the per capita income of $2,300 remains among the world's lowest.

In an ugly twist of fate, the region of Nigeria with the greatest oil reserves, the Niger Delta area, is also one of the poorest and most heavily degraded. For the Ogoni people in particular, oil has been a curse. Ogoni lands have six oilfields, a fertilizer plant, and a petrochemical plant and yet as of 2001, no single all-season road. They also have the highest child mortality in the country, approximately 80% of the population is illiterate, 85% are unemployed, and few homes have electricity. In addition, this once fertile delta area has become an environmental nightmare. There have been frequent oil spills that have degraded the land and water, and flares from the unused (or wasted) natural gas burn 24 hours a day.

The plight of the Ogoni people was taken up by the Nigerian Ken Saro-Wiwa, a relatively wealthy and successful writer and member of the Ogoni. Saro-Wiwa fought for environmental justice in the delta region and organized rallies in protest. He led a nonviolent campaign against environmental degradation of the land and waters of Ogoniland by the operations of the petroleum industries, especially Shell. He was also an outspoken critic of the Nigerian government, which he viewed as reluctant to enforce environmental regulations on the foreign petroleum companies operating in the area. He demanded that some of the proceeds from oil money be allocated to the development of the delta region. Tragically, at the peak of his nonviolent campaign, Saro-Wiwa was arrested, hastily tried in military court, and executed in 1995 by the military government. The charges against him were widely viewed as entirely politically motivated and completely unfounded.

In the aftermath of Saro-Wiwa's death, Shell Oil has made some efforts to reinvest in the most negatively impacted delta communities. But the potential degradation and inequalities created by of oil wealth continue to be the cause of sometimes violent unrest in the region. Unrest in the Nigerian delta impacts us all, as oil markets respond to unrest and violence in the region with increased prices for oil. The region remains one of the most polluted in the world. (See Watts 2008 for more.)

EJ and Healthy Environments

As the description of Long Beach, California, in the introduction to this chapter illustrates, environmental injustices can take multiple forms. While the bulk of EJ research and activism has examined environmental hazards by focusing attention on the needs of disenfranchised populations, a recent shift has placed more emphasis on environmental amenities such as parks, green spaces, and the built urban environment in general.

Several factors have contributed to this shift. For one, there have been architectural changes following the oil shocks of the 1970s, especially the construction of airtight buildings. These buildings can become "sick," and many are now known to contain air contaminated with chemicals and other toxins. While wealthier populations have been able to pay to modify their buildings and remove some of the hazards such as asbestos, the poor continue to occupy many such dwellings with hazardous conditions. In addition, rapid urbanization around the world, the sprawling expansion of cities, and the failure to set aside adequate space for recreation have given new meaning – and urgency – to the idea of environmental health. The obesity epidemic in developed nations has also called attention to land use and transportation as determinants of physical activity.

Disparities in the built environment can be identified in at least five arenas: housing, transportation, food, green spaces, and squalor. As described above, the location of green spaces, such as parks, in Long Beach tends to be disproportionately proximate to the wealthier and whiter population of the city. Parks and green spaces represent critically important environmental amenities, because contact with nature is highly valued and offers a range of health benefits. In cities and towns, parks are the principal venue for regular public access to nature. Parks also offer settings for physical activity and social interaction.

Race and ethnicity considerations continue to matter for EJ issues in at least two ways. First, members of minority groups in some cities may lack access to parks, trails, and other green spaces. Second, it has been shown that minority populations tend to prefer recreational uses while dominant groups will tend to favor land conservation. These attitudes are reflective of environmental politics in the United States, but similar dispositions can be found world-wide. The relationship between parks and urban areas is thus a complex one: unlike the case of hazards, which are universally unwanted, different communities might have very different preferences for the types of parks or green spaces they want. These differences call for culturally sensitive park design.

An interesting example can be found in the city of Baltimore, where researchers found that African American and white populations had differing access to parks. The study found that that parks in the African American areas were small and close by, while those in the white suburbs were larger and further from where people lived. The Baltimore study highlights how community preferences matter in terms of EJ of amenity environments. What about you? Would you rather live within walking distance of a small park or within driving distance of a large one?

Also, a worrisome irony is that urban green space may, in some cases, increase adjacent residential property values, while in others it can result in property value decreases due to unsafe conditions: poorly maintained or poorly policed parks can become magnets for crime and other activities. This reduces the benefits of living near an open space.

Transportation is another area where EJ scholars and activists are beginning to work. According to Frumkin (2005), equity concerns in transportation take at least two forms. First, certain elements of transportation infrastructure, such as highways and bus depots, are "locally undesirable land uses." Poor people and people of color live disproportionately near these locations and suffer associated health consequences – the effects of diesel air pollution, noise, injury risks, and ugliness. Second, transportation systems that do not necessarily provide poor people with convenient, practical access to employment, medical care, and other necessities, and undermine their health in numerous ways. As Jocoy and Del Casino's (2010) study of homelessness and transportation in Long Beach demonstrated, services available to the homeless require that they have access to transportation, yet transportation systems in southern California are organized in such a way so as to restrict homeless people's access to services such as health care.

More recently, there has been an increasing recognition that the built environment may affect what people eat. In poor neighborhoods where members of minority groups disproportionately live, junk food, soda, and cigarettes are readily available in small markets. Scholars are now using the term **food deserts** to describe areas or neighborhoods with meager (and distant) access to fresh foods. Meanwhile, grocery stores that sell fresh foods are scarce and/or expensive. These urban environmental factors matter as they help explain why people who live in poor neighborhoods eat less healthy diets. They have also spawned numerous urban garden movements to counteract these effects.

Is Climate Change an Environmental Justice Issue?

Climate change is surely one of the greatest environmental challenges of our generation (see Chapter 8), but is it also an EJ issue? According to climate justice scholars, climate change touches on two main ethical concerns. The first is that global warming will likely hurt the poorest and least able to adapt to a changing environment (for example, people living in low-lying areas such as the Pacific islands or coastal Bangladesh which will likely be flooded). The second ethical issue is that efforts to curb carbon emissions and slow global warming will fail unless all parties involved feel they are being treated justly. For example, if China and India, two emerging industrial powers, feel that they are being "cheated" out of a chance to use fossil fuels to power their economic growth, then a global climate change agreement will not be reached.

Others, however, argue that climate change is an environmental justice issue of another sort. Climate justice scholars argue that some populations will suffer disproportionately from policies designed to counter climate change. In general, climate justice research focuses on the disproportionate impacts of climate change and climate policies (Adger 2004). Three basic assertions of climate justice are:

1. Climate change and the policies to address climate change will have numerous economic and social consequences. Some have called this a problem of "double exposure," in that some societies will be faced with a squeeze from both biophysical changes (climate), and at the same time, the pressures of economic globalization (Leichenko and O'Brien 2008).
2. The impacts of both climate change and related policies will not be felt uniformly by all members of society; some groups may face more severe and adverse effects than others.
3. Low-income groups and people of color will likely face more severe impacts than other groups from climate change and related policies.

In short, the consequences of climate change will likely have a disproportionate impact on vulnerable communities, and without proper planning climate policy itself may impose unequal burdens. As is the case with most EJ issues, many of these differences reflect long-standing systemic and societal inequalities. For a simple example, one possible climate change policy is a carbon tax (a tax on all forms of fossil fuel burning including gasoline). Such a tax would reduce fossil fuel consumption and would result in many environmental benefits. Do you think a carbon tax will impact all members of society equally? Surely as a student on a low budget, an increase in gas prices would have a greater impact on you than it would on a wealthier individual. What would be the impact on a family with low income?

A primary goal of the climate justice movement is to ensure that no racial, ethnic, or socioeconomic group bears a disproportionate share of the negative consequences of climate change and related policies. Climate justice advocates seek to protect vulnerable communities from the impacts of climate change and to implement policies that provide a just transition. Climate justice also places great emphasis on community participation in state decision-making, research, and policy development.

A final aspect of climate justice is to document the extent to which the major economic and health costs of climate change will be disproportionately borne by low-income communities and people of color. Disproportionate effects of climate change fall into four primary health and economic categories: heatwave-related mortality, increases in ambient ozone, employment risk, and impacts on prices of basic necessities. One of the few ways to make this tangible is to think critically about what aspects of climate change would most affect your own community.

Chapter Summary

Environmental justice is a distinctive human–environmental perspective in that it focuses on the inequitable distribution of environmental qualities among various human groups. It is also closely associated with environmental activism. Most commonly EJ scholars have documented how disadvantaged groups (especially the poor and minority populations) often bear the least responsibility for causing environmental degradation while they bear the brunt of its negative consequences. EJ is concerned with documenting the unequal patterns of exposure of different groups to hazards (especially hazardous waste and other industrial pollutants) and explaining the causes of these. EJ also takes an explicitly activist stance aimed at mitigating or removing the hazards and the powerful forces that result in disproportionate exposure to them. Although the field of EJ began with a focus on industrial hazards, EJ scholars are now documenting how environmental amenities, such as green spaces, are unequally distributed among different social and economic groups. A major challenge for the future will be to address issues of climate justice that will emerge as our society attempts to mitigate and adapt to climate change.

Critical Questions

1 Can you think of local examples, in your community, where citizens have cited "environmental justice" as a term in their activist pursuits? How were they using the term?

2. How is the issue of "scale" useful or challenging in pursuing environmental justice claims?
3. What are the problems in trying to prove environmental racism?
4. What aspects of human–environment thinking are neglected in the environmental justice literature or activist groups that might benefit EJ in the long run?
5. Does everyone have a "right" to a clean place, clean water, and healthy air? How do our basic biological needs parallel our political solutions for pollution; or is there a disconnect?

Key Vocabulary

activism
air quality
Bhopal (incident)
Chernobyl
climate justice
environmental burden
environmental equity
environmental justice
environmental racism
food deserts
Fukushima (nuclear power plant)
generational equity
geographic equity
Kuznets Curve
landscapes of risk
Minamata Bay (and disease)
NIMBY
nuclear waste
open space
procedural equity
remediation
social equity
social movement
toxic waste
toxins
unit-hazard coincidence

References

Adger, W.N. (2004) The right to keep cold. *Environment and Planning A*, 36(10), pp. 1711–15.

Anderton, D.L., Anderson, A.B., Oakes, J.M., and Fraser, M.R. (1994) Environmental equity: the demographics of dumping. *Demography*, 31, pp. 229–48.

Been, V., and Gupta, F. (1997) Coming to the nuisance or going to the barrios? A longitudinal analysis of environmental justice claims. *Ecology Law Quarterly*, 24, pp. 1–56.

Bowen, W. (2002) An analytical review of environmental justice research: what do we really know? *Environmental Management*, 29(1), pp. 3–15.

Bryant, B.I. (1995) *Environmental Justice: Issues, Policies, and Solutions* (Washington: Island Press).

Bullard, R.D. (1983) Solid waste sites and the black Houston community. *Sociological Inequity*, 53, pp. 273–88.

Bullard, R.D. (1990) *Dumping in Dixie: Race, Class, and Environmental Quality* (Boulder, CO: Westview Press).

Bullard, R.D. (ed.) (1994) *Unequal Protection: Environmental Justice and Communities of Color* (San Francisco: Sierra Club Books).

CNRA (2009) California Climate Adaptation Strategy. Document available at http://resources.ca.gov/climate_adaptation/docs/Statewide_Adaptation_Strategy.pdf

Chavis, B., Jr. (1994) Preface. In R.D. Bullard, ed., *Unequal Protection: Environmental Justice and*

Communities of Color, pp. xi–xii (San Francisco : Sierra Club Books).

Cutter, S. (1995) Race, class and environmental justice. *Progress in Human Geography*, 19(1), pp. 111–22.

Durlin, M. (2010) The shot heard round the West. *High Country News*, 42(2, February).

Faber, D. (ed.) (1998) *The Struggle for Ecological Democracy: Environmental Justice Movements in the United States* (New York: Guilford Press).

Foreman, C.H. (1998) *The Promise and Peril of Environmental Justice* (Washington, DC: Brookings Institution Press).

Frumkin, H. (2005) Guest editorial: health, equity, and the built environment. *Environmental Health Perspectives*, 113(5), pp. A290–A291.

Greenberg, M. (1993) Proving environmental inequity in siting locally unwanted land uses. *Risk: Issues in Health & Safety*, 4, pp. 235–52.

Heiman, M. (1996) Race, waste, and class: new perspectives on environmental justice. *Antipode*, 28(2), pp. 111–21.

Hollander, J.M. (2004) *The Real Environmental Crisis: Why Poverty, Not Affluence, Is the Environment's Number One Enemy* (Berkeley, CA: University of California Press).

Jocoy, C.L. and Del Casino Jr., V.J. (2010) Homelessness, travel behavior, and the politics of transportation mobilities in Long Beach, California. *Environment and Planning A*, 42, pp. 1943–63.

Kosek, J. (2006) *Understories: The Political Life of Forests in Northern New Mexico*. Durham, N.C.: Duke University Press.

Kwong, J. (2004) *Globalizations Effects on the Environment: Boon or Bane?* (St. Charles, MO: Lindenwood University's Economic Policy Lecture Series).

Lee, K. and Ehrenfeucht, R. (2001) The origins and future of the environmental justice movement: a conversation with Laura Pulido. *Critical Planning*, 8 (Summer), pp. 16–21. http://www.spa.ucla.edu/critplan/past/volume008/003%20Pulido%20interview.pdf (accessed July 3, 2012).

Leichenko, R. and O'Brien, K. (2008) *Environmental Change and Globalization: Double Exposures* (New York: Oxford University Press).

Masco, J. (2006) *The Nuclear Borderlands: The Manhattan Project in Post-Cold War New Mexico* (Princeton: Princeton University Press).

McGurty, E.M. (1997) From NIMBY to civil rights: the origins of the environmental justice movement. *Environmental History*, 2, 301–23.

Mohai, P. and Saha, R. (2006) Reassessing racial and socioeconomic disparities in environmental justice research. *Demography*, 43(2), pp. 383–99.

Mohai, P. and Saha, R. (2007) Racial inequality in the distribution of hazardous waste: a national-level reassessment. *Social Problems*, 54(3), pp. 343–70.

Peet, R., Robbins, P., and Watts, M. (eds.) (2011) *Global Political Ecology* (New York: Routledge).

Walker, B. (2010) *Toxic Archipelago: A History of Industrial Disease in Japan* (Seattle: University of Washington Press).

Watts, M. (ed.) (2008) *The Curse of the Black Gold: 50 Years of Oil in the Niger Delta* (Brooklyn: PowerHouse Books), http://www.curseoftheblackgoldbook.com/.

Part III

Thematic Issues in Human–Environment Geography

8

Climate, Atmosphere, and Energy

Icebreaker: The Perils of a Micronesian Island State

In many ways, the Micronesian island state of Kirabati (Figure 8.01) might be considered a development success story. Kirabati has a population of about 100,000 people spread over 34 islands with a per capita income of US$950. While most of the population (80%) is engaged in a subsistence-based farming and fishing economy, the country also has significant

An Introduction to Human–Environment Geography: Local Dynamics and Global Processes,
First Edition. William G. Moseley, Eric Perramond, Holly M. Hapke and Paul Laris.

Published 2014 by John Wiley & Sons, Ltd.

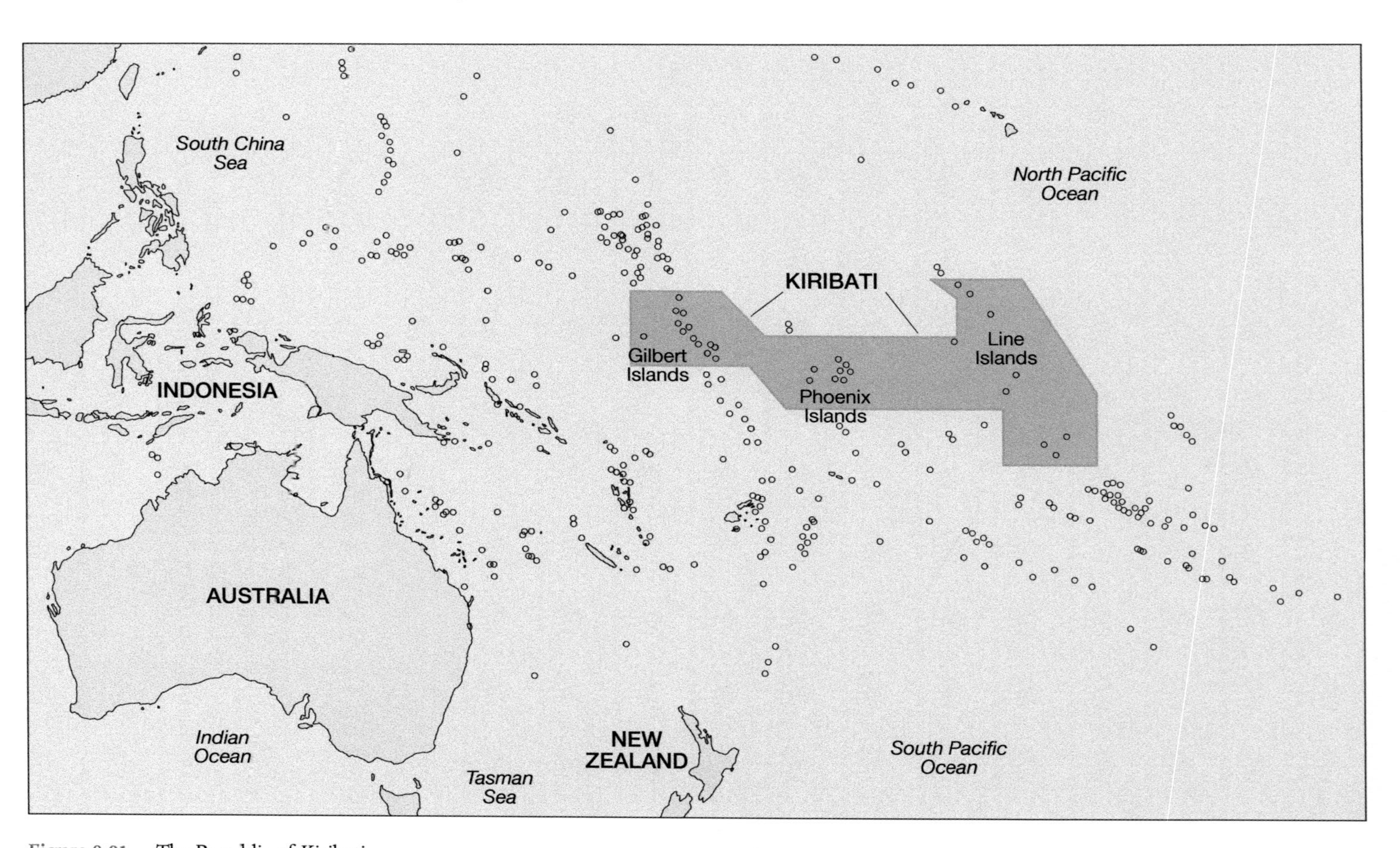

Figure 8.01 The Republic of Kiribati.

phosphate deposits which it is mining and exporting. Unlike other countries which have unwisely used their resources (such as oil in Nigeria), the Kirabati government has carefully used its non-renewable resources to care for its people. Since 1956, the proceeds from phosphate extraction have been placed in a trust fund which is invested offshore by two London-based account managers (Gibson-Graham 2004). The returns on this fund are used to finance government services, including health care and the development of a communication and transportation infrastructure between the islands. What this means is that most residents in Kirabati are free to continue living a subsistence lifestyle, yet still have access to sustainably financed government services. This case represents a situation where a non-renewable resource (phosphate) has effectively been converted to a renewable resource (a self-sustaining trust fund) which provides for ongoing investments in a country's human capital.

The tragedy of Kirabati is that it may soon (literally) be under water. Much of Kiribati is no higher than 5 meters above sea level and the ocean around its islands has been rising around 5 millimeters a year since 1991. This sea-level rise is a direct result of global climate change largely driven by carbon emissions from fossil fuel combustion in the world's most industrialized countries. The country's president, Anote Tong, has said that increased flooding has forced some villagers to move inland. The ability to move, however, is limited as many of the country's islands are so narrow that there really is no place to go. The reality is that many of the country's citizens will eventually need to leave the country and resettle elsewhere. Does the international community have an obligation to help Kiribatins settle elsewhere and purchase land? Is it fair for those in the Global North to consume energy at unprecedented levels if it means that others lose their homes to sea-level rise?

Chapter Objectives

1. To understand the nature of acid deposition as well as its causes, consequences, and remedies.
2. To comprehend the nature of stratospheric ozone depletion as well as its causes, consequences, and remedies.
3. To grasp the nature of global warming as well as its causes, consequences, and remedies.
4. To review the causes and consequences of local-level atmospheric problems as well as the initiatives and policies to manage these problems.
5. To examine energy production and consumption in relation to atmospheric problems.

Introduction

Air is the classic **open-access resource**, that is, it is a resource with no owner and no one can prevent us from using it. Unlike **private property**[1] or **common property**,[2] the beauty of an open-access resource is that it is available to everyone no matter their wealth, income, or group affiliation. The problem with air, however (and other open-access resources), is that not only may it be used by anyone at no cost, but it may also be abused by anyone at limited cost. As such, the air is not only a **resource**[3] which is essential for life but it is a **sink**[4] which absorbs pollution. When one individual, group, or organization pollutes the air, the cost of such pollution is not wholly felt by that same entity, but by the entire population of an area. As such, there is a mismatch as the benefits of using the air or atmosphere as a sink for pollution accrue entirely to one entity but the costs of abusing such a sink are diffused to a larger group. This divergence of benefits and costs is sometimes referred to as the **tragedy of the commons** (Harden 1968) (see Chapter 12 for a discussion of this problem in the context of fisheries). According to this parable, individuals abuse resources in cases where access is not restricted and they are not bearing the full cost of such use. To be clear, the term 'commons' is misapplied in this now classic scenario (as the use of commons is regulated by a group) and open-access resources is the more apt descriptor. In spite of the misapplication of the term commons, the tragedy of the commons scenario does help us understand why air pollution has long been a problem that is typically only solved via government regulation. Air pollution is also a problem that manifests itself at different scales, from localized problems to those which are truly global in scope.

Harnessing energy to do work has been an age-old preoccupation of human beings. At the simplest level, humans consume vegetable and animal tissue (both forms of stored energy) to secure the sustenance needed to perform work. But humans can only perform so much labor in a day, and we have long sought to harness other sources of energy to enhance the impacts of our labor (such as using draft animals to pull plows) or to power machines that perform labor on our behalf (such as using water currents to drive a flour mill or windmills to pump water). Wood was the other common source of energy, used to supply the fires that cooked food, heated homes, and smelted iron. But at some point wood started to run out. The Chinese are likely to have been the first to use a fossil fuel (coal) as a source of energy to smelt iron some 1000 years ago. In the early 1600s, the Dutch were running out of wood and shifted to locally abundant peat (another fossil fuel) to cook their meals, heat homes, and process manufactured products. But it was in Great Britain during the 17th and 18th centuries that fossil fuels (coal in this case) would be forever linked to industrialization. Here wood had grown scarce, rising some 700% in price between 1500 and 1630, and coal was abundant. Coal was burned to produce the steam to power all sorts of mechanical processes. Along with the burning of fossil fuels came air pollution, such that by the end of the 17th century London had already gained a reputation for its smoke (Crosby 2006).

Fossil fuels (oil, natural gas, coal) are essentially the remains of anaerobically decomposed organisms (often phytoplankton and zooplankton) that settled to the bottoms of seas and lakes millions of years ago. By exploiting such stockpiles of energy, we are essentially digging into the energy of the sun that was stored in plants and animals in bygone eras. This treasure trove of energy has powered globalization and allowed humanity to accomplish levels of work never before imagined. But the unprecedented use of such resources has also altered the atmosphere at scales never before seen, leading to new problems and the need for different solutions.

This chapter examines three major global atmospheric issues: acid deposition, stratospheric ozone depletion and global warming. It then reviews a handful of more localized atmospheric issues. The chapter concludes with a review of energy production and consumption around the world, an activity which often – but not always – has some bearing on atmospheric conditions.

Global Atmospheric Issues

Acid Deposition

Acid deposition, also known as acid rain, refers to the deposition of acids and substances by rainfall, snow, and dust particles falling from the atmosphere that contribute to soil acidification. The increasing acidity of precipitation and dust is related to emissions of sulfur dioxide and nitrogen oxides. Sulfur dioxide (SO_2) combines with oxygen (O) and water (H_2O) in the atmosphere to form sulfuric acid (H_2SO_4). Similarly, nitrogen dioxide (NO_2) combines with water (H_20) to form nitric acid (HNO_3).

Most sulfur dioxide emissions result from the combustion of impure fossil fuels such as coal and fuel oil which are often burned to generate electricity. Nitrogen oxides (NO_x) are generated by motor vehicles, industrial processes, and thermal power plants. Acid deposition tends to be a problem in those areas downwind of major source areas around the world (see Figure 8.02), including significant urban and manufacturing areas as well as sites of coal-fired power generation.

The pH of normal rainfall is around 5.5. More acidic rainfall falls below 4.6 and extremely acidic rainfall would measures under 4.3. Acid deposition may lead to a number of environmental and economic costs. Lakes may become so acidic that fish and other aquatic life can no longer survive in them. At first glance, such lakes may appear crystal clear, but further inspection reveals that they are unnaturally sterile. Forests exposed to acid deposition may also experience slower growth and increased susceptibility to disease, leading to the iconic dead forests of northern Europe. Acid rain weakens trees by damaging their leaves, limiting the supply of nutrients to their roots, or poisoning them with toxic substances released from the soil (such as aluminum). The problem may also impact human health by affecting drinking water supplies as more acidic pH levels may lead to the leaching of lead and aluminum into such waters. The leading economic impact of acid deposition is related to the corrosive effect of such precipitation. Exterior paint and statuary may all corrode as a result of acid rain. In fact, one of the authors of this text recalls a huge sale at a local used car dealer when he was in graduate school which was triggered by a significant acid rain event. The rain (registering acidity levels in the 3s!) corroded the paint on cars for sale in an open-air lot, forcing the dealer to sell the used cars at a steep discount. While the event was great for those potential car buyers on limited budgets, it represented a significant loss of value for the dealer.

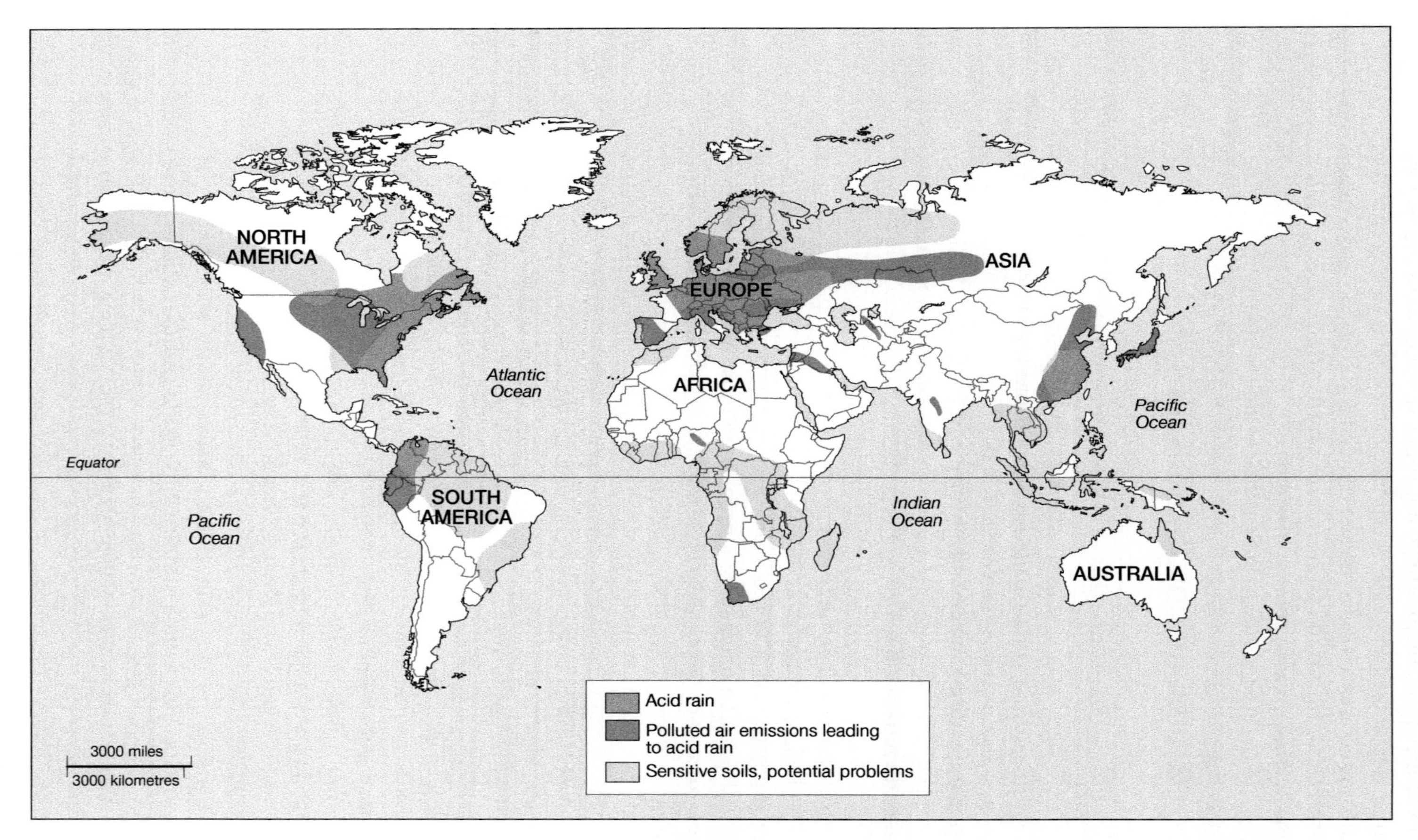

Figure 8.02 The global problem of acid rain. *Source*: ST9 Acid Rain Map from http://go.hrw.com. Copyright © by Holt, Rinehart and Winston.

Acid rain is reduced by controlling the emissions of sulfur dioxide and nitrogen oxides. Sulfur dioxide emissions may be limited by burning low-sulfur coal, installing sulfur scrubbers, or reducing combustion temperatures in power plants. A key way to reduce the emissions of nitrogen oxides is to drive less. Given that acid deposition does not respect national borders, effective solutions must be regional or international in scope. The Convention on Long Range Transboundary Air Pollution was signed in 1979. It was signed by most European Countries, the USA and Canada. The USA launched its own Acid Rain Program in 1990.

Stratospheric Ozone Depletion

Ozone naturally occurs in the upper atmosphere (stratosphere) and is important because it intercepts ultraviolet radiation. Stratospheric ozone is different than ground-level ozone, which is a form of pollution that may be dangerous to human health. Stratospheric ozone depletion refers to the accumulation of ozone-depleting chemicals such as CFCs which contribute to the loss of ozone in the stratosphere. The depletion of stratospheric ozone leads to more ultraviolet radiation reaching the Earth's surface, resulting in human health as well as environmental problems.

In 1985, scientists discovered a 40% reduction in stratospheric ozone over Antarctica (subsequently named the **ozone hole**). They have also discovered a much less significant reduction of stratospheric ozone depletion over the Arctic region. Polar stratospheric clouds form at both the North and South Poles. Chemical reactions on ice crystals convert chlorine (from synthetic chlorofluorocarbons or CFCs) from non-reactive to reactive forms which are sensitive to sunlight. Warming in springtime leads to chlorine chain reactions that transform ozone into oxygen.

Increased ultraviolet radiation as a result of stratospheric ozone depletion is a problem for plants and animals alike. It causes biological damage to plants by decreasing photosynthesis and water-use efficiency. It may depress crop yields by as much as 30%. Amongst animals, humans are particularly susceptible to ultraviolet radiation, given their exposed skin. It has been estimated that every 1% decrease in stratospheric ozone may lead to a 4–6% increase in skin cancer. Ultraviolet radiation may also depress the human immune system.

The leading causes of stratospheric ozone depletion are two synthetic chemicals: chlorofluorocarbons (or CFCs) and bromine. CFCs are a class of synthetic substances originally developed for refrigeration. The trademark name for CFC-12, a very common CFC, is Freon. One third of CFCs were used for refrigeration and cooling (including cars) in the US Another third were used as blowing agents for rigid foams (styrofoam) and flexible foams (foam pillows). They were also used in industrial solvents and in aerosol sprays. Bromine (in halon) is a more potent destroyer of ozone than CFCs (although used in much smaller quantities). Halons are used in fire extinguishers.

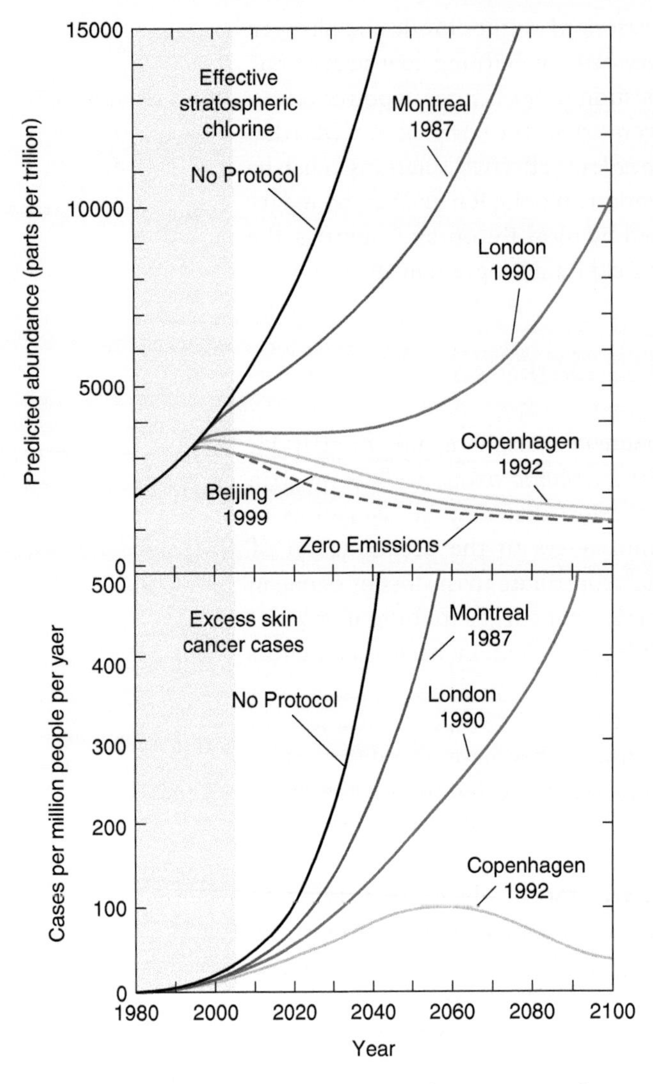

Figure 8.03 Effect of the Montreal Protocol. *Source*: D.W. Fahey, "Twenty questions and answers about the ozone layer," 2006 update. Panel Review Meeting for the 2006 Ozone Assessment; www.usa.gov.

The international community has been relatively successful at reducing the production and use of CFCs. The first major international agreement on CFCs was the **Montreal Protocol**, which was spurred by discovery of the ozone hole in 1985. The treaty was ratified in 1987 by over 125 countries and called for a 50% reduction in emissions by 1999. It also limited the use of halons as fire retardants. The signatories to the protocol subsequently agreed on total elimination of the use of CFCs in North America and Europe by 2000. All production of CFCs in developed countries had ceased by 1996. The Montreal Protocol has undergone seven revisions, in 1990 (London), 1991 (Nairobi), 1992 (Copenhagen), 1993 (Bangkok), 1995 (Vienna), 1997 (Montreal), and 1999 (Beijing). It has now been ratified by 196 countries, or all member states of the United Nations. Scientists estimate that the ozone layer could recover to 1979 levels by 2050. One of the problems with CFCs is that they persist in the atmosphere for a very long time before breaking down, meaning that this problem will continue for some time after the complete discontinuance of production. In general, it is probably fair to say that international efforts to address stratospheric ozone depletion are considered to have been quite successful (see Figure 8.03). The reasons for success are related to a few factors. First, the ozone hole was a very tangible problem, with clear human health impacts, that was easy to understand by the public. Second, the most used CFCs, such as Freon, had substitutes. As such, while the cost was a little higher, it was not a great inconvenience to switch from one refrigerant to another.

Climate Change and Global Warming

When discussing global climate change it is important to make a distinction between climate and weather (see Chapter 3). **Climate** refers to long-term environmental conditions in a particular area, such as average temperature and precipitation patterns over decades. In contrast, **weather** is the subject of the evening news, that is, the day-to-day fluctuations in meteorological conditions. The distinction between climate and weather is important because commentators sometimes remark on a particularly warm day, or even a warmer than average season, as evidence of global warming. In contrast, the study of climate change focuses on trends that occur over decades or centuries rather than days, weeks, or months.

It is important to remember that the global climate has fluctuated throughout history between warmer and cooler periods. These natural fluctuations in climate may be distinguished from human-induced or **anthropogenic climate change**. For many years, scientists struggled to distinguish between natural and anthropogenic climate change. In 2001 (and in subsequent meetings), the **International Panel on Climate Change (IPCC)**[5] concluded that "The balance of evidence suggests that there is a discernible human influence on global climate."

The central problem is that humans are producing certain types of gases which are amplifying the **greenhouse effect**. As depicted in Figure 8.04, the atmosphere regulates the flow of energy between space and the Earth's surface, much like a greenhouse regulates temperatures for plants. This greenhouse effect occurs because sunlight passes through the atmosphere relatively unimpeded, yet much of the outgoing radiation is temporarily trapped in the atmosphere, keeping the Earth warmer. This happens because energy returning to space from Earth has longer wavelengths than sunlight and the atmosphere is only partially transparent

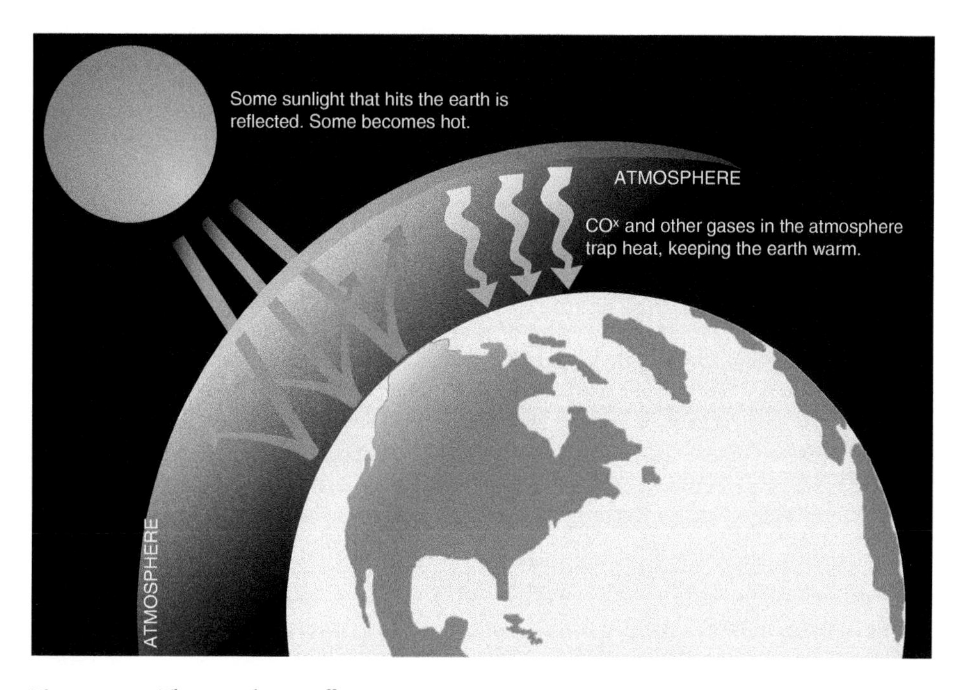

Figure 8.04 The greenhouse effect.

to these wavelengths. We would all freeze to death in the absence of the naturally occurring greenhouse effect. However, this effect becomes a problem when it is amplified by additional greenhouse gases produced by humans.

Greenhouse gases are those components in the atmosphere which are responsible for the greenhouse effect, including water vapor, methane, ozone, nitrous oxides, carbon dioxide, and trace gases. Many of these have natural as well as human sources. As discussed, human caused increases in greenhouse gas concentrations, lead to amplification of the greenhouse effect, and induce global warming. The leading greenhouse gases and their sources may be found in Table 8.01. Increasing emissions have amplified the greenhouse effect and led to increasing temperatures since the 19th century (Figure 8.05).

Estimates of future temperature increases vary. According to the IPCC, the average surface temperature of the Earth is likely to increase by 1.1° to 6.4° C (2° to 11.5° F) by 2100, relative to 1980–90, with a best estimate of 1.8° to 4.0° C (3.2° to 7.2° F) (IPCC 2007). This is a temperature increase which is about double that experienced in the 20th century. The warming will occur unevenly across

Table 8.01 Major greenhouse gases

Greenhouse gas	*Chemical formula*	*Anthropogenic sources*
Carbon dioxide	CO_2	Fossil fuel combustion, land-use conversion, cement production
Methane	CH^4	Fossil fuels, rice paddies, waste dumps, livestock
Nitrous oxides	N_2O	Fertilizers, industrial processes, fossil fuel combustion
CFC-12	CCI_2F_2	Liquid coolants, foams

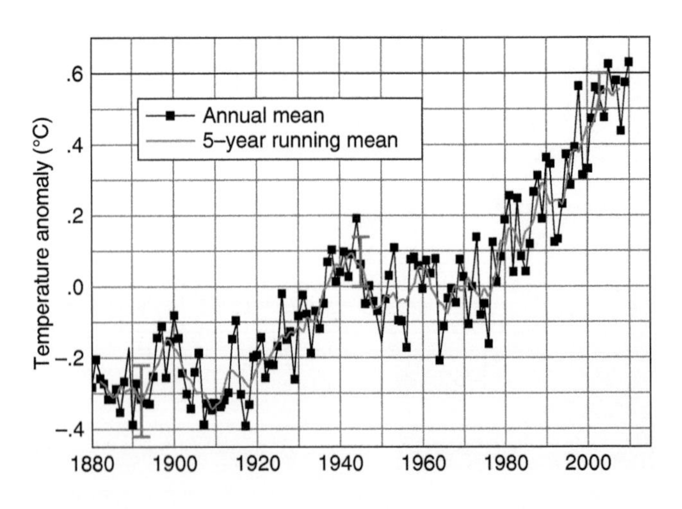

Figure 8.05 Global temperature trends.
Source: J. Hansen, M. Sato, R. Ruedy, K. Lo, D.W. Lea, and M. Medina-Elizade, "Global temperature change," *Proceedings of the National Academy of Science*, 103, pp. 14288–93 (doi:10.1073/pnas.0606291103) (2006). http://data.giss.nasa.gov/gistemp/graphs/. Reprinted by permission of the authors.

the planet (see Figure 8.06). At a general level, warming will be more pronounced over land masses than oceans, and winters will warm more than summers in most areas.

As noted in Table 8.01, the major human-produced greenhouse gases are carbon dioxide, methane, nitrous oxide, and CFCs. While it is important to limit the emission of all of these gases, carbon dioxide has garnered the most attention because it is responsible for about 84% of human-caused global warming (IPCC 2007). Emissions of greenhouse gases, and carbon dioxide in particular, vary greatly from country to country depending on levels of industrial activity, consumption and efficiency of energy use, spatial living patterns (dispersed or concentrated), and transportation choices and infrastructure. Figure 8.07 depicts three different views of carbons emissions by country or block of countries.

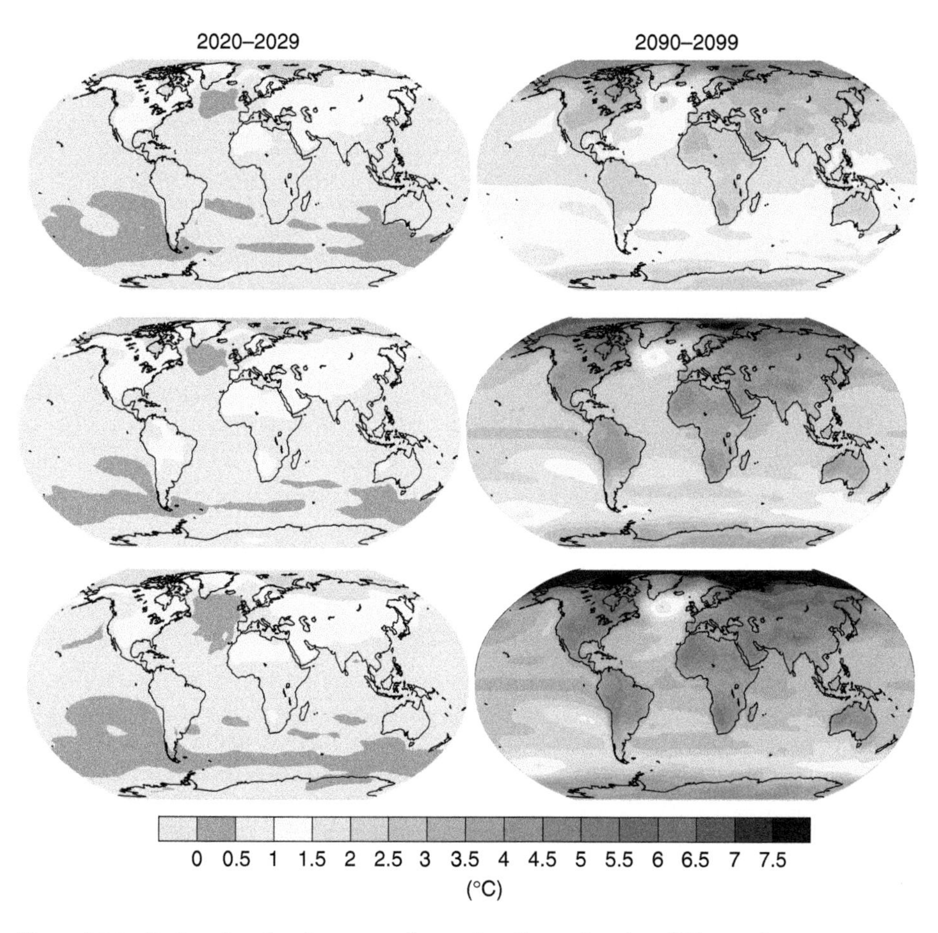

Figure 8.06 Projected regional patterns of warming. *Source*: Based on "Climate change 2007: the physical basis," Working Group I Contribution to the *Fourth Assessment Report of the Intergovernmental Panel on Climate Change* (Cambridge: Cambridge University Press), figure SPM 6. Reprinted by permission of IPCC.

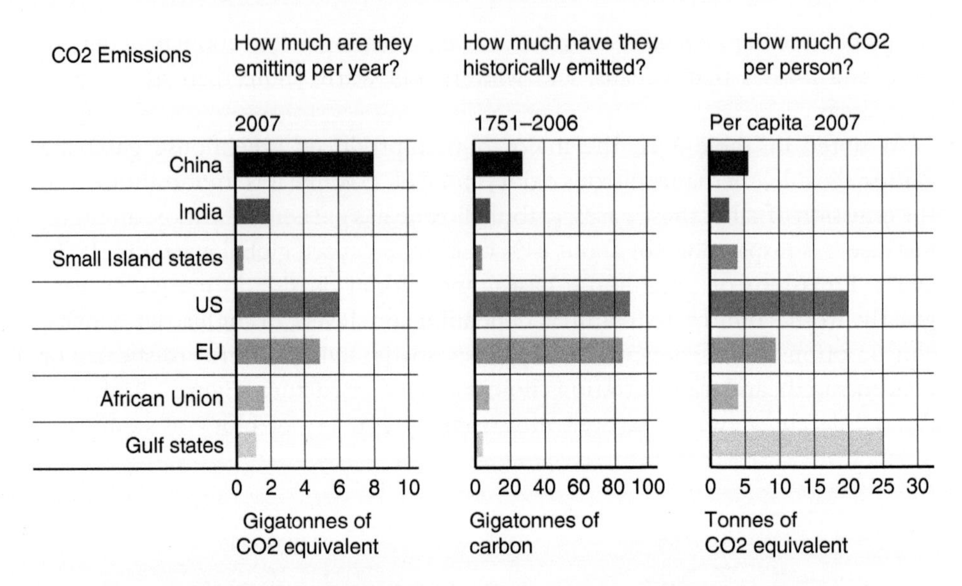

Figure 8.07 Three different ways to see carbon emissions. *Source*: CDIAC, Potsdam Institute for Climate Impact Research and Economics of Adaptation to Climate Change. *Source*: http://news.bbc.co.uk/2/hi/asia-pacific/8411768.stm.

These measures include total emissions per annum in the present, cumulative emissions since the 18th century, and emissions per capita in the present. The top emitters of greenhouse gases differ depending on which one of these measures is employed. The leader in terms of total emissions is China, having recently surpassed the United States in 2006. China's very large population and industrial economy largely explain this ranking. In terms of cumulative historical emissions, the US slightly surpasses the European Union. This reflects the historically large US economy as well as inefficiencies in transportation and urban development patterns (e.g., limited public transportation, low-density urban development). Finally, the Gulf States have the highest emissions on a per capita basis. The significant oil-extraction business in many of these countries, coupled with relatively small national populations, explain these high per capita emissions.

Not only may emissions of greenhouse gases be measured in different ways on a per country basis, but this is also a highly political question because it involves assigning responsibility or blame. In the run-up to the Bali climate talks in December 2007, the geographer Joshua Muldavin (see Box 8.01) published an op-ed in the *Boston Globe* critiquing the tendency to increasingly scapegoat China as the top emitter of greenhouse gases. As he adeptly argued in his editorial, China had essentially become the factory for the world, emitting greenhouse gases to largely produce goods for consumers in the West. The ultimate cause of the problem is the unrelenting, and highly unequal, levels of consumption.

Box 8.01 Op-ed in the *Boston Globe* (Wednesday, December 19, 2007)

China's not alone in environmental crisis
by Joshua Muldavin

Leaders from around the world gathered these past two weeks at the Bali climate change talks to chart our collective future. Looking out my window in Beijing through the dense haze that envelops this powerful city with world-record levels of smog, dust, and deadly pollution, it is easy to understand why many there perceived China as the Godzilla of global warming. As a country choking on its own "success," now producing over 20 percent of global greenhouse gases, China makes for easy scapegoating. However, targeting China does little to address the fundamental causes of climate change, mitigate its consequences, or provide lasting solutions.

The West has worked long and hard to transform China into what it is today: an industrial platform for the world where some of the most noxious, occupationally hazardous production processes are concentrated. Western governments and corporations have not only benefited, but have helped lead China down this road of energy-intensive, environmentally destructive development with resulting rapid increases in greenhouse gas emissions. In addition, Western consumers have directly profited from the inexpensive products that pour from China's factories. Fundamental to the rise of China's emissions is the rapacious growth of consumption, and the championing of it – especially in the West. The carbon dioxide embedded in China's exports to the United States in 2004 alone is estimated at 1.8 billion tons, equivalent to 30 percent of the US total.

The World Bank, Japan, and Western donor countries have provided more than $200 billion in loans to China since the early 1980s – the largest global flow of development aid during this period – to create the infrastructure that has enabled China to become the world's factory. Multinational companies received contracts to help build China's infrastructure – the power plants, electrical grids, railways for coal transport, natural gas pipelines, highways, ports, and airports. Combined with its large, mobile, low-cost workforce of rural peasants, China became highly attractive to globalizing companies.

Simultaneously, Western leaders have promoted neoliberal economic policies increasing capital mobility. For 25 years, corporations moved factories to China, often partnering with local companies and subcontractors to take advantage of lax environmental and occupational conditions and achieve higher profits. In moving manufacturing jobs to China, footloose corporations have de-industrialized other parts of the world.

China's global integration was further enabled by Beijing's own devotion to rapid growth at any cost, averaging more than 10 percent per year for over two decades. Paradoxically, the resulting environmental destruction threatens that very growth, with hundreds of protests around the country every day reflecting the big divide between those who reap the profits and those who suffer the consequences of China's far-flung production networks. While the greatest benefits fill corporate coffers in China and abroad, the real costs are imposed upon local environments and Chinese workers' bodies.

The long-term destructive environmental consequences of China's development path are well known to the country's leadership and citizens. Official statistics point to pollution as the primary cause of death. And global warming's catastrophic consequences for China provide strong incentives for action. The rapidly shrinking Himalayan water tower foretells a dire future for billions in China, India, and Southeast Asia as Asia's rivers dry up. This helps explain China's increasing engagement with the international community at the Bali talks.

But China's global integration means its footprint of environmental destruction does not stop at its borders. The world's companies pull global resources through China from far-flung corners of the planet – timber from Siberia, Mozambique, and Burma; petrochemicals and minerals from Sudan, Indonesia, and Bolivia. The impacts on global warming through deforestation, as just one example, are magnified far beyond China itself.

The West must acknowledge its own role in shaping and benefiting from China's global integration and rapid increase in consumption of resources. Instead of being diverted by the relatively easy and therefore attractive answer of blaming China or any single country for rising greenhouse emissions, we must focus on the real root of the problem: a highly unequal and unsustainable international system of production, distribution, and consumption that insulates winners from losers, and delivers the greatest share of the benefits to a lucky few while jeopardizing the future for everyone else.

Source: http://www.boston.com/bostonglobe/editorial_opinion/oped/articles/2007/12/19/chinas_not_alone_in_environmental_crisis/. Used by permission of the author. Joshua Muldavin is a Professor of Geography at Sarah Lawrence College.

Another geographer, Luke Bergman, has taken this thinking a step further and determined the quantity of carbon emissions embedded in imports and exports. By undertaking such analysis, Bergman has been able to determine a country's total carbon emissions (related to consumption within its borders) which are the result of emissions produced within a country's borders, plus emissions embedded in imports, minus emissions embedded in exports. Based on analysis of import and export data for 2004, Bergman found, for example, that the United States has another 360,000 10^9g in net virtual carbon emissions

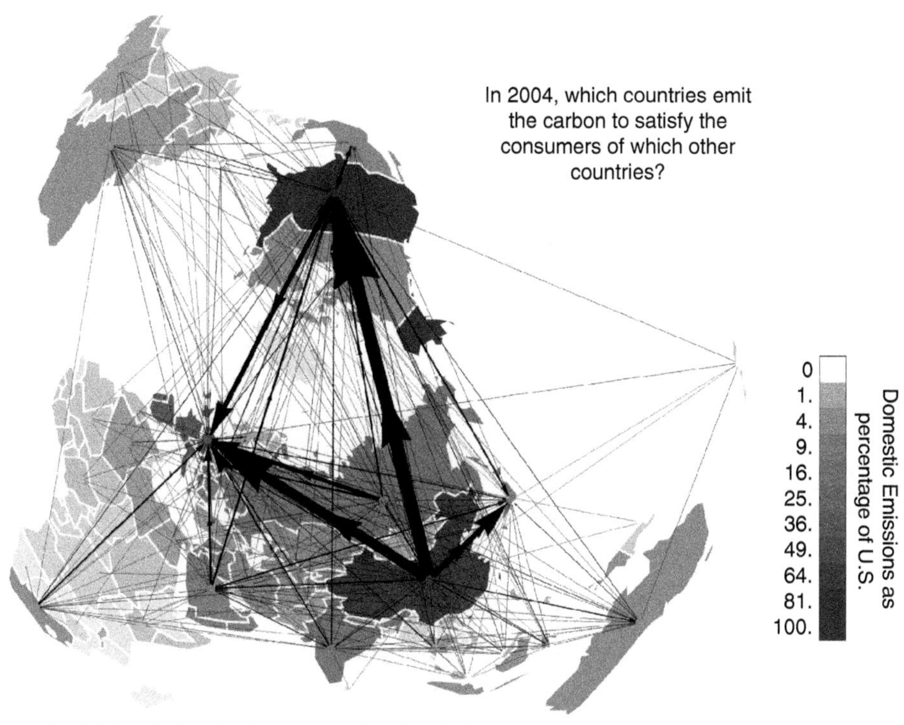

Figure 8.08 Carbon emissions embedded in imports and exports. *Source*: Used by permission of Luke Bergmann, Department of Geography, University of Washington, Seattle.

from China as a result of its consumption of Chinese goods (after subtracting out the carbon content of US goods exported to China). Similarly, western Europe and Japan also have significant net virtual emissions from China (280,000 10^9g and 140,000 10^9g respectively) (Bergmann 2010). Figure 8.08 cartographically displays such transfers.

Another important element to consider when calculating net carbon emissions is the environment's ability to absorb carbon, also known as **carbon sequestration**. Vegetative matter, soils, and oceans all have the ability to bind up carbon. Some of the more practical ways to increase carbon sequestration include planting more trees, preventing deforestation, and encouraging no-till agriculture (see Chapter 10 on agriculture). For example, the Amazon rainforest (the world's largest such forest), which was already viewed as important for biodiversity, is now also seen as important for carbon sequestration. Here, as in other forested areas, the loss of a tree both releases carbon into the atmosphere and represents the end of a sink for future CO_2.

As a result, many multilateral and non-governmental environmental organizations have doubled their efforts to protect the tropical rainforest. Reducing Emissions from Deforestation and Forest Degradation (or **REDD**) is a framework which emerged from the 2007 climate summit in Bali. It is a set of steps designed to use market incentives in order to reduce the emission of greenhouse gases from deforestation and forest degradation. While its original objective was to reduce greenhouse gases, it later sought to deliver other benefits such as biodiversity conservation and poverty alleviation. REDD has been criticized for ignoring the rights of indigenous peoples, for relying on failing carbon markets for its success, and for problematic methodologies in setting levels of compensation (McDermott et al. 2011). Under REDD, some environmental organizations have developed carbon offset programs which finance forest conservation efforts in the world's tropical areas. In such cases, donors (often airline customers, given their carbon-intensive form of travel) support projects to compensate for their carbon emissions. To be more specific, travelers may make charitable contributions to organizations which undertake reforestation efforts to partially offset the carbon emissions from their flights. Efforts to protect or expand the world's tropical rainforests are sometimes viewed with cynicism, especially when environmental organizations from the Global North are involved, as the majority of the globe's greenhouse gas emissions are produced in the industrial nations.

Global warming is of concern to the international community for a number of reasons. The first concern is sea-level rise. Sea levels are likely to rise as global temperatures increase because water naturally expands as it warms. The melting of glaciers and the polar ice cap will further contribute to the sea-level increase. While the estimates of sea-level rise vary greatly, the IPCC anticipates a rise of 48 to 79 cm by 2100. Such a rise will be particularly problematic for the 25% of the global population that live within only 1.1 meters of sea level. As discussed in the opening story about Kiribati, some island nations may almost entirely disappear. While less dramatic than submersion, sea-level rise will present a number of other highly problematic challenges for many coastal residents, including increased coastal erosion, flooding and storm damage, the loss of coastal wetlands, and the increasing salinity of groundwater (which will threaten drinking water supplies).

Droughts and poorer conditions for agriculture are another area of concern, particularly in those regions of the world which are predicted to experience higher levels of warming (see Figure 8.06). To the extent that increasing heat augments evapotranspiration, temperature increases in already warm areas – particularly those with more marginal rainfall – are likely to make crop agriculture more challenging. As discussed by Kates (2000), increasing irrigation is one response to this type of challenge. However, even if one ignores the environmental consequences of such a response, this is often a socially problematic option at two different scales. First, at the scale of the nation-state, wealthier countries are more likely to be able to afford to adapt to climate change through dams and irrigation schemes than poorer countries. Second, even within poorer countries, such a response often leads to very uneven outcomes for rich and poor households as the redistribution of land in flood-irrigated zones typically favors wealthier or more politically connected households.

Others have discussed how a rise in temperatures will lead to a shift in vegetation communities. For example, Figure 8.09 depicts current and future beech tree distribution in North America based on two different climate change scenarios. In

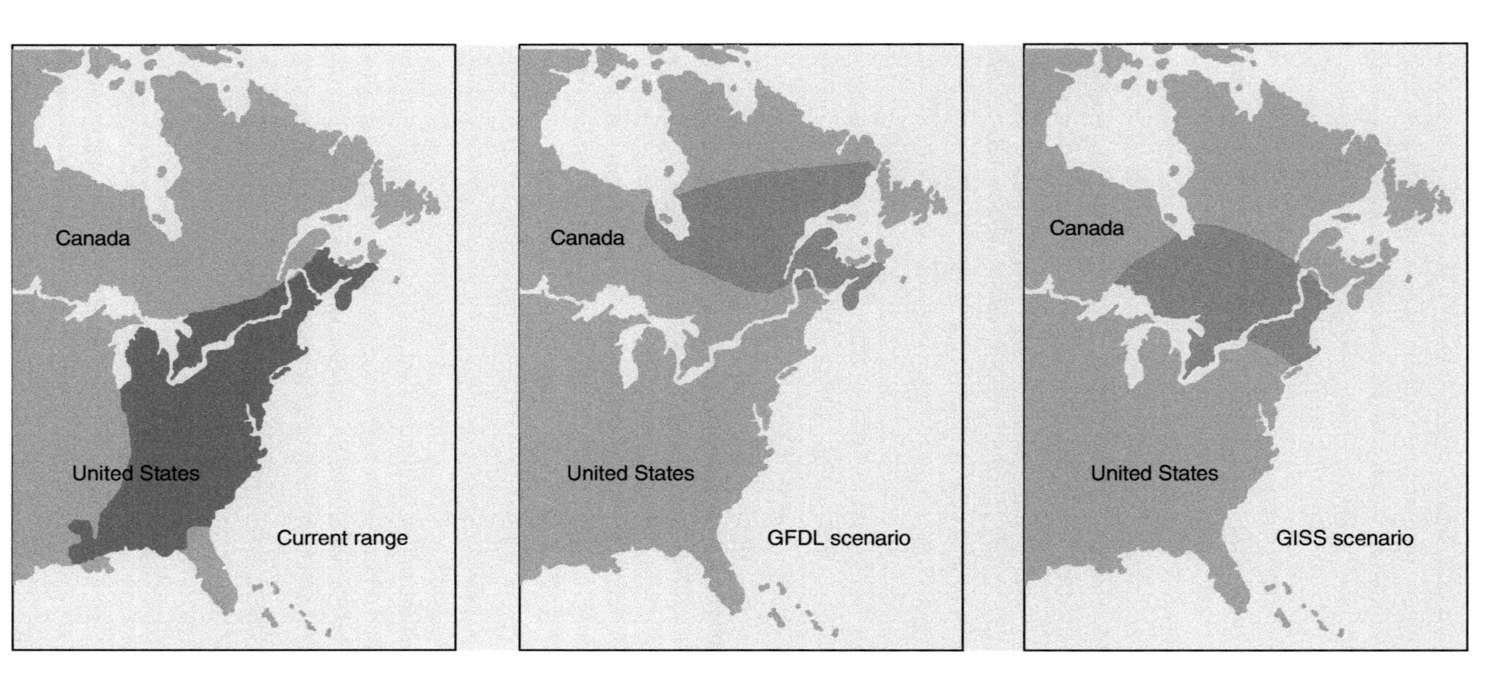

Figure 8.09 Current and projected ranges for beech trees in North America. *Source*: Philippe Rekacewicz, UNEP/GRID-Arendal (http://www.grida.no/graphicslib/detail/forest-composition-case-study-in-north-america_2def). Data: US Environmental Protection Agency.

other words, species (beech trees in this case) have a certain range of tolerance (see Chapter 3 for a discussion of this concept). When an environmental condition permanently shifts, such as temperature or rainfall, then the species will attempt to migrate in order to survive (with tree species doing this as a population, not as individuals, over generations). At first glance, such migration over time may not necessarily appear to be problematic. However, these shifts may prove to be problematic for human economic activity which has come to rely on such trees, or on animal communities which depend on this vegetation for habitat.

Finally, climate change is also likely to impact disease distribution. The most likely impact will be on vector-born diseases, that is, diseases spread or transmitted by insects or parasites, such as malaria by mosquitoes. As many disease vectors are sensitive to environmental conditions, changes in temperature and precipitation often impact their range. Figure 8.10 depicts the potential change in the range of malaria with climate change, with shading indicating the current distribution of malaria-transmitting mosquitoes and additional new areas by 2050.

In contrast to acid deposition or stratospheric ozone depletion, the international community has only more recently begun to deal with global warming, and the effectiveness of our efforts in this regard has been mixed at best. The first major international agreement forged on climate change was the **Kyoto Protocol** signed in December of 1997 (a protocol to the United Nations Framework Convention on Climate Change). The key aspects of this agreement are outlined in Box 8.02. You will notice that the levels of cuts vary somewhat by country or region. Japan, for example, was obligated to make fewer cuts as it was already thought to be extremely efficient in terms of energy use. In contrast, Europe was thought to have an easier task to cut emissions given the deindustrialization that had occurred in eastern Europe following the break-up of the Soviet Union. Developing nations were allowed to voluntarily sign onto the Protocol, and a number subsequently did.

A major concession to the United States at the time of the original agreement was a provision to allow for **emissions trading**. Emissions trading was controversial because experts disagreed sharply on its effectiveness. Some of this disagreement was muted in the case of the climate change agreement because of the specific nature of the pollutants under consideration.[6] Proponents of emissions trading argued that it was a cost-effective approach for reducing overall (or global) levels of certain pollutants. In the case of the Kyoto agreement, the basic premise was that the cost of reducing CO_2 emissions varied by locality. In general, it was assumed that it was cheaper to reduce emissions in places with outdated technology than it was in those with more state-of-the-art, efficient technology. Emissions trading involves a market for carbon credits. Those who can meet or exceed their goals (set by a regulatory body) more cheaply do so and potentially sell their excess credits on the market. Those for whom it is more expensive to reduce emissions may opt to buy these credits (if they are cheap enough) as a way of reaching their goal, rather than reducing their own emissions. As such, the system permits the community to reach an overall goal in the most efficient manner, allowing it to stretch its investment further. One twist on such trades is a

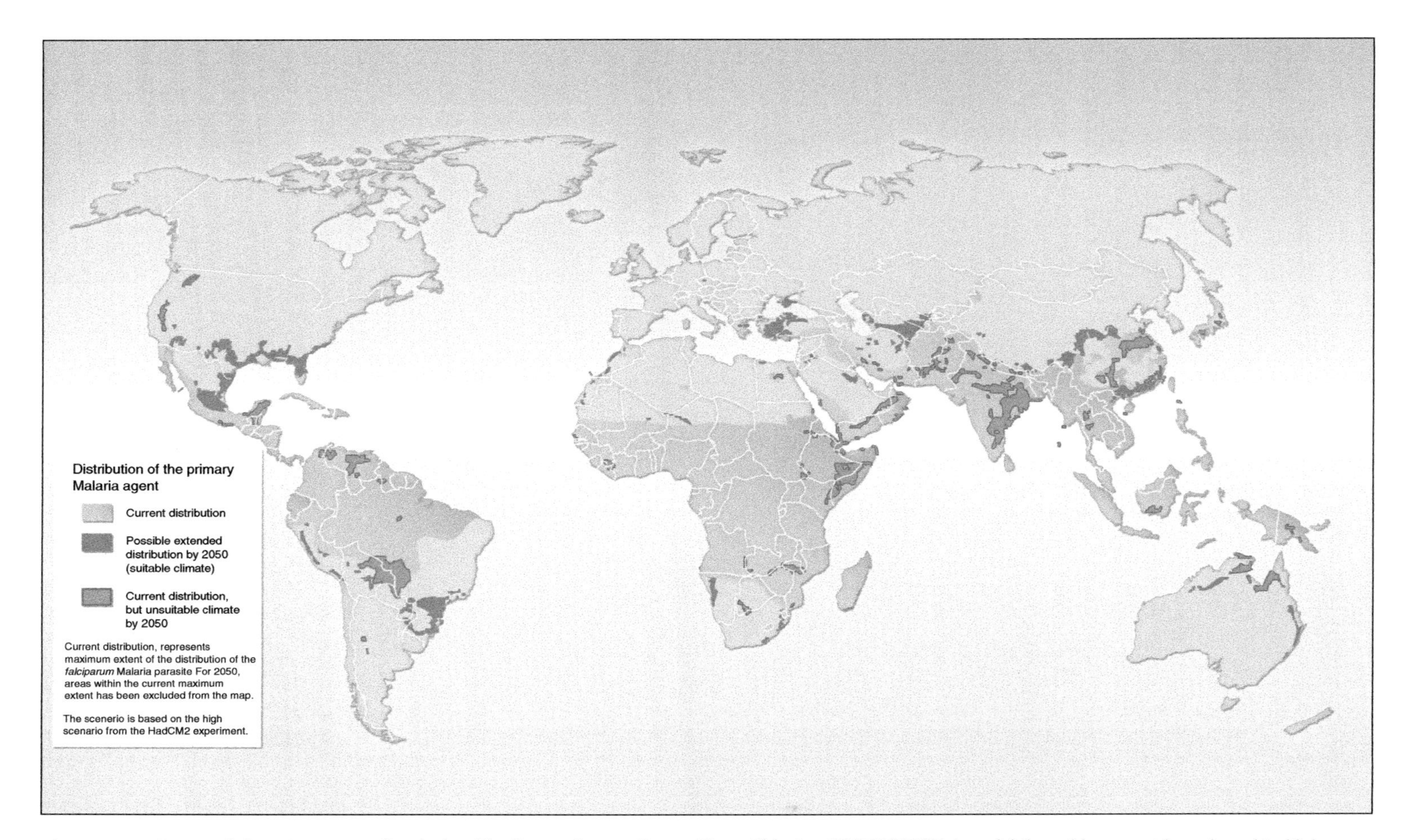

Figure 8.10 Potential changing range of malaria with climate change. *Source*: Hugo Ahlenius, UNEP/GRID-Arendal (http://www.grida.no/graphicslib/detail/climate-change-and-malaria-scenario-for-2050_bffe). Data: Rogers and Randolph, "The Global Spread of Malaria in a Future, Warmer World," *Science* (2000), pp. 1763–6.

process known as **joint implementation**. Under joint implementation a company in one country (typically a wealthier one) will build, for example, a more efficient power plant in another one (typically a poorer country). The company in the wealthier country will then receive the carbon credits for the reduced emissions in the other country. Companies or governments are motivated to undertake such investments if the cost per reduction in emissions is lower in the other country. One of the major concerns with the emissions trading component of the Kyoto Protocol was that there were carbon credits on the market that were not a result of efficiency investments but a byproduct of deindustrialization. These type of credits, also known as **hot air**, were at play as Russia had experienced a significant degree of deindustrialization in the post-1990 period (the benchmark year for the Kyoto Protocol).

Box 8.02 Core elements of the Kyoto Climate Change Protocol (December 1997)

1. US agreed to cut greenhouse gas emissions by 7% below 1990 levels by 2008–2012.
2. Japan agreed to cut greenhouse gas emissions by 6% below 1990 levels by 2008–2012.
3. Europe agreed to cut greenhouse gas emissions by 8% below 1990 levels by 2008–2012.
4. Emissions trading among signatory nations.
5. Developing countries may sign on voluntarily.
6. 37 countries, representing 55% of global greenhouse gas emissions, signed on.[7]

While then Vice President Al Gore had signed the Kyoto Protocol on behalf of the United States, his country's commitment would not be binding until the agreement was ratified by the US Congress. One of the complaints of some Congress members (among others) was that the agreement did not include major carbon emitters in the developing world. Then President Bill Clinton held off submitting the agreement to Congress in hopes that developing countries would voluntarily sign on. While a few such countries did so (e.g., Argentina and Chile signed on at a follow-up meeting in Rio de Janeiro in 1999), this was never enough to sway Congress, and the agreement was never ratified by the USA. Then a new US President, George W. Bush, came to power in January of 2001 with a different set of priorities. Under Bush, the United States formally pulled out of climate change negotiations at a meeting in Bonn, Germany, in 2001. The withdrawal of the then largest emitter of greenhouse gases in the world dealt a huge blow to the Kyoto process.

With another change of US administrations with Barack Obama assuming the presidency in 2009, the United States has re-engaged with global climate change negotiations. The most recent United Nations climate change conference took place in Copenhagen, Denmark, in December 2009. The conference produced a succinct, non-binding agreement. Over 100 countries, representing 80% of global emissions, are listed as parties to the document. Most significantly, this list not only includes the US, but China and India as well. The agreement calls for limiting the rise in global temperatures to no more than 2° C (or 3.6° F), beyond pre-industrial levels. Very significantly, this was the first such agreement to address adaptation (long resisted due to a fear that this would signify an acceptance of climate change). The agreement calls for spending as much as $100 billion a year to help emerging countries adapt to climate change and develop low-carbon energy systems. It seeks to accelerate energy technology transfers to the developing world and take steps to protect tropical forests from destruction. There was great hope and momentum on climate change leading up to this conference, but a global economic downturn has limited much of that drive.

Local and Urban Air Pollution

While this chapter is largely about broader-scale atmospheric issues (and related levels of energy consumption), it is important to remember that local-level air pollution is, in some ways, a more serious and pressing concern for most of the world's citizens. Furthermore, in some instances, particular pollutants are locally problematic in one way, and then troublesome in another manner at broader scales.

Major urban air pollutants include particulate matter (soot, fly, ash, dust, pollen), sulfur dioxide, nitrogen dioxide, carbon monoxide, ground-level ozone, and lead. Particulate matter emerges from a variety of sources (mainly as the result of fossil fuel combustion and wood-burning) and can lead to cardio-respiratory disease. While sulfur dioxide and nitrogen dioxide both lead to the regional atmospheric problem of acid deposition (discussed earlier), these are also local-level pollutants. High concentrations of sulfur dioxide emissions, resulting from the combustion of impure fossil fuels such as coal and fuel oil which are often burned to generate electricity, can result in breathing problems for asthmatic children and adults who are active outdoors. Other effects associated with longer-term exposure to sulfur dioxide, in conjunction with high levels of particulate soot, include respiratory illness, alterations in the lungs' defenses, and aggravation of existing cardiovascular disease. Similarly, prolonged exposure to nitrogen dioxide, generated by motor vehicles, industrial processes, and thermal power plants, may decrease lung function and increase the risk of respiratory symptoms such as acute bronchitis and cough and phlegm, particularly in children.

Other local air pollutants include carbon monoxide, ozone, and lead. All of these tend to be related to automobile emissions. Carbon monoxide is particularly harmful in confined spaces where concentrations may climb (and it enters the blood and limits oxygen supply). At lower levels it creates breathing troubles and is particularly problematic for those who suffer from cardiovascular disease. Ground-level

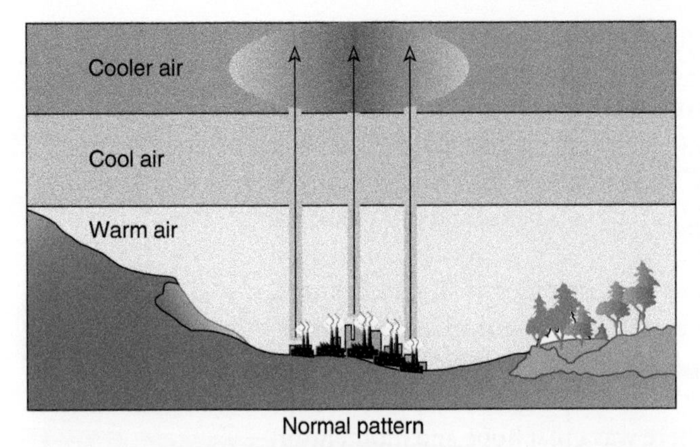

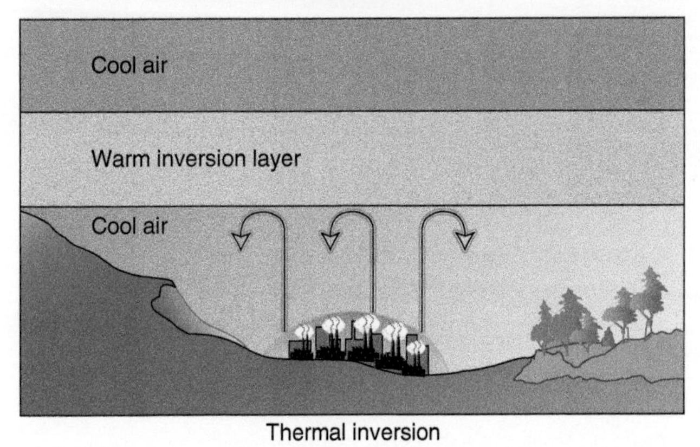

Figure 8.11 Temperature inversion.

ozone (O3) forms when nitrogen oxides (NOx) combine with volatile organic compounds. Ozone contributes to respiratory difficulties. Finally, airborne or ambient lead is most common in those areas of the world where leaded gasoline for motorized vehicles is still employed. Lead is a neurotoxin and is particularly problematic for the developing brains of young children. Airborne lead concentrations in the congested areas of some cities have been found to be alarmingly high. For example, Cairo, Karachi, Bangkok, Mexico City, and Manila all have recorded ambient lead concentrations which exceed World Health Organization (WHO) standards.

Temperature (or thermal) inversion involves air masses interacting with the landscape and anthropogenic air pollution (and such air–land–human interactions are a long-standing concern of physical and human–environment geographers). It is a local atmospheric phenomenon which can exacerbate urban air pollution problems, sometimes to the point of being extremely hazardous. Under normal circumstances, temperature decreases with increases in altitude. This temperature gradient allows for warm air to rise, and this natural mixing dissipates pollution, thereby limiting the harm it may cause to local inhabitants. Inversion refers to a situation where the temperature gradient is reversed, with warm air sitting on top of cooler air (see Figure 8.11). This situation is problematic because the warm air on top of the air column acts like a lid which prevents air from mixing, and air pollution from dissipating. A number of circumstances may lead to inversion.

Radiation inversion is the most common type of inversion and occurs when the Earth's surface cools rapidly. This is common on cool, clear nights or in higher-altitude enclosed valleys where cool, down-slope air movement occurs at night. The inversion is typically most pronounced in the morning and then burns off. It may be prolonged (and more problematic) if there is a thick fog which blocks sunlight and prevents warming of the lower air. This is a frequent problem in higher-altitude cities like Mexico City or Denver, USA. **Subsidence inversion** is associated with slow-moving, high-pressure systems. Under such circumstances, a high-pressure cell may descend, compressing and warming the air below it. This becomes the warmer lid over a surface layer that typically cools at night. Subsidence

Figure 8.12 Advective inversion during the winter in Cape Town, South Africa. *Source*: W.G. Moseley. Used with permission.

inversions are common over the northern continents in the winter. **Advective** (or advection) **inversion** is mostly likely to occur in coastal cities, and is even more probable in such cities which are located in valleys. This is most likely in the winter months when warm ocean air rolls in over cooler surface air over the land. Such inversions are not uncommon in cities like Cape Town, South Africa (see Figure 8.12) and Los Angeles, USA.

There have been a few famously tragic cases of inversion, many in the 19th and 20th centuries, when coal-burning produced massive concentrations of sulfur dioxides, particulates, and carbon monoxide in urban or industrial areas. In December 1952, for example, such an event occurred in London. Because of the cold December weather at the time, Londoners began to burn more coal, which increased air pollution in the city. An inversion event that month trapped and increased London's air pollution. The result was the Great Smog of 1952 that was blamed for thousands of deaths. Another famously tragic case occurred in 1948 in the eastern United States in the Monongahela River Valley of Pennsylvania. Here, the strong inversion occurred over the small industrial town of Donora that prevented the flushing of industrial pollution (sulfur dioxide, particulate, and carbon monoxide). Half of the town's 12,000 people became ill with respiratory difficulties and 20 people died.

Energy Consumption

The world is now more dependent on fossil fuel consumption than it has ever been in human history and the byproducts of this reliance (i.e., the emissions of fossil fuel combustion) explain many of our contemporary and historic atmospheric and air pollution problems. Emblematic of humanity's energy addiction is the night-time view of the world featured in Figure 8.13. Such is the intensity of our energy consumption that it is even visible from outer space. This image also hints at the unevenness of energy consumption across the planet. The map shows the geography of night-time electricity consumption for outdoor lighting (rather than the geography of population). It clearly depicts that cities are concentrated in Europe, the eastern United States, Japan, China, and India. However, the eastern United States is far brighter than the more densely populated areas of China and India.

As discussed earlier, CO_2 emissions (from fossil fuel combustion) are responsible for 82% of global warming from greenhouse gases that are produced by humans. World energy consumption trends have increased steadily over time, with an occasional pause or downturn in demand. In some instances, demand has slackened because of a supply shock, and related rising prices, which have cut into demand (such as the formation of the OPEC oil cartel in 1973–5 or the Iran–Iraq war in 1979–82). In other instances, a global economic downturn has led to slackening demand and falling prices (such as in 1990–3 and 2008–10). What these trends show is that energy consumption is fairly price-sensitive (or elastic) as consumption tends to fall when prices are high and rise when prices fall or stay low for an extended period of time. People adapt to rising energy prices by changing their behavior, making do with less, or (over time) investing in more efficient technology. The converse is also true with lower prices. The impact of price is also detectable across countries as different national-scale energy policies (and taxation schemes) have led to different behaviors and consumption patterns.

Energy efficiency, or the amount of energy it takes to produce one unit of GNP, varies tremendously across countries. On the one hand, energy intensity reflects the efficiency with which a country uses energy. Efficiency has tended to improve when energy prices are high, or be better in those countries where fuel prices are higher. On the other hand, energy intensity reflects the nature and composition of a country's economy. As evidenced in Figure 8.14, energy intensity has tended to decline in some of the wealthiest economies over time as manufacturing has fallen off (and moved to other countries) and the service sector has become more prominent. This is noticeable in the trends for the US and UK. Energy intensity increased for Korea as it became a more prominent global manufacturer from the 1980s. The case of Japan is interesting as it has maintained a strong manufacturing sector and has long been known for having some of the most energy-efficient production methods in the world. While Japan has found it difficult keep improving its efficiency levels (and has lost the edge it held over the competition in the 1980s), it still has the most energy-efficient economy in the world if one considers the

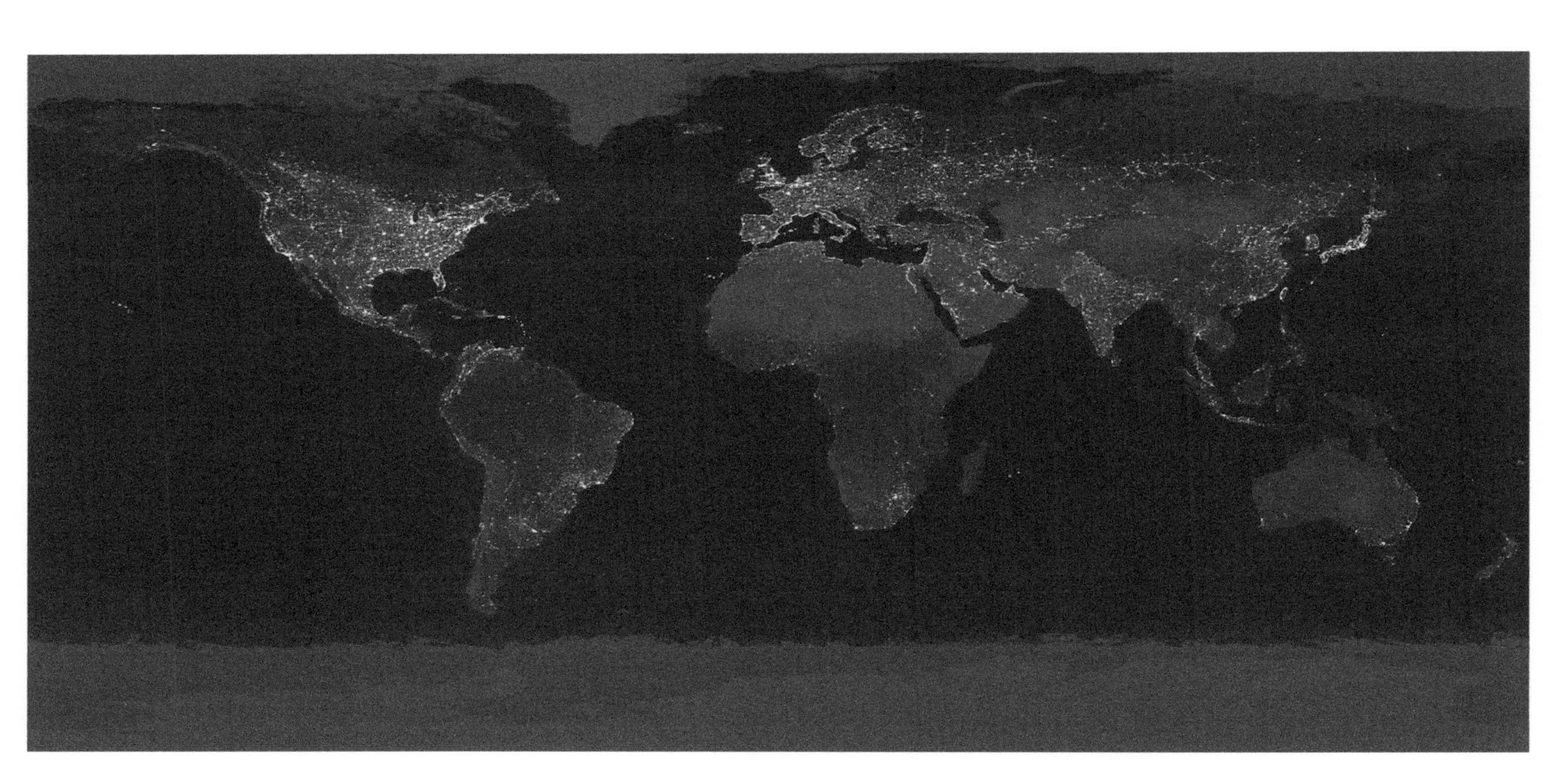

Figure 8.13 The world at night. *Source*: http://eoimages.gsfc.nasa.gov/images/imagerecords/55000/55167/land_ocean_ice_lights_8192.tif.

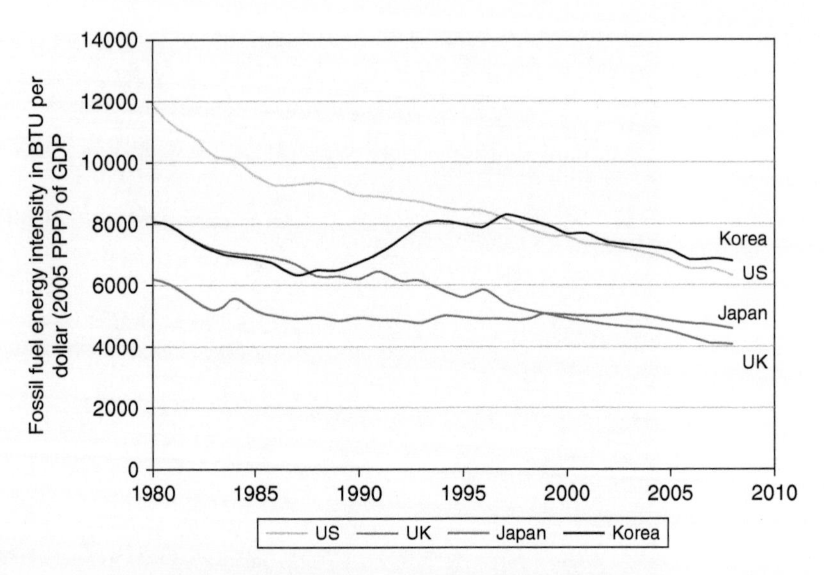

Figure 8.14 Fossil fuel energy intensity for selected countries, 1980–2008. *Source*: Ivo Šlaus and Garry Jacobs, "Human capital and sustainability," *Sustainability*, 3(1) (2011), pp. 97–154. Reprinted with permission.

nature of its economic activities. According to one study undertaken in the mid-1990s, if the whole world had the energy efficiency of Japan, the GDP of the world could be doubled and global energy consumption cut by 25%.

Conventional and Alternative Sources of Energy

Conventional fossil fuels (oil, coal, and natural gas) dominate the world's energy markets. Despite the growing importance of some of these fuel sources over time, such as oil, many experts are concerned about their long-term viability for at least two reasons. The first concern is that these are non-renewable resources and they will eventually run out. To build the global economy on the back of a resource that will eventually run out seems problematic, especially if the search for alternatives is not well under way. Furthermore, some scholars believe we have reached a situation known as **peak oil**, an assertion that we have reached or surpassed the globe's peak oil-production capacity and that this will decline as fewer and fewer exploitable new reserves come under production. The second concern is that our fossil fuel dependence is the major factor driving greenhouse emissions and that we must reduce our consumption of these by using less energy or switching to alternative fuel sources (or both). This has led to an increased drive to discover and develop alternative energy sources.

The major alternative sources of energy include: nuclear, hydroelectric, geothermal, tidal, solar, wind, and biofuels. Nuclear power generated 14% of the world's electricity in 2009. The US is the largest nuclear power producer in the world (2010 data), yet France produces the largest share of its electricity from this source (at 74% in 2010). Global interest in nuclear power has waxed and waned, with public perception of this energy source being heavily influenced by some prominent nuclear accidents, including the Three Mile Island plant accident near Pittsburgh, USA in 1979, the Chernobyl accident in Russia in 1986, and the Fukushima accident created by an earthquake in Japan in 2011. Perhaps even more serious than the risk of an accident is the problem of storing spent nuclear fuel, which remains dangerous for thousands of years. The appeal of nuclear power is based on the world's ample reserves of uranium, and the low levels of greenhouse gas emissions produced by this source of power.

Hydropower provides about one-fifth of the world's electricity. The biggest producers (in ranked order) are China, Canada, Brazil, the US, and Russia. Norway has the highest percentage of its electricity generated by hydropower. Hydroelectric power generation works by building dams and running water through tubes to turn turbines and generate electricity. In previous eras, water was also used to turn wheels to power grain mills and factories. While hydropower produces zero greenhouse gas emissions, it creates upstream and downstream effects in rivers which are problematic for sediment flows and aquatic life. Given these problems, there is actually an active program to remove small dams in the US A subset of hydropower is tidal power, which accounts for a very small proportion of electricity generated around the world. It works best at locations with sufficiently high tidal ranges or flow velocities (limiting the number of potential sites). It includes a variety of systems to harness tidal energy. These may be permanent structures which traverse the estuary of a river (such as on the Rance River in France), also known as tidal barrages, or tidal turbines which function like windmills under the surface of the water. Tidal barrages are not without their problems as they may lead to siltation and have harmful effects on some aquatic species.

Geothermal energy refers to heat trapped within the Earth which may be used to warm or cool homes via small passive systems, or generate electricity with larger and deeper systems (by heating water to produce steam to turn turbines). The first people to attempt to harness geothermal energy were arguably early humans who lived in caves that were warm in the winter and cool in the summer. In the contemporary period, geothermal or ground source heat pumps are used for heating homes. By piping water to relatively shallow depths, and circulating this back through homes, such systems may be used to heat and cool houses. Some 24 countries generate electrical power from geothermal sources. Historically, most such power generation has occurred in tectonically active areas, but new technology has allowed this to become more common in other regions. The United States is the largest geothermal energy producer in the world, followed by the Philippines. The country with the largest share of energy from geothermal sources is Iceland, which generates 24% of its electricity and heats 87% of its buildings in this way.

Most solar power is generated and consumed in a decentralized manner, making it very attractive in areas with underdeveloped power grids. This also means that comparative international statistics on solar energy are difficult to assemble.

Solar power may be divided into passive and active forms. Passive solar energy involves the proper design of structures, building materials that insulate or store energy, careful planting of vegetation, and correct orientation of buildings to provide proper heating and cooling. Active solar power generation uses mechanical devices to collect and store solar radiation for heating and cooling. The simplest type of collector consists of a flat black plate encased in an insulated, glass-covered box. Solar heat is used to warm air or water as it moves across the plate, and this is then circulated through homes. Photovoltaic cells have become increasingly common as their price has declined. These thin silicon wafers convert sunlight directly to electricity.

Two final alternative sources of power remain to be discussed, wind and biofuels, and both have grown tremendously in the past decade. Wind power accounted for 2.5% of the electricity generated in the world in 2010. China is the largest wind energy producer in the world, followed by the US, Germany, Spain, and India. Wind accounts for the largest share of electricity generated in Denmark (21%), followed by Portugal (18%), Spain (16%), and Germany (9%). World wind energy generating capacity grew tenfold between 2001 and 2011 (World Wind Energy Association 2011). This growth has largely been driven by improvements in turbine technology and higher prices for conventional fuels.

Biofuels refers to fuels produced from plant matter or biomass, including corn (or maize), sugar cane, vegetable oil, switch grass, etc. Biofuels accounted for 1.8% of the world's transport fuels in 2008 and their production has grown steeply in recent years. Ethanol is a particular type of biofuel which is typically (but not always) made from corn. The US is the largest producer of ethanol in the world and it has seen its production of this fuel increase steeply in recent years, with 25% of the US corn crop being used for ethanol production in 2008. This level of ethanol production has been controversial for at least three reasons. First, the amount of energy required to produce corn-based ethanol is almost as high as the energy content of the fuel generated (in terms of the energy it takes to grow the corn and then process it into ethanol). This means that there is very little energy to be gained (or greenhouse gas emissions curtailed) from producing ethanol. Second, given very high global food prices in recent years, many are concerned about diverting significant amounts of grain to produce fuel. Finally, ethanol production in the US has been supported by substantial government subsidies, and many find this to be problematic.

Energy and Transport

The transport sector accounts for a significant proportion of energy use and air pollution. It is also a sector where urban design (or the layout of cities), networks within and connecting cities, and modes of transportation have a significant impact on energy consumption. Emissions per capita (not to mention energy use) for transport vary tremendously amongst the most developed countries in the world. Figure 8.15 shows how CO_2 emissions per capita related to transport are much higher

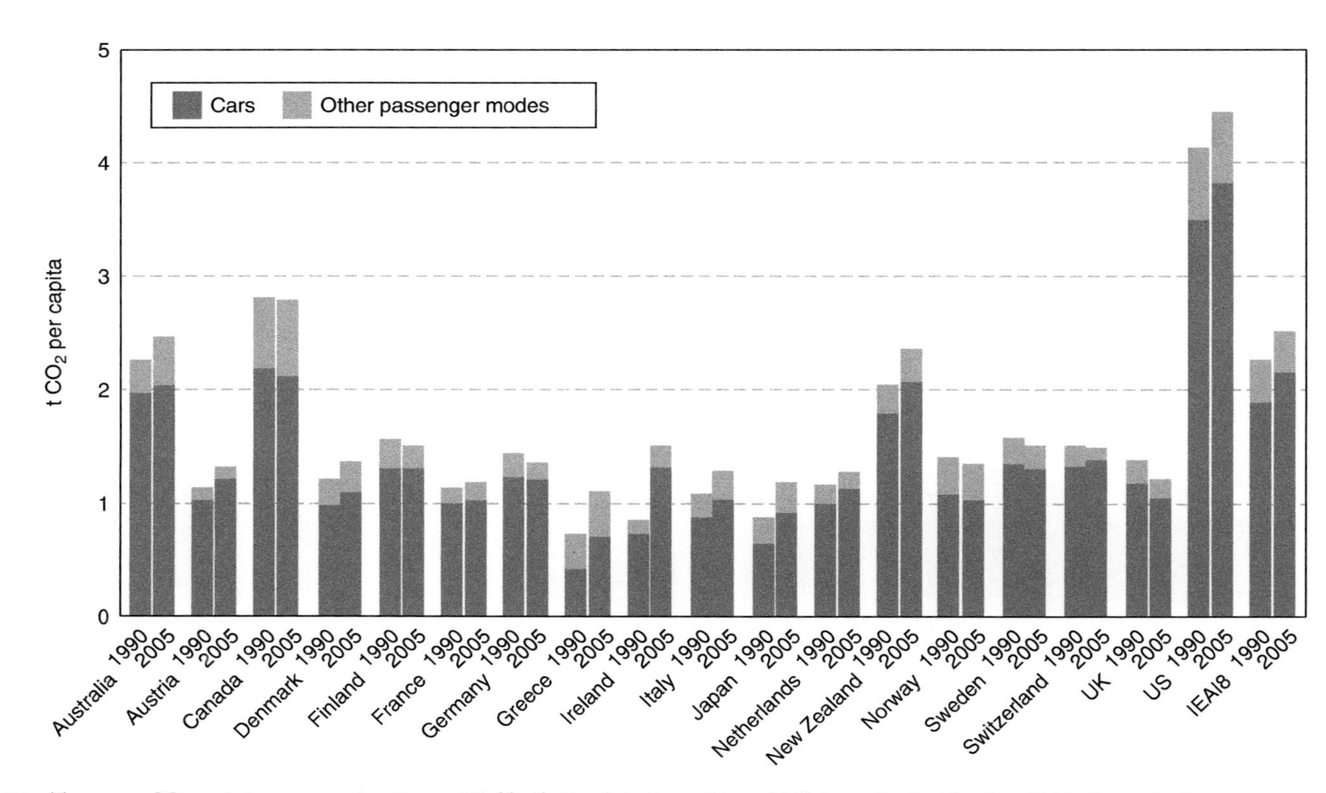

Figure 8.15 Transport CO_2 emissions per capita. *Source*: *Worldwide Trends in Energy Use and Efficiency: Key Insights from IEA Indicator Analysis* © OECD/IEA 2008, figure 6.3 (p. 60).

Figure 8.16 Hong Kong's pedestrian walkways facilitate foot traffic. *Source*: W.G. Moseley. Used with permission.

in some wealthy countries than in others, and that these emissions generally rose over the 1990–2005 period. Some of this has to do with population density, with those in high-density countries (Japan and the Netherlands) traveling much less than those in low-density countries such as the US, Canada, and Australia. That said, the differences seem to extend beyond population density. US residents, for example, drive more than those in other countries and are much less likely to use alternative form of transportation. While automobiles account for the majority of travel in all countries to varying degrees, secondary modes of transit also differ substantially. For example, train travel is the second most common transit mode in Japan (at 30% in 2005) as well as in many European countries. This compares to air travel being the second most common form of transit (in terms of share of distance traveled) in Australia, Canada, and the US (at 10% or more in 2005).

At the level of the city, urban form and transportation infrastructure have a tremendous impact on individual transit decisions. In those cities where public transit is affordable, extensive, and convenient, public transit ridership tends to be very high. This type of interaction between the built environment and individual behavior is consistent with a key principle in human–environment geography: that humans interact synergistically or dialectically with their environments. In other words, humans shape their environments which, in turn, influence human behavior. In the city of Hong Kong, for example, more than 90% of all travel is done on public transit or on foot. Not only is there a subway system, but an affordable network of water taxis and an extensive pedestrian walkway system (see Figure 8.16).

Hong Kong is not alone in its ability to create an environment where cars are not necessary. In the cities listed in Table 8.02 the majority of residents do not own a car. The high cost of driving, and the comparatively low cost of transit, militate against such a choice even when one has the money to purchase a car. As suggested by the data in Figure 8.15, residents of the US drive automobiles more than any other people on the planet. That said, public transit does flourish in a handful of cities in the US (as listed in Table 8.03). Many of these top US transit cities initially evolved before the automobile era and their higher urban density and distinct **central business districts** (CBDs) tend to be more amenable to **hub and spoke transit networks**. Transit critics argue that cities which boomed in the automobile era have lower population densities and no single distinct core, making them harder to service through public transit. That said, urban form is constantly evolving, meaning that it can change and adapt to transit corridors. Washington, DC,

Table 8.02 Top metro systems in the world (serving over 1 billion passengers per annum)

System	*Passengers per annum in 2009 or 2010*
Tokyo Subway	3.16 billion
Moscow Metro	2.392 billion
Shanghai Metro	2 billion
Beijing Subway	1.84 billion
New York City Subway	1.604 billion
Paris Metro	1.479 billion
Mexico City Metro	1.414 billion
Hong Kong Metro	1.41 billion
Guangzhou Metro	1.18 billion
London Underground	1.107 billion

Table 8.03 Top 10 US cities for public transit ridership in 2006

City	*Percentage who take public transit to work*[a]
New York City	54.24
Jersey City	46.62
Washington, D.C.	38.97
Boston	31.6
San Francisco	30.29
Philadelphia	26.43
Arlington	26.28
Yonkers	25.47
Chicago	25.38
Newark	24.04

[a] As opposed to driving, walking or biking.
Source: Based on data from US Census Bureau (2006) *American Community Survey* (Washington, DC).

is a good example of a city which did not have a well-developed metro system in place until the late 1970s / early 1980s. After the system was introduced, high-density housing developments sprung up along the city's metro lines, making it one of the most heavily used systems in the US today.

The two maps in Figure 8.17 nicely contrast emissions per unit area versus emissions per household in the city of Chicago in central USA. Given relatively high population densities in urban areas, it is no surprise that carbon emissions are high in terms of area. What is interesting is that emissions on a per household basis in the city are much lower than in surrounding suburban areas. This is largely related to the fact that urban residents tend to drive less and use more public transportation. Zones of lower automobile use also tend to follow commuter rail lines which radiate out from the city. Please note that Chicago sits on one of the largest freshwater lakes in the world, Lake Michigan, which is just east of the central business district.

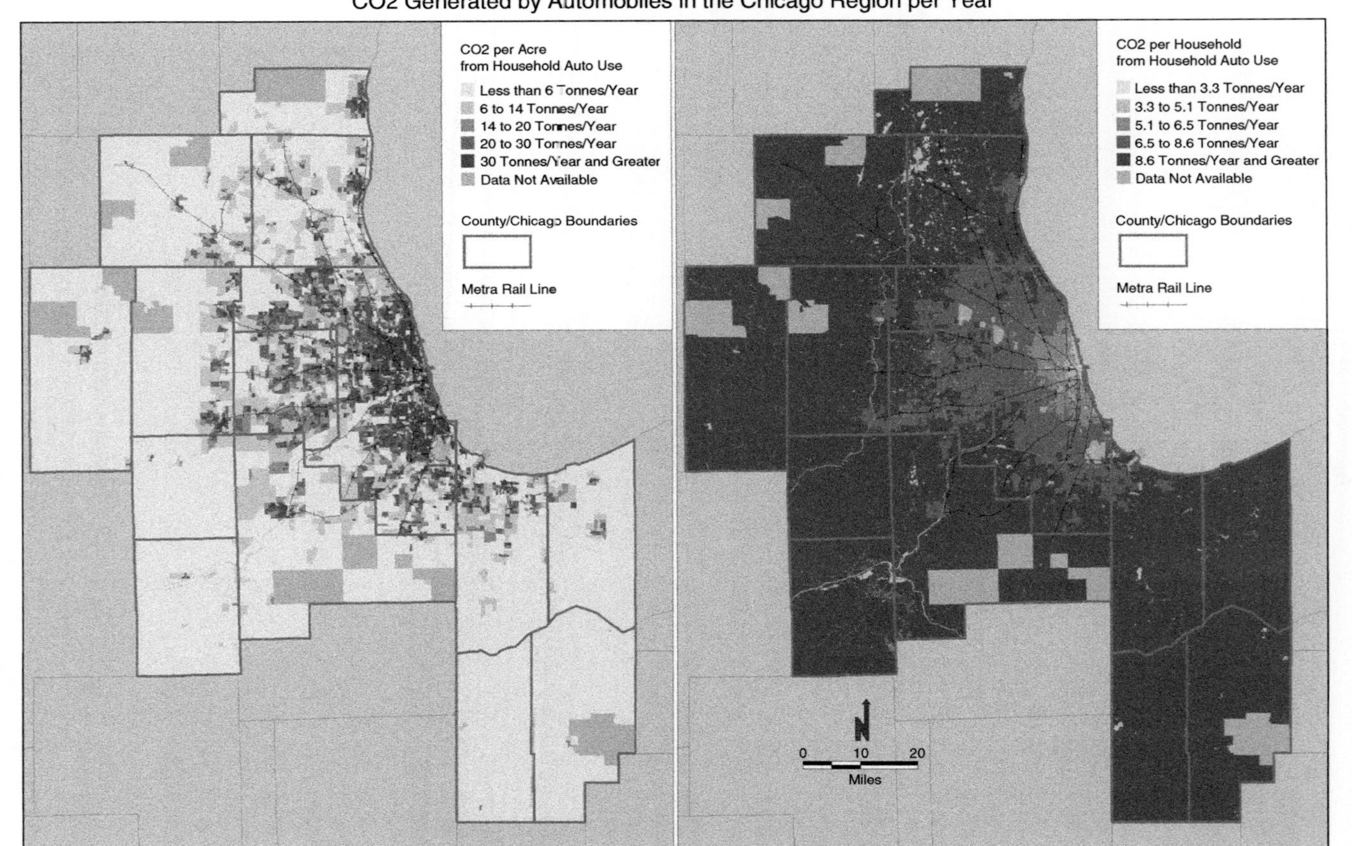

Figure 8.17 Automobile-related CO_2 emissions per square mile versus per mile. *Source*: Center for Neighborhood Technology. Reprinted with permission.

Bicycles are another significant form of alternative transportation in the world, accounting for 50% of urban trips in China, 30% in the Netherlands, 18% in Japan, and 1% in the USA. Many consider Amsterdam, the Netherlands, to be the biking capital of the world where 40% of all travel takes place by bicycle. Another major biking city is Copenhagen, Denmark, where 37% of commuters bike to work. Like mass transit, urban form and infrastructure seem to foster this mode of transportation. Both cities have designated bike lanes, free or rentable public bicycles, safe bicycle storage, and robust bike cultures. While statistics are hard to come by, studies suggest that the percentage of urban transportation undertaken on bicycles may be much higher in some other cities, such as 77% and 65% respectively in Tianjin and Shenyang, China.

One might be inclined to think that cities in colder climes would have fewer bike commuters than in those municipalities with more temperate conditions. The evidence does not support this. For example, the US city of Minneapolis (known for its long, harsh winters) has the second highest percentage of commuters who bike to work (at 4.3%) in that country. Despite a potentially adverse climate, Minneapolis has built an infrastructure that promotes bicycling on many fronts (from bike lockers and designated street lanes to recreational trails and snowplows dedicated to clearing off-street paths). While only 1% of Americans bike to work (even though half of all commutes are under 5 miles or 8 km), surveys suggest that 20% would do so if there were safe bike lanes, secure bike storage, and showers at work.

Box 8.03 Imagining a less energy-intensive United States

Some experts have estimated that the true cost of gasoline would actually be about $70 per gallon (or 12.77 euros per liter) if one accounted for the environmental damage created by energy consumption, the health effects of fossil fuel-related pollution, subsidies provided to the oil industry, and the wars fought to secure access to cheap fossil fuels. Use your geography concepts, and the field's propensity to make connections, to imagine what the United States would look like if Americans paid even one-fifth of this estimated true cost for a gallon of gasoline, say $14 per gallon (or 2.55 euros per liter). How might higher energy prices influence the form of cities, the way people move around, the manner in which food is produced, or the country's foreign policy?

Figure 8.18 is one image of what the landscape of a less energy-intensive USA might look like (from Portland, where light rail has been promoted and urban sprawl discouraged). Perhaps higher energy prices would mean that low-density urban environments with sprawling suburbs would have to change, as well as the dominant modes of transport. No longer would it be possible to live in car-oriented suburbs with no sidewalks and limited access to public transportation, or to commute long distances to work alone in an

Figure 8.18 Light rail and pedestrian-friendly streets in Portland, Oregon. *Source*: Courtesy of the Project for Public Spaces (www.pps.org).

automobile. Perhaps cities would become denser, demand for more and diverse forms of public transportation would rise, and the market for alternative-energy homes and energy-efficient modes of private transportation would skyrocket.

Would not the way food is produced also have to change? The current agricultural system is extremely energy-intensive, requiring fossil fuel-based fertilizers and pesticides, and energy to run tractors, transport agricultural goods around the world, and process foods. In order to be cost-effective, new ways of producing food would have to be more localized, capitalize on natural plant relationships, encourage agro-biodiversity, and be less processed.

An energy revolution in the United States would also change the world outside of its borders as we know it. Decreasing demand for fossil fuels in the US, driven by higher prices, would allow the country to stop fighting foreign wars to secure cheap access to oil. This would mean an end to its support for despotic and undemocratic leaders. Declining fossil fuel use would significantly reduce greenhouse gas emissions, reducing the threat of global warming to the poorest nations, not to mention America's decreasing use of energy (the most significant driver of increasing food prices in 2007–8) would also reduce hunger around the world as global food prices would begin to decline.

The biggest rub to such a change would be how to best assist the poorest Americans in dealing with higher energy costs. A simple subsidy to this group (funded at a fraction of the costs of the foreign wars the US is fighting to secure cheap energy) would help in this transition. Other anticipated savings could be used to build a more robust public transportation network. Government policy helped Americans make the transition to high energy use after World War II; couldn't policy help the country reverse the trend?

Chapter Summary

This chapter examined three major global atmospheric issues: acid deposition, stratospheric ozone depletion, and global warming. It then reviewed a handful of more localized atmospheric issues. The second half of chapter reviewed energy production and consumption around the world, an activity which often – but not always – has some bearing on atmospheric conditions. This review included looking at different sources of alternative energy. The chapter concluded by examining energy consumption in the transportation sector as it is affected by modes of transit and urban form.

Critical Questions

1. Consider the three major global atmospheric problems (acid deposition, global warming, and stratospheric ozone depletion) and the varying degrees to which the international community has been successful at addressing these issues. Why has there been such a difference in levels of success vis-à-vis these problems?
2. Discuss how some air pollutants act to create local as well as regional or global scale problems.
3. Discuss which forms of alternative energy production you see as the most promising and why.
4. One of the basic principles of human–environment geography is that humans interact synergistically or dialectically with their environments. Describe how this principle is realized in some of the world's major built environments (i.e., cities).
5. How do you travel to school or college every day? What type of factors influence your personal transit decisions? What would it take for you to switch to a more energy efficient mode of transportation?

Key Vocabulary

acid deposition
anthropogenic climate change
carbon sequestration
central business district
climate vs. weather
common property
emissions trading
fossil fuels
global warming
greenhouse effect
greenhouse gases
hot air
hub and spoke transit network
International Panel on Climate Change (IPCC)
joint implementation
Kyoto Protocol
Montreal Protocol
open-access resource
ozone hole

peak oil
private property
REDD
resource
sink
stratospheric ozone depletion
tragedy of the commons
temperature inversion (advective, radiation, subsidence)

Notes

1. Private property is property which is owned by an individual, group, or organization. The owner of the property has the right to exclude others from using it.
2. Common property, which is sometimes confused with open-access resources, is property which is controlled and managed by a group. There are often rules in place regarding use by members of the group. Non-members are often excluded from using such resources.
3. A resource is something found in the environment which is useful to humans. Until such a use is found, such resources are just neutral stuff. Oil is a classic example of a substance which was just black goo in the ground until humans discovered that they could burn it to create energy.
4. A sink is a particular type of resource which is useful because of its ability to process waste or pollution.
5. The International Panel on Climate Change (IPCC) is a group of the world's leading climate and atmospheric scientists.
6. In contrast, in the case of a pollutant like sulfur dioxide, there are often equity (or environmental justice) concerns with the emissions-trading approach which may reduce total emissions at the national level but allow for high levels of pollutants in certain areas. These concentrations of air pollution then create health risks.
7. With 191 countries signing and ratifying the treaty by 2010.

References

Bergmann, L. (2010) Global capitalism, carbon emissions, and China: re-narrating the relations. *International Conference on China and the Future of Human Geography* (Sun Yat-sen University, Guangzhou, China).

Crosby, A. (2006) *Children of the Sun: A History of Humanity's Unappeasable Appetite for Energy* (New York: W.W. Norton).

Gibson-Graham, J.K. (2004) Surplus possibilities: re-presenting development and post-development. Conference on Economic Representations: Academic and Everyday, April, University of California Riverside.

Harden, G. (1968) Tragedy of the commons. *Science*, 168(3859), pp. 1243–8.

IPCC (2007) *Climate Change 2007: The Physical Science Basis* [ed. S. Solomon, D. Qin, and M. Manning]. Contribution of Working Group I to the Fourth Assessment Report of the Intergovernmental Panel on Climate Change.

Kates, R. (2000) Cautionary tales: adaptation and the global poor. *Climate Change*, 45(1), pp. 5–17.

McDermott, C.L., Levin, K., and Cashore, B. (2011) Building the forest-climate bandwagon: REDD+ and the logic of problem amelioration. *Global Environmental Politics*, 11(3), pp. 85–103.

US Census Bureau (2006) *American Community Survey* (Washington, DC).

World Wind Energy Association (2011) *World Wind Energy Report 2010* (Bonn).

9

The Population–Consumption–Technology Nexus

Icebreaker: Complicating Overpopulation

Consider two countries in the world, India and the Netherlands. Now think about your perceptions of population dynamics in each of these countries. Which one of these countries is thought to have too many people and which is considered not to have an overpopulation problem? Without giving it too much thought, many students would likely indicate that the Netherlands is a stable, prosperous country that does not suffer from an overpopulation problem. In contrast, perceptions of India may be of a poorer, albeit up-and-coming,

An Introduction to Human–Environment Geography: Local Dynamics and Global Processes, First Edition. William G. Moseley, Eric Perramond, Holly M. Hapke and Paul Laris.

Published 2014 by John Wiley & Sons, Ltd.

country that continues to struggle with population issues. As the world's second most populous country (with 1.14 billion people in 2007), India is sometimes considered to be the poster child of overpopulation. Reflect, for example, on the words of Paul Ehrlich in the introductory chapter of his 1968 classic, The Population Bomb:

> *I have understood the population explosion intellectually for a long time. I came to understand it emotionally one stinking hot night in Delhi a couple of years ago. My wife and daughter and I were returning to our hotel in an ancient taxi. The seats were hopping with fleas. The only functional gear was third. As we crawled through the city, we entered a crowded slum area. The temperature was well over 100, and the air was a haze of dust and smoke. The streets seemed alive with people. People eating, people washing, people sleeping, people visiting, arguing and screaming. People thrusting their hands through the taxi window, begging. People defecating and urinating. People clinging to buses. People herding animals. People, people, people, people. ... since that night I've known the feel of overpopulation. (Ehrlich 1968: 15–16)*

This emotive, visceral sense of overpopulation is not uncommon. In this passage, Ehrlich is clearly articulating the overwhelming sense of material poverty that some visitors from the Global North feel when they first visit a country in the Global South. While not explicitly stated, a causal connection is also being made between too many people and poverty. Although India has changed a lot since Ehrlich wrote this passage over 40 years ago, the perception that overpopulation is driving poverty in India persists. It is interesting to note that the population density of India (360 persons per square kilometer) is actually slightly less than that of the Netherlands (400 persons per square kilometer). As such, is it really population density (one measure of overpopulation) that is driving poverty in India? Or, perhaps, do we see poverty and think backwards – assuming that this poverty must be driven by overpopulation?

An alternative perspective would be to set population aside as one of many potential causal variables of poverty, and to allow other potential explanations to be considered. If we consider the history of these two countries, we would no doubt note that the Netherlands is a former colonial power, having once controlled territories in Southeast Asia, South Africa, and South America (not to mention several islands). India, in contrast, was a colony of Britain (to varying degrees, between 1613 and 1947). If you accept that colonialism was largely about resource extraction, then these two very different histories (one as colonizer and the other as colony) may help explain contemporary differences in wealth. Another explanatory variable to consider is trade, which is not unrelated to historical colonial relationships. While agriculture is still a significant component of the Dutch economy (especially cut flowers, bulbs, and dairy products), the country is not food self-sufficient. That said, hunger or poverty are not significant problems in the Netherlands because it has a major industrial export economy and is able to bring in more than enough resources to purchase sufficient food. While the Indian economy is now rapidly changing (with a vibrant manufacturing and high-tech sector) and has been self-sufficient in food production since the 1970s, until recently it exported few industrial and manufactured goods. Instead, agricultural items such as tea, which fetch relatively lower earnings, made up the bulk of its exports. These trade patterns, combined with its history as a colony (as well as other economic and political factors), may better explain the wealth gap between India and the Netherlands than demographic factors.

Chapter Objectives

The objectives of this chapter are for the student to be able to:

1. Understand how human population patterns vary in space and time.
2. Critically assess the relationship between population, consumption, and technology.
3. Evaluate the major perspectives on the population–resource question.
4. Analyze the major models concerning population change and development.
5. Review the factors that influence fertility and birth rates.

Introduction

In this chapter we consider a variety of human population issues in relationship to resources, the environment, and economic development. As human population numbers alone do not determine resource impacts, we will consider this variable in association with rates of consumption, prevailing technologies, and other factors. We will begin by examining human population patterns over time and in space. We will then examine the nexus of population numbers, per capita consumption rates, and technology. We will next examine the major theoretical perspectives on the population–resource question, as well as models concerning the relationship between population change and development. We will end with a review of factors that influence fertility and birth rates.

Population Change in Time and Space

Most of the history of human population on Earth is one of gradual increase in both absolute numbers and rate of population growth. More recently, both growth rates and absolute numbers have increased rapidly (see Figure 9.01). While such descriptions of world population growth tell us an important part of human population history, this broad-scale interpretation also masks a great deal of spatial and temporal variation. The global trend obscures enormous declines in population suffered by some regions, even as others experienced relatively rapid growth. It also misses periods in some regions when population doubled in as little as a century, while the world population as a whole was taking 600 to 1000 years to double. Much of the apparent stability in world population size before 1750 represented a balance of regional stability, rapid growth, and decline in population.

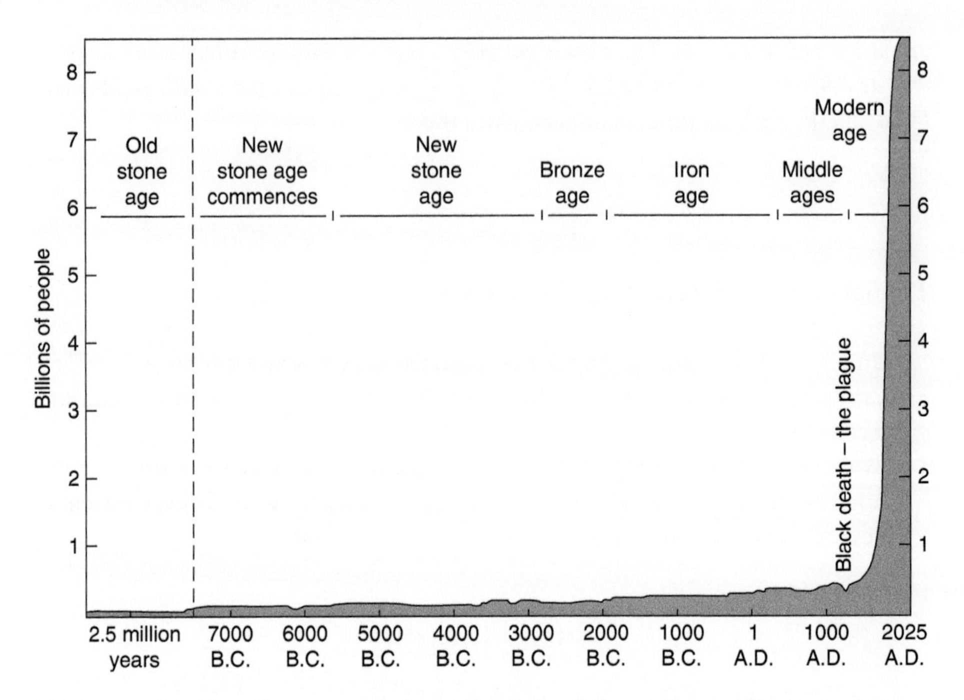

Figure 9.01 World population growth. *Source*: Population Reference Bureau, "World population: towards the next century" (1994). Reprinted with permission.

At the global scale, human population is often divided into four **population eras**:

- *The First Era*. This era occurred during the Paleolithic or Stone Age. It began about 1 million years ago (when human first began using tools) and ended about 10,000 years before the present. During this period, humans largely lived by hunting and gathering. As migrants left tropical environments (see Figure 9.02), they generally tended to escape areas with high burdens of parasites and disease organisms. Migration to new territory also meant that resources may have been less constrained. As such, this period is generally considered to be one of population growth. There is some disagreement amongst scientists about whether population gradually increased throughout the entire period or if it increased more rapidly in last 20,000 years of this era.
- *The Second Era*. This period begins with the first agricultural revolution (the intentional cultivation of plants and domestication of animals) about 10,000 years ago. Population growth began slowly with the advent of crop farming (about 8000 BC) as food sources became more regular. However, some scholars (e.g., Diamond 1987) have argued that population growth likely declined in the early stages of the first agricultural revolution as diets became less varied and nutritious. Population growth likely peaked at the beginning of the Iron Age (1000 BC) and then leveled off by about AD 500. As the first agricultural revolution began at different times around the world (although largely

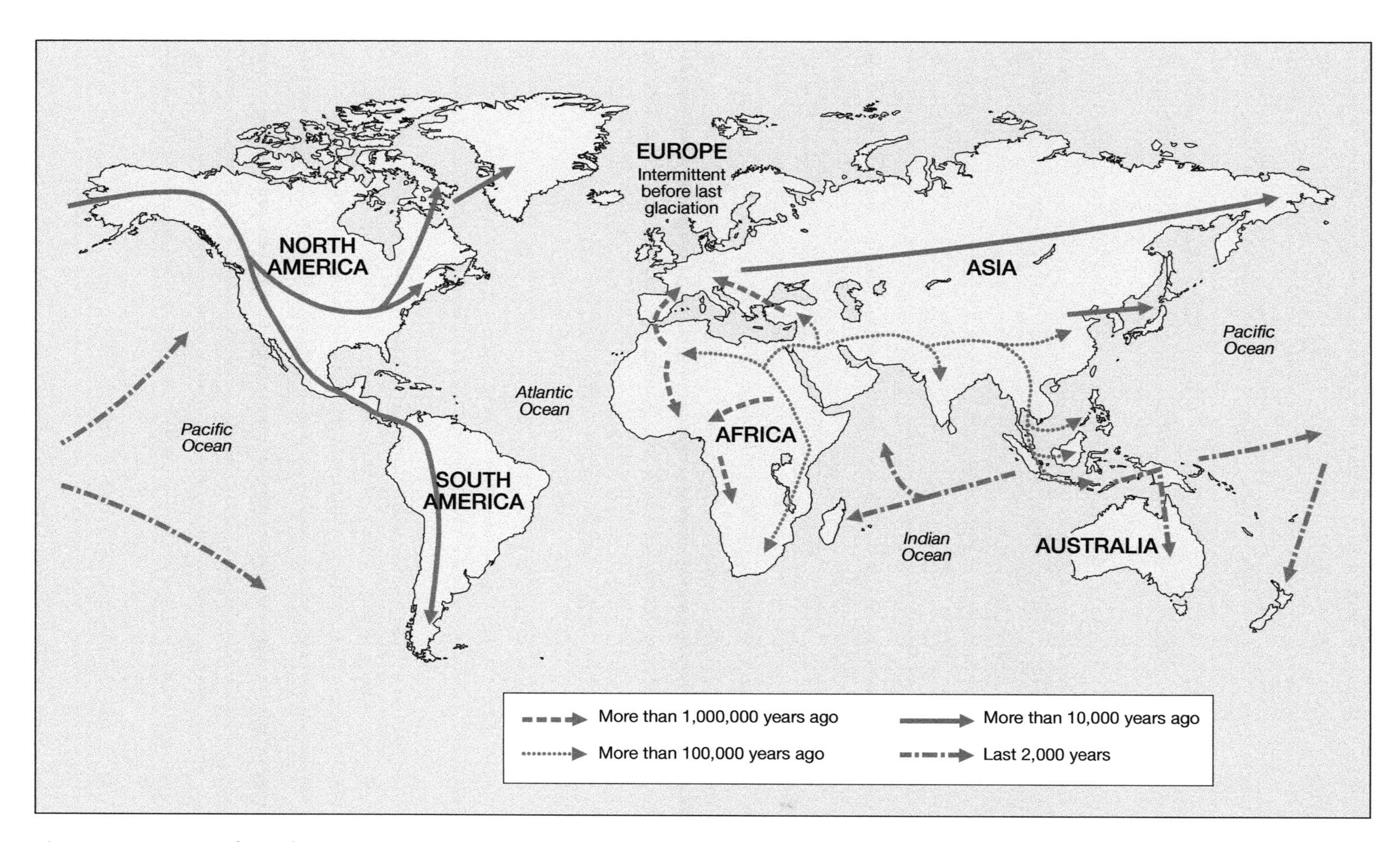

Figure 9.02 Historical population migration.

considered to be complete by AD 1000), there would have been considerable spatial variation in population growth trends during this period. Areas of political stability and instability would also have influenced these trends. For example, the relatively stable situation created by the Roman Empire in the Mediterranean basin during part of this period led to rapid population growth within this region.

- *The Third Era.* The third era occurs between the culmination of the first agricultural revolution (around AD 1000) and the beginning of industrial revolution in Europe (around 1750). Population growth during this period tends to be more rapid but less consistent than in previous eras. For example, in Europe there is a general upward population trend, but with severe declines during the Black Death (second half of 14th century) and during the Thirty Years War (1600–50).
- *The Fourth Era.* The fourth era begins with the start of European industrial revolution in 1750 and runs until the present. During the first century of this period we see a rapid upturn in population in Europe, Anglo-America, the Caribbean, and China. Central to the explanation for this upturn is the industrial revolution, which brought improved living conditions, particularly better nutrition and public health. The latter part of this period differs from previous eras in terms of the sources of population growth and decline. In previous eras, population grew because of political stability and economic prosperity (e.g., growth under the Roman Empire) and declined as a result of famine and pestilence during times of political instability and economic decline (e.g., the Thirty Years War). This relationship is reversed in the fourth era. More politically powerful and stable areas now have more stable populations, while less developed countries have more rapid growth.

In the contemporary era, we see the global population growth rate slowing, with most demographers predicting that the world population will likely peak around 2050 at between 9 billion and 10 billion people (see Figure 9.03). Perhaps more significant is the increasingly urban nature of the global population. In 2010, the world population became majority urban and it is predicted to be 70% urban by 2050.

The increasingly urban nature of the world's population does have resource implications because, as people move from rural to urban areas, their diets typically begin to change. One of the most significant changes is that urban dwellers generally consume more meat. As it takes 7 to 10 kilograms of grain to produce 1 kilogram of meat, this consumption shift has very significant resource implications. Urban dwellers also typically begin to consume more processed foods and foods that have traveled a greater distance from field to fork.

Just as human population numbers have not increased uniformly over time, it is important to remember that the human population is also not distributed evenly across the surface of the Earth. Approximately 90% of the world's population lives north of the equator where the largest portion of total land area

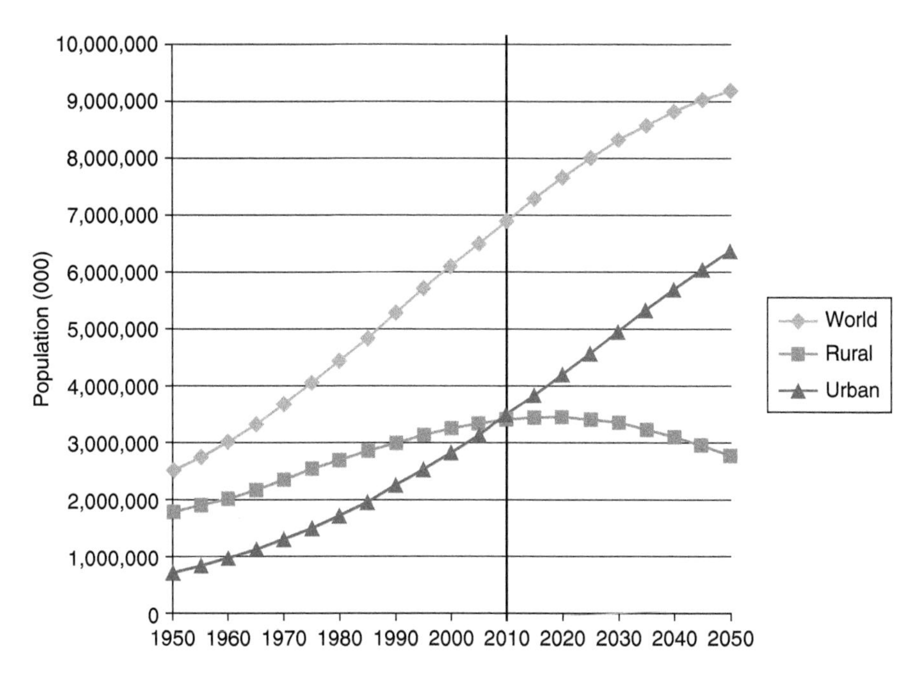

Figure 9.03 Global population trends. *Source*: *World Urbanization Prospects: The 2011 Revision* (New York: United Nations Department of Economic and Social Affairs, Population Division).

(63%) is found. As can be seen from Figure 9.04, some areas of the world are much more densely populated than others. In fact, the entire global population only occupies about 10% of the total land area. Most people live near the edges of land masses, near the oceans and seas, or along rivers, with easy access to navigable waterways. Furthermore, most of the world's population lives in temperate, low-lying areas with fertile soils. So, clearly, physical geography explains many of the population distribution patterns we observe. That said, sometimes people live in undesirable locations because the situation is advantageous vis-à-vis regional or global trading networks. Both Chicago and Venice are examples of cities with unfavorable site characteristics (both were built on swamps) but good situations, the former because it served as a node connecting the shipping routes of the NorthAmerican Great Lakes to 19th-century railroad networks and the latter because it was positioned as key trading point between Europe and the Near East on the Mediterranean Sea.

Population, Consumption, and Technology in Relation to the Resource Base

Biologists have long posited that there is a relationship between the sustenance produced in a particular environment and the number of individuals of a particular species that may be supported. Let us start this discussion with

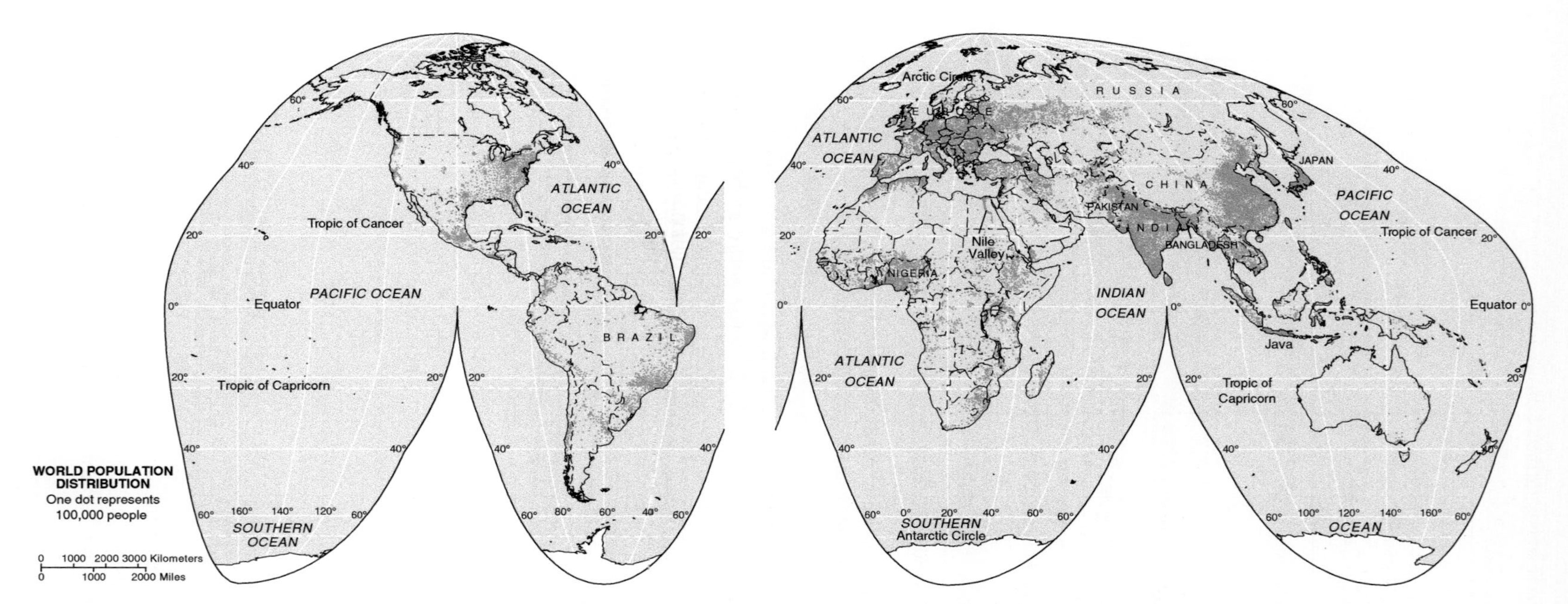

Figure 9.04 World population distribution. *Source*: Erin H. Fouberg, Alexander B. Murphy, and H. J. de Blij, *Human Geography: People, Place, and Culture*, 10th edn. (Hoboken: John Wiley & Sons, 2012), fig. 2.5.

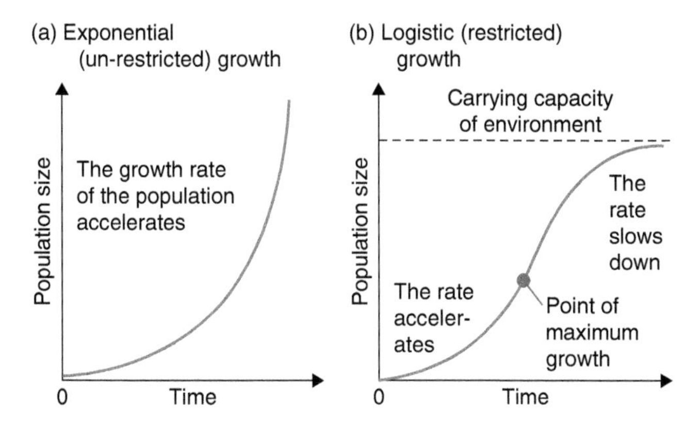

Figure 9.05 J and S population curves.
Source: http://www.cdli.ca/courses/biol2201/unit04_org01_ilo01/b_activity.html.

two basic curves which depict potential population trends vis-à-vis the resource base (see Figure 9.05).

The first type of curve is commonly referred to as a J curve, which depicts unrestricted or **exponential growth**. Exponential growth refers to a population that grows at a consistent rate (say 3% per annum) over time, but because of the effect of compounding, growth in absolute numbers occurs very slowly at first but then increases very rapidly. Such a growth pattern may occur when a new species enters a region with limited predators and bountiful food supplies. It is, however, unlikely that exponential growth will occur indefinitely as most populations must, sooner or later, deal with resource constraints. This leads to the second type of curve, commonly referred to as an S curve, which depicts restricted growth in the latter stages. To be more precise, the rate of growth tapers off as population approaches the maximum number of individuals which a particular environment may support.

Carrying Capacity

The maximum threshold population that a region may support is also referred to as the **carrying capacity**. A classic study of carrying capacity occurred on Isle Royale in Lake Superior (see Figure 9.06). Isle Royale, now a national park, is the second-largest natural island in the North American Great Lakes (at 535.43 sq. km). The island is located 24 km (or 15 miles) from the Canadian and Minnesota shores. In 1905 moose swam to the island, and their numbers soon exploded because of relatively abundant food and the absence of predators. However, their population soon exceeded the carrying capacity of the island, crashing because of food scarcity in 1930 and then again in 1945. Such a fluctuation around the carrying capacity is not unusual for some populations. Interestingly, wolves walked to the island in 1948 on an ice bridge following a colder than normal winter. The wolves (natural predators of moose) culled the moose numbers and moderated the population swings around the carrying capacity threshold.

Whether or not biology's carrying capacity concept may be applied to human populations is a question open for serious debate. Yet some scholars have applied this

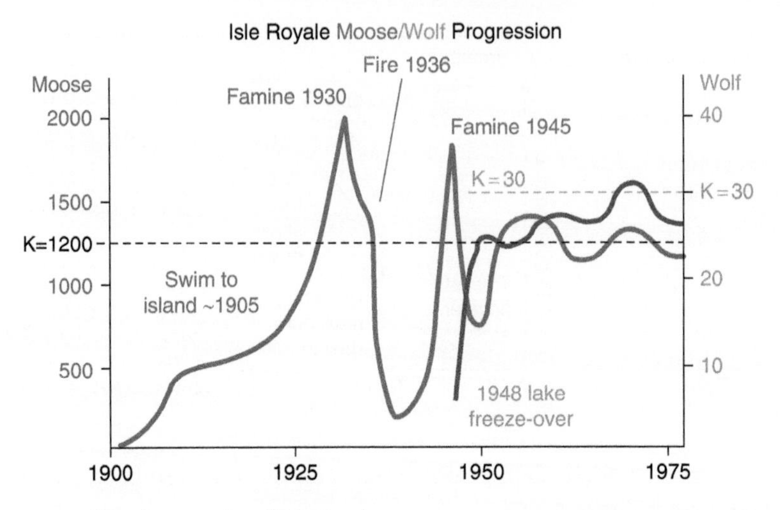

Figure 9.06 Carrying capacity of Isle Royale.

idea to human populations. In reviewing predictions of the Earth's human carrying capacity, Joel Cohen (1995) found that estimates ranged from 1 billion to 1000 billion, with most in the range of 4 billion to 16 billion. The mere fact that estimates of this capacity vary so widely suggests that these calculations may be based on different types of assumptions. A crude (if not humorous) way to think about potential problems involved in the transfer of this concept to human populations might be to consider ways in which humans differ from moose in their use of resources:

1 After one accounts for differences in size and age, moose (which are herbivores in the deer family) have fairly standard rates of consumption. One study of moose populations in North America and Europe calculated that the average moose consumes about 7200–9000 kg of fresh mass, or an average of 2700 kg of dry matter per annum. While moose may vary their consumption from year to year based on environmental conditions, differences in consumption rates between moose in similarly productive environments are minimal. In contrast, humans have highly variable rates of consumption. Consider the fact that the typical American consumes 70 times as much as the typical Ugandan in any given year. As consumption (intake of all material items, energy, and food) is the main way that humans impact the environment, assumptions about per capita consumption would have a huge impact on any calculation of carrying capacity. A graphic depiction of different rates of consumption may be found in Figure 9.07.
2 Use of the environment by moose is generally not modified by technology. In contrast, human use of the environment is often modified by technology in at least two different ways. First, technology and management strategies may influence the productivity of a particular environment. For example, certain approaches to farming or forestry (and related technologies) are much more productive than others. Technology may also influence how efficiently humans use resources. To wit, the Japanese economy is, on average, about twice as efficient as

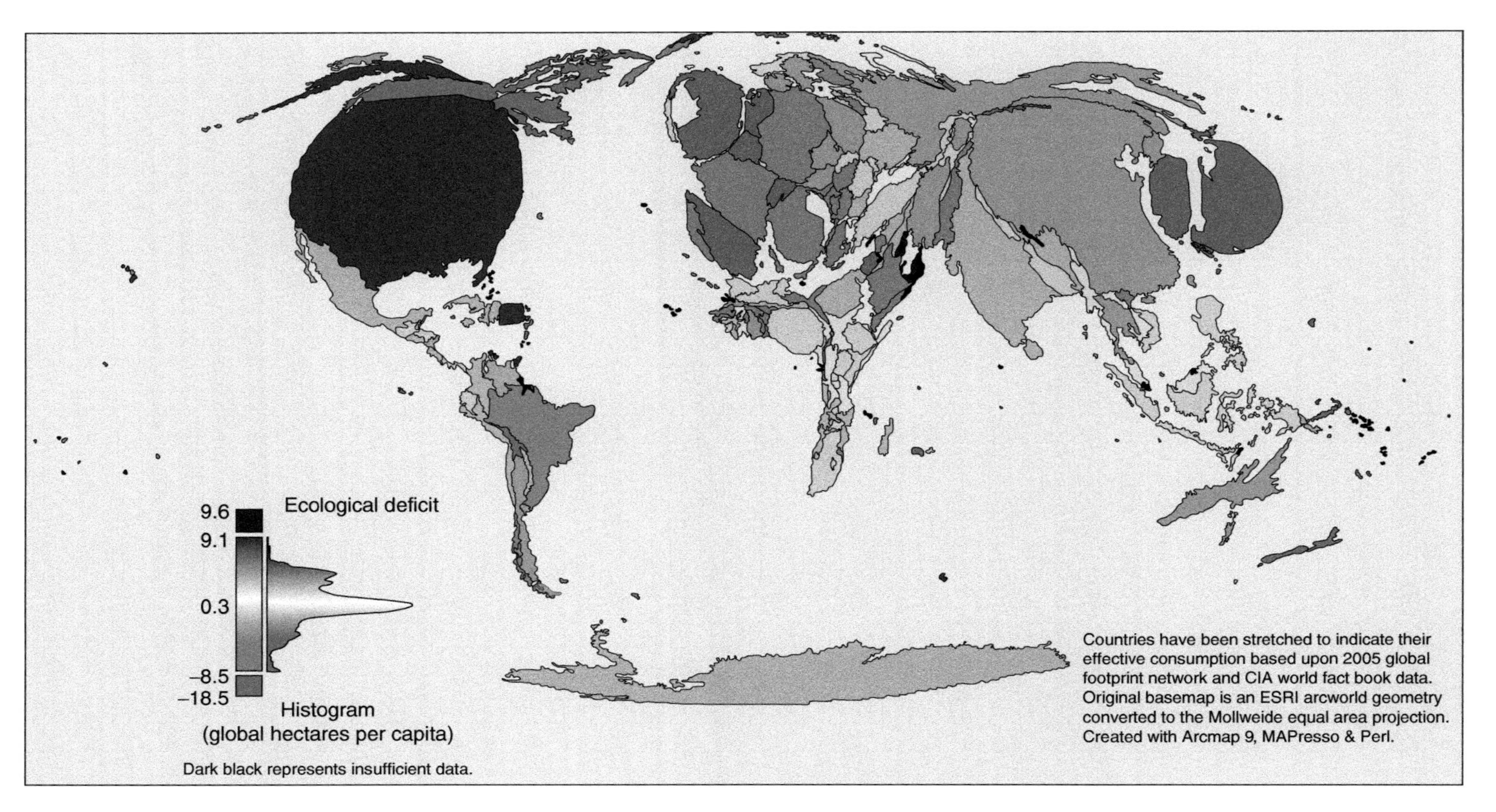

Figure 9.07 Cartogram of relative rates of consumption by country (2005 data). *Source*: http://pthbb.org/natural/footprint/. Reprinted with permission.

the American one in terms of the amount of energy it takes to produce a unit of Gross National Product (GNP). A more specific example of technology influencing resource use might be the fuel economy of an automobile. As such, assumptions about technology would have a significant influence on carrying capacity estimates.

3 Finally, given that carrying capacity is often calculated for a particular area, country, or region, it may be important to consider resource transfers between places. While moose are largely dependent on potential food supplies within a certain range of mobility, humans in any given location may rely on vastly different-sized **food** or **resource sheds** – the area from which a population draws its food and other material supplies. While some humans may largely draw their foodstuffs and material supplies from a particular area, others may source these items from around the world. Consider, for example, the wealthy Asian city-state of Singapore. With 4.8 million people and 624 sq. km of land, Singapore cannot possibly rely solely on its national territory for all of its food and material needs. As such, it trades for much of its supplies. Carrying capacity calculations that treat an area as a stand-alone entity may therefore be problematic.

Major Perspectives on the Population–Resource Question

The relationship between human population numbers, food supply, and the environment has long been debated. In this section we will explore three major perspectives on this relationship: (1) **Malthusian** and **neo-Malthusian**, (2) **structuralist** and **neo-structuralist**, and (3) **technocratic** or **cornucopian**.

Malthusian and Neo-Malthusian Perspectives

The British clergyman Thomas Malthus has arguably been the most influential thinker on population–resource issues in modern human history. Malthus' 1798 paper, entitled "Essay on the Principle of Population," suggested that human population grew at an exponential rate while agriculture grew at an **arithmetic rate** – that is, by a fixed quantity each year (see Figure 9.08). His assertion was that the human urge to breed was uncontrollable and, therefore, it was inevitable that

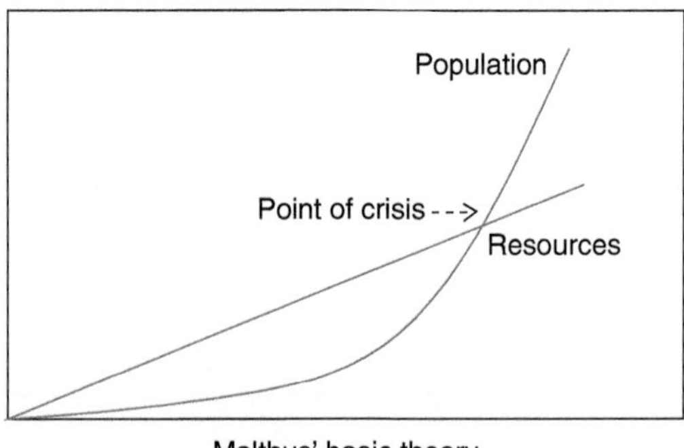

Figure 9.08 Malthus' understanding of the relationship between population and agricultural growth.

population would outstrip food supply. For Malthus, starvation and famine were natural checks that brought the population back into line with food supply.

It is important to understand the historical context in which Malthus was making his arguments (Lohmann 2003). First, Malthus was using his particular understanding of the relationship between population and food production to argue against the Poor Laws, a bare-bones welfare system in late 18th-century England. For Malthus, any assistance to the poor would militate against natural checks on population growth. Second, as was not uncommon at the time, Malthus had a very classist understanding of human behavior, wherein the wealthy were seen as able to control their fertility while the poor could not. He further argued that having more children is what impoverished families. Third, it is important to understand that Malthus was writing around the high point of the enclosure movement in England, in which private landlords were closing off much of the land that had previously been available to commoners (or villagers) for the grazing of their animals or hunting and gathering. The elimination of such safety nets for the poor was putting pressure on the poor in local communities (and increasing the ire of the wealthy classes of which Malthus was a member).

Malthus' ideas have been critiqued on a number of fronts. Many find his support of famine as an inevitable natural check and his class-based arguments to be ethically problematic. Furthermore, his projections for population and agricultural change were simply assertions rather than trends supported by empirical evidence. Actual trends in agricultural production have been much more positive than those projected by Malthus. Population growth rates have also not persisted at an exponential rate indefinitely, but have tended to moderate as economies became less rural and more urban. Others have questioned Malthus' assertion that poverty is driven by overpopulation, suggesting that poverty has more to do with a capitalist economic system that does not pay workers a fare wage than it does with numbers of people.

Although periodically discussed throughout the 19th and early 20th centuries, Malthus' ideas lay dormant until 1968 when **Paul Ehrlich** published his well-known text, *The Population Bomb*. In it, Ehrlich resurrected Malthus' central concern that unabated population growth would outstrip food supplies. Unlike Malthus, his assertions were based on actual population and agricultural production figures. He also argued for modern contraception methods, rather than famine, as a means to control population numbers. Perhaps most significant, Ehrlich essentially made population control an environmental issue in a way that it had never been before. Part of the reason his ideas took hold was timing, as his book was published shortly before the first Earth Day in the United States (in 1970). Because Ehrlich and others revitalized and updated Malthus' ideas, they are referred to as neo-Malthusians.

Structuralist and Neo-Structuralist Perspectives

Structuralist and neo-structuralist ideas essentially grow out of a Marxist critique of Malthus' original 1798 paper. Friedrich Engels (a close associate of Karl Marx) published a short response to that paper, entitled "Outlines of a Critique of

Political Economy," in 1844, 10 years after the death of Malthus. In it, Engels argued that each additional person was a net benefit to society as the value of their labor exceeded the cost of their sustenance. The problem was that employers extracted the "surplus value of labor" from workers by paying low wages. This, Engels argued, was what drove poverty, rather than overpopulation.

Building on the ideas of Marx and Engels, neo-structuralists have emphasized how scarcity is largely a constructed problem based on the hoarding of resources by some countries and segments of society, as well as grossly unequal levels of consumption. Betsy Hartmann has written "It is true that population growth can put additional pressure on resources in specific regions. But the threat to livelihoods, democracy and the global environment posed by the fertility of poor women is hardly comparable to that posed by the consumption patterns of the rich" (Hartmann 1994).

An example of how population pressure may actually be a good thing is documented and theorized in the work of late Ester Boserup (1965). Boserup critiqued the causal logic of Malthus. While Malthus suggested that the level of food production determined the level of population (with food production being the independent variable), Boserup argued that this logic was backwards, based on her experiences and research in Asia. According to Boserup, as populations increase, they farm more and more land. Once they have used up all the land, people will farm existing land more often. In order to combat declining soil fertility, farmers will supply more and more labor (to intensify tillage, build irrigation and terracing infrastructure, etc.) and other inputs (such as manure) to increase the productivity of the land. This intensification leads to higher production levels that keep up with population growth. As such, the "problem" of an increasing population creates its own solution.

One important caveat, and problem for Boserup, was a recognition that markets could interfere with the process of agricultural intensification. If people left rural areas to work in the city, or in other countries, then there would not be enough able-bodied laborers for intensification to occur. Lack of intensification could lead to unaddressed soil fertility declines and eventual underproduction of food.

Another well-known scholar on this issue from the neo-structuralist perspective is David Harvey. In his 1974 article, entitled "Population, Resources and the Ideology of Science," Harvey provocatively asked why policymakers and scientists tend to focus on population numbers to the exclusion of other variables when analyzing the population–resource question. A simplified version of his answer is that opinion leaders characterize consumption levels as immutable (changing these would largely impact the most developed countries) and choose instead to see fertility decisions as more easily changed (a modification that would disproportionately impact poorer, developing countries).

The Cornucopian or Technocratic Perspective

The third and final population viewpoint we examine is the cornucopian or technocratic perspective. The essence of this perspective is that technological innovation will overcome resource constraints. One of the best-known cornucopian scholars in this debate is the late Julian Simon. Simon argued that more people meant more brains to solve the world's problems. Yet, unlike Boserup, he saw free markets as crucial in this process, as rising prices (an indicator of scarcity) would encourage people to innovate. In his 1981 book, *The Ultimate Resource*, Simon looked at major categories of resources and found that their prices had fallen over time (when adjusted for inflation and wage-earning power). In some instances, the economic concept of a substitute was critical for understanding how the cost of a resource like energy could fall over time. Human societies have worked through a variety of energy resources (wood, peat, coal, oil, solar power) and each time a substitute has been found as the older, dominant, source of energy becomes more scarce and expensive. In this sense, nothing ever runs out as we keep conceptually creating new resources. That said, one might reasonably ask:

1 Does every resource have a substitute?
2 Do rising prices give us enough advance warning to find substitutes?
3 Are some resource prices artificially low (because of unaccounted costs) and is this a problem in the cornucopian model?

Another well-known cornucopian thinker is the Danish statistician Bjorn Lomborg, who wrote a book entitled *The Skeptical Environmentalist* (2001). Similar to Simon, he argued that concerns about overpopulation, declining energy resources, deforestation, species loss, water shortages, certain aspects of global warming, and a variety of other global environmental issues were unsupported by analysis of the relevant data. For Lomborg, many environmental concerns were based on analysis of short-term trends rather than more relevant long-term trends.

I=PAT

One model to keep in mind for balancing the concerns of the neo-Malthusians, neo-structuralists, and technocrats/cornucopians is the **I=PAT model**, which was originally proposed by Paul Ehrlich and John Holdren (1971) and subsequently refined by Barry Commoner (Chertow 2001). The acronym IPAT represents the following: environmental impact (I) = Population (P) × Affluence (A) × Technology (T). While this formula has largely been used as a conceptual device, one could calculate this using national population statistics, per capita Gross National Income (GNI) (or a similar measure), and a measure of energy efficiency such as the ratio of per capita GNI over per capita energy consumption. The result would be an index (rather than a real number) that one could compare across countries.

Although the I=PAT formula acknowledges the multiple causes of environmental impact, it has received considerable criticism for a number of reasons. First the model assumes a multiplicative relationship among the variables that does not necessarily exist. That is, doubling population does not necessarily lead to a doubling of impact. Second, the model sees humans only as parasites and predators on the natural environment so the I (impact) is fundamentally only a negative measure of degradation and destruction from pollution (Hynes 1999: 51). I = PAT does not account for positive impacts through conservation and restoration initiatives or the sustainable use of resources that characterizes the behavior of many indigenous peoples and rural peasants around the world. Third, different scholars have been able to use I = PAT to support different sides of the argument about which variable (P, A, T) may be most significant for impact. Ehrlich and Holdren used I = PAT to argue the greater importance of population growth for environmental impact while Commoner demonstrated that technology was more determinative. This suggests that the model may be too broad and general to adequately capture the interrelationships between the different variables in environmental impact (Chertow 2001).

Finally, some have criticized I = PAT for not only simplifying a very complex situation, but also looking at the various variables in isolation from a political context (Maniates 2002). For example, the word "population" implies an unspecified universal that evades the issue of *who* among the P of I = PAT is responsible for the high fertility of poor women (Hynes 1999) or the environmental damage caused by strip mining, the burning of fossil fuels, or industrial waste. Population policy and resource use decisions are the outcome of political processes in which different countries and groups of people possess different degrees of power and influence to assert their interests. Social structures thus influence the impact of the formula's variables in complex ways that are completely unaccounted for in the model. Furthermore, in the politics of the debate about population and environment, blame has tended to fall on the relatively high fertility rates of the poorest P (who are the least powerful) instead of on the affluence (A) and technology (T) of the wealthy and most industrialized countries who consume the greatest amount of resources. The I = PAT formula also does not consider the reasons poor people have children and the complex relationship that exists between population change and economic development.

Box 9.01 Op-ed in the *Philadelphia Inquirer* (July 11, 2007)

A population remedy is right here at home: US overconsumption is a bigger issue than fertility

By William G. Moseley

Today is World Population Day, a time when the United Nations calls on us to reflect on global fertility. Environmentalists have often framed the size and growth of the world's population as a problem. But if the question is defined as how many people the world can sustain, then facts suggest that American overconsumption is the real culprit.

Ever since the British philosopher Thomas Malthus wrote "An Essay on the Principle of Population" in 1798, we have been concerned that human population growth will outstrip available food supply. In the early 1970s, the burgeoning American environmental movement came to see overpopulation as one of its key issues.

But human impact on the environment is not just a question of population numbers. At least two other factors – efficiency of technology and levels of consumption per capita – have as much, or more, influence in determining the environmental impact of a given population.

Conventional wisdom used to suggest that the wealthiest countries would employ the most efficient technologies. This assumption was shattered by the proliferation of gas-guzzling SUVs in the 1990s. In fact, the average fuel economy of American cars and trucks grew worse after the mid-1980s, even as per capita wealth grew.

In fact, no other population on Earth consumes at the same rate as the United States. With 4.6 percent of global population, Americans consume 24 percent of its energy. While China, Brazil and Ethiopia may have population growth rates that are, respectively, the same and 2.3 and 4 times higher than our own, Americans consume, on average, 6.8 times as much energy as the Chinese, 7.3 times as much as Brazilians and 28 times as much as Ethiopians.

In other words, in terms of environmental impact, our already high and exponentially growing per capita energy consumption far outweighs any population growth in the developing world.

So why focus on controlling population numbers when environmental impact is the result of three factors, not one?

As a college professor, I have watched students debate this issue for years. While students recognize the importance of all three factors, they invariably argue that it just isn't practical to try to control overconsumption. They suggest that the pragmatist must focus on what can be done – for instance, developing energy-efficient technology worldwide and supporting education and distribution of family planning methods in the developing world.

I am perplexed by the assumption that encouraging families in the developing world to have fewer children is more doable than reducing US consumption. Having fewer or no children may be easy for a middle-class person in the United States, where raising children is expensive and most of us expect no economic return from children as they grow older. In fact, one could argue that having children in the American context is economically irrational.

It's true that millions of families in the developing world desire access to modern contraceptives, and filling this unmet need is important. However, for millions of others, children are crucial sources of farm labor or important wage earners who help sustain the family. Children often act as the old-age social security system for their parents. For these families, having fewer children is not an easy decision.

We also have misconceptions about overconsumption – that it's synonymous with human well-being and development. But booming rates of childhood obesity, depression and environmental degradation contradict those connections.

While individuals can and should be encouraged to reduce consumption on their own, patterns won't change unless we address the underlying causes of interrelated consumption and development patterns. For years, government policy has promoted inefficient vehicles and auto-friendly suburban development to the detriment of mass transit. Subsidies also favor energy-intensive industrial agriculture over more efficient local farms.

It's time population control came off the top of the environmental agenda. While we should help those who want access to better family planning abroad, the real focus should be on controlling wasteful consumption at home.

Population Change and Development

Probably the most prevalent framework for thinking about population change and its relationship to economic development is the **demographic transition** model (or theory). As discussed earlier in this chapter, for most of human history, population growth proceeded very slowly and remained low until the 19th and 20th centuries, during which the world's population increased sevenfold from 1 billion in 1804 to 7 billion-plus today (see Figure 9.01 above). Although population growth remains relatively high in most developing countries, it has slowed considerably, and even declined in some countries in the industrialized Global North. This change in population growth over the past century has been described as the demographic transition – a model based on the observations of population change primarily in Europe by the American demographer Warren Thompson in 1929.

The demographic transition model consists of four stages or phases (see Figure 9.09). The first phase coincides with pre-industrial period and is characterized by high birth rates and high death rates. Modern medical knowledge was non-existent, hygiene was poor, and food scarcities sometimes occurred. Infant and child mortality was quite high and adult life expectancy low. Because high birth rates were offset by high death rates, overall population growth was stable and low. Currently no country is in this stage. Phase II represents a transitional stage of rapid growth in population. Improved sanitation, hygiene, and modern medicine drive death rates down as people began to live longer and children survive into adulthood, but birth rates remain high. This is because agriculture still dominates local and national economies and families need children to work, and because there is a lag time between the increased survival rates of children and the adjustment in people's fertility behavior in response. During this phase it is the decrease in death rates while birth rates remain high that accounts for the rapid growth in population. While Europe passed through this stage relatively quickly, it

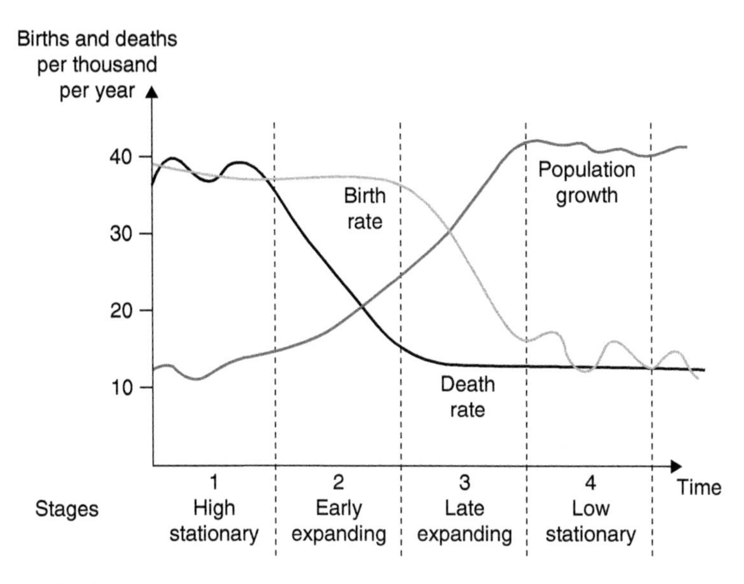

Figure 9.09 The demographic transition model. *Source*: http: // geographyfieldwork.com / Demographic Transition.htm. Reprinted with permission of Barcelona Field Studies Centre SL under the Creative Commons License.

appears that this phase may be longer for less developed countries for reasons discussed below – a situation that is sometimes referred to as the **demographic trap**.

Phase III is a second transitional stage. Death rates stay low and birth rates begin to fall in response to a number of socioeconomic changes that took place during the period of industrialization. First, in rural areas as more children survive into adulthood, couples begin to have fewer children. Urbanization further decreases the incentive to have large families. Mandatory education and child labor laws enacted in the latter part of the 19th century changed the economic role of children within the family from that of economic assets to that of costly dependants. As household incomes increase and the middle class expands, the economic necessity of children further decreases. The education and entrance of women into the paid workforce also decreases the incentive to have several children. In Phase IV of the demographic transition both birth rates and death rates are low and population growth is again stable; however, the total size of the population is significantly higher than in Phase I. Industrialized countries in the Global North are currently in Phase IV of the transition. Newly industrialized and semi-peripheral countries in Asia and Latin America are generally in Phase III and approaching Phase IV, while a number of less developed countries in sub-Saharan Africa and other parts of the world are still in Phase II. In recent decades some have posited a possible Phase V, in which population actually declines. Japan and a number of countries in western Europe are experiencing declining populations as fertility rates have dropped below the replacement rate of 2.1 children per couple. This has prompted the need to import workers from other countries to maintain economic growth.

The latter stages of the demographic transition model correspond to the period when countries in the Global North achieve a fully industrialized economy and a well-developed middle class. This suggests a strong connection between population dynamics and economic development in which one could argue that economic development is a critical factor in decreasing population growth. If that is the case, one could then logically reason that if economic development decreases population growth, it is poverty that likely drives "overpopulation" rather than the reverse as implied in neo-Malthusian perspectives on population dynamics.

However, the demographic transition model is based on the particular history and experience of Europe and other industrialized countries in the Global North. It is unclear that the model is very relevant to the situation of developing countries for a number of reasons. First, the decline in mortality in Europe took place gradually over a period of 150 years, in contrast to 50 years in less developed countries. In addition, fertility rates in the Global South were higher at the outset than in Europe. In Britain, the Phase I–II fertility rate was about 35 / 1000 people/ year. In some developing countries, the rate was 45+/ 1000. Thus, Phase II represents much more dramatic change for developing countries than what Europe experienced. Compounding this is the fact, that Europe was able to "export" tens of millions of people to the Americas during its period of population growth, and a significant number of people moved from rural to urban areas. Urban economies offer a different set of incentives to have children than do rural economies. Furthermore, the possession of colonies around the world allowed Europe to extract enormous resources that fueled its industrialization and economic development. As former colonies of Europe, less developed countries are still struggling to industrialize. Urbanization and economic development have been slow to unfold, and these countries possess limited options for migration. The particular economic context in which population dynamics are situated in many developing countries requires us to explore further factors driving persistently high birth and fertility rates in these countries.

Factors that Influence Birth and Fertility Rates

Demographers use two measures of fertility. The **crude birth rate** is the number of live births per year per 1000 people in a given population. The **total fertility rate** is the average number of children a woman would have over her lifetime assuming she survives to the end of her reproductive life and bears children in accordance with current fertility rates for her age cohort. A number of factors influence birth and fertility rates. First is education and affluence. Birth and fertility rates are usually lower in developed countries where levels of education and affluence are higher. Female literacy and access to education are particularly influential. Women with only a couple of years of education tend to have fewer children than women who have none. As women's educational levels increase, their involvement in family fertility decisions increases and they tend to desire

Figure 9.10 Children's work plays an important role in household and farm economies. *Source:* (a) The Museum of English Rural Life, University of Reading; (b) © Mike Goldwater/Alamy.

fewer children. Related to education is access to employment. As women enter the paid workforce the opportunity cost of having children increases, thereby impeding their ability to earn an income, and becomes a disincentive. Average age at marriage (or onset of sexual activity) is also influential. The earlier a woman marries, the longer her reproductive years and the tendency to have more children. Women who marry at age 25 or older tend to have fewer children than women who marry at a younger age.

Another set of very influential factors in fertility rates is related to the economic role of children in the family. In rural households children perform important tasks critical to the survival of their households. They work in fields, tend livestock, gather fuel and water, help take care of younger children and elderly family members, and perform a wide range of other work that supports their families (see Figure 9.10). Girls often assist with household chores such as cooking and cleaning, which frees their mothers to earn income from work outside the home. In developed countries the high cost of raising and educating children serves as a disincentive to have large families. But for poor families in developing countries children are an economic asset, so having more children actually makes economic sense. Furthermore, in the absence of pension and social security programs, children provide economic security for parents in old age. In such a context and where infant and child mortality rates are high, couples need to have several children in order to ensure that a least one or two survive into adulthood to be able to take care of them.

Social traditions, religious beliefs, and cultural norms present another set of important factors in fertility behavior. In many societies children are highly valued culturally, and having large families is a source of social status. In strongly Christian countries, religious beliefs may oppose abortion and discourage the use of birth control. In strongly patriarchal societies, male children are important both socially and economically so couples will continue having children until a son is born. Finally, access to reliable methods of birth control is necessary. The desire to limit family size cannot be realized if people do not have access to the means to do so. The lack of access to affordable birth control methods is particularly acute in rural areas and in poor countries.

Population and Consumption Politics and Policies

Although population matters have been of concern to rulers dating back to the ancient empires, population policy and programs are essentially a modern phenomenon emerging with the development of the liberal state in the 19th century. **Population policy** may be defined as deliberately constructed or modified institutional arrangements through which governments influence, directly or indirectly, demographic change (Demeny 2003), including the size, composition, and distribution of population. Population policies articulate general goals and objectives, and **population programs** consist of the specific strategies and instruments for achieving them. Demographic change is affected by fertility, mortality, **immigration** (in-migration) and **emigration** (out-migration). Since mortality is typically viewed as a matter of health, population policies generally focus on fertility and migration (Demeny 2003).

Population policies may have either a pro-natalist or an anti-natalist orientation. "**Natalism**" comes from the Latin word for "birth" and is a belief that supports human reproduction. **Pro-natalist** population policies are promoted by governments that wish to expand their populations or are concerned about declining populations. Rapid population decline can lead to severe economic problems. Countries with population decline may have an increase in the proportion of elderly people, which leads to large medical and social security expenses. Population decline also shrinks the workforce, which can lead to economic slowdown. In subsistence economies agricultural de-intensification often occurs, which threatens the stability of food production. Pro-natalist population programs offer financial and social incentives to couples to have more children (e.g., subsidized daycare) and often limit access to abortion and contraception. An example of a country that has pursued pro-natalist policies at different times in the past century is Japan. In the 1930s, the Japanese government urged women to have as many children as possible to fuel the war effort. Abortion clinics closed, and contraceptives were made unavailable (White 2002 by Chapple 2004). In the years following World War II, when population was growing rapidly and women were joining the workforce, abortion was again legalized, but the birth control pill only became legal in 1999. Today, Japan is again facing a potential crisis from population decline and the government is exploring ways to encourage higher fertility rates. Other countries pursuing pro-natalist population policies include Singapore, Taiwan, South Korea, and Iran.

Apart from the few exceptions discussed above, population policies and programs since the 1950s have tended to be **anti-natalist** in orientation – that is, concerned with lowering birth and fertility rates. The decade following 1950 is significant for several reasons. The world's population in 1950 was 2.5 billion, more than half (1.7 billion) of which resided in developing countries that were beginning to gain political independence with the demise of European colonialism. Death rates had dropped considerably in these countries, but birth rates remained quite high – on average 44/1000. The prospect of rapid global population growth loomed large, and there was tremendous concern about the implications of such growth for economic development. According to Demeny, "the large and widening differential between the more developed and the less developed countries in terms of population size and average income levels was seen as holding out the prospect of major dislocations and long-term instabilities within the international system" (2003: 12). As a result, an intense debate about what policies might reduce fertility in the less developed world began in the 1950s. Population policy became a matter of international concern. US foreign development assistance began including funding for family-planning programs, and a series of international population conferences began to be held. Early conferences in Rome in 1954 and Belgrade in 1965 essentially consisted of technical sessions that focused on the exchange and improvement of scientific information about population variables, determinants, and consequences. The United Nations sponsored conferences in 1974 (Bucharest), 1984 (Mexico City), 1994 (Cairo), and 1999 (New York), and these were the first that government representatives attended. They were much more political in nature and tone.

From the outset international population policy discussions were fraught with controversy. Considerable debate ensued about the extent to which rapid population growth constrained economic development and what the appropriate role of the state should be with respect to population matters. Concerns about national sovereignty also came into play (De Jong 2000). The idea that a population development "problem" might exist and should be addressed in developing countries was essentially a diagnosis made by the Global North – the United States in particular – and this was met with resentment by the developing countries of the Global South. Family-planning programs funded by the Global North under the guise of development were also suspected to be motivated by underlying concerns about international migration and environmental degradation. In recent decades women's organizations have criticized the way family-planning programs have tended to victimize women and disregard basic human rights (De Jong 2000).

The first UN conference considered the "Draft World Population Plan of Action" and debate ensued over two opposing approaches to lowering fertility. The Global North argued that the Global South was producing too many babies, and this posed a threat to economic development and environmental integrity. Suggested policies included the imposition of population targets and programs to aggressively reduce family size. The developing countries challenged the idea that a population problem in fact existed by pointing out that the industrialized countries consumed many more resources and produced much more waste in both absolute and per capita terms than the developing countries. They argued that economic development would be a much more effective route to lowering fertility than family-planning programs. By the time of the next UN conference in Mexico City in 1984, viewpoints had reversed. Developing countries had become persuaded of the need to reduce population independently and alongside efforts to foster economic development – many had already formulated family-planning programs (see Box 9.02) – while the developed countries advanced the "development as contraceptive" argument. This perspective was influenced by the turn toward free market principles (neoliberalism) in the US and UK and the assessment that population problems represented failures of development for which excessive state intervention were to blame (see De Jong 2000).

The Cairo International Conference on Population and Development in 1994 is understood to have achieved an unprecedented international consensus and is credited with introducing a completely new approach to population policy that shifted focus away from a macro-level preoccupation with the impact of population growth on economic development to a concern for individual sexuality and reproductive rights. The consensus that emerged agreed on the need to bring down population growth rates, but it also advocated deliberate steps to reduce poverty and disease, foster sustainable development, improve educational opportunities for girls and women, and improve reproductive and maternal health care. Freedom of choice, recognition of diverse cultural practices, and respect for human dignity also figure prominently in the action plan adopted. That the action plan recognized the connection between gender relations and reproductive decision-making and sexual behavior was in large part because of the influence numerous women's organizations had gained at the international level.

In part as a result in the paradigm shift emanating from the Cairo conference, best practices in family planning recommend programs that: provide information on birth spacing, birth control, and pre-natal care; include teenagers and sexually active unmarried people; expand access to family planning services; and expand programs to educate men about the importance of having fewer children. According to UN studies, an estimated 300 million women in developing countries want to limit the number and determine the spacing of their children, but they lack access to services. This could prevent an estimated 5.8 million births a year and over 130,000 abortions per day. Furthermore, in many societies, women have no control or autonomy over their own sexuality and reproductive decision-making – this lies in the hands of their husbands. In fact, the average male is responsible for producing more children during his lifetime than the average female.

Box 9.02 Family planning programs in less developed countries: China and Thailand

Most developing countries have developed some type of family-planning program. China and Thailand each present interesting case studies of such programs. Both are anti-natalist in orientation but the approach each country has adopted differs profoundly from that of the other. China currently has the world's largest population – 1.22 billion. In the 1960s the Chinese government determined that the only alternative to mass starvation was strict population control. Since China's government has considerable control over its people, it was able to impose a consistent population policy throughout the country. The fact that China has a fairly homogeneous population and a widespread common written language helped. China established the most extensive, intrusive, and strict population policy in the world. Couples are strongly urged to postpone the age of marriage and to have no more than one child – the "one child policy." Married couples have ready access to free sterilization, contraceptives, and abortion. Paramedics and mobile units ensure access in rural areas. Couples who pledge to have no more than one child are given extra food, larger pensions, better housing, free medical care, and salary bonuses. The child is given free education and preferential treatment in employment. Couples who break the pledge lose all benefits.

Between 1972 and 1996 China's crude birth rate dropped from 32 to 17 per 1000 people. Total fertility rate dropped from 5.7 to 1.8. China's current population growth rate is 1.1%. Eighty-one percent of women in China use modern contraceptives as compared to 57% in developed countries and 35% in other developing countries. Making the program locally available (rather than asking people to go to distant centers) has been seen as one of the program's strong points. Although the apparent positive outcomes of the policy are striking, China's approach has been criticized as being incompatible with democratic values and basic human rights. These type of policies are often ineffective in the long run because people eventually resist. In addition, because China is a patriarchal society with a strong cultural preference for boys, female infanticide and the abortion of female fetuses have increased. This has had a long-term effect on society in that adult males now considerably outnumber females, and men face a "shortage" of women to marry.

Thailand's family-planning program presents a different approach. It adopted a very imaginative and high-profile advertising and public information campaign to persuade families to have fewer children. This campaign focused actively promoted the use of condoms, which were made widely available, recruiting local people such as shopkeepers and schoolteachers to assist in the effort. At the same time, resources went into training local health-care workers to prescribe contraceptives and to provide better health care for mothers and children. It also made efforts to advance women's rights. Thailand benefited from a high literacy rate among women and the support of the country's (Buddhist) religious leaders. The government furthermore was willing to provide financial support and work with non-government organizations to promote its programs. Its family-planning initiative was also coupled with an effective economic development program that doubled per capita income between 1971 and 1996. Part of the program made available small loans to households that participated in family-planning measures. Between 1971 and 1996, the country's population growth rate dropped from 3.2% to 1.4%. The total fertility rate dropped from 6.4 to 2.2. In 1996, 64% of married women were using contraceptives (as compared to 57% in developed countries and 48% in developing countries). This program is considered to be very successful.

Figure 9.11 A billboard promotes use of condoms as part of Thailand's family-planning program. *Source*: © Peter Ashton/flickr.com/peamasher.

Chapter Summary

This chapter has addressed a number of issues related to human populations and their impact on the environment. Of particular interest has been the question about the relationship between population change, technology, and environmental degradation. One perspective on this relationship tends to see population as the determinative factor, while another perspective sees consumption and production and resource extraction technologies as being the more significant. The other question we have considered is the relationship between population growth, poverty, and economic development. Does overpopulation cause poverty, or do poverty and lack of development drive overpopulation? What exactly do we mean by overpopulation in the context of modern industrial society and its consumption habits? Finally, population policies and programs have been quite politically contentious both nationally and internationally. How many people is too many, and who gets to decide?

Critical Questions

1. In what ways do human populations vary over space and time?
2. What factors do you think are most important in the population-consumption-technology nexus?
3. What role should population policies and programs play in economic development?

Key Vocabulary

anti-natalist
arithmetic rate
carrying capacity
crude birth rate
crude death (mortality) rate
demographic transition
demographic trap
emigration
exponential growth
fertility rate
food sheds/resource sheds
immigration
I=PAT model
Malthusian and neo-Malthusian perspective
natalism
population eras (1st, 2nd, 3rd and 4th)
population policy
population program
pro-natalist policies
structuralist/neo-structuralist perspective
technocratic and cornucopian perspective
total fertility rate

References

Boserup, E. (1965) *The Conditions of Agricultural Growth: The Economics of Agrarian Change Under Population Pressure* (Chicago: Aldine).

Chapple, J. (2004) The dilemma posed by Japan's population decline. *Electronic Journal of Japanese Studies*, 4(1) (Discussion Paper #5). Posted October18. http://www.japanesestudies.org.uk/discussionpapers/Chapple.html

Chertow, M.R. (2001) The IPAT equation and its variants: changing views of technology and environmental impact. *Journal of Industrial Technology*, 4(4), pp. 13–29.

Cohen, J. (1995) *How Many People Can the Earth Support?* (New York: W.W. Norton).

De Jong, J. (2000) The role and limitations of the Cairo International Conference on Population and Development. *Social Science and Medicine*, 51, pp. 941–53.

Demeny, P. (2003) *Population Policy: A Concise Summary*, Working Paper 173. (New York: Population Council).

Diamond, J. (1987) The worst mistake in the history of the human race. *Discover* (May).

Ehrlich, P. (1968) *The Population Bomb* (New York: Ballantine Books).

Ehrlich, P.R. and Holdren, J.P. (1971) Impact of population growth. *Science*, 171, pp. 1212–17.

Hartmann, B. (1994) Population fictions: the Malthusians are back in town. *Dollars & Sense* (Boston, Sept./Oct.). Repr. in T. Goldfarb (ed.), *Taking Sides: Clashing Views on Controversial Environmental Issues*, 6th edn. (Guilford, CT: Dushkin, 1995).

Harvey, D. (1974) Population, resources and the ideology of science. *Economic Geography*, 50(3), pp. 256–77.

Hynes, H.P. (1999) Taking population out of the equation: reformulating I=PAT. In J. Silliman and Y. King (eds.), *Dangerous Intersections: Feminist Perspectives on Population, Environment, and Development* (Cambridge, MA: South End Press).

Lohmann, L. (2003) Re-imagining the population debate. The Corner House. http://www.thecornerhouse.org.uk/resource/re-imagining-population-debate.

Lomborg, B. (2001) *The Skeptical Environmentalist: Measuring the Real State of the World* (New York: Cambridge University Press).

Maniates, M. (2002) Individualization: plant a tree, buy a bike, save the world? In T. Princen, M. Maniates, and K. Conca (eds.), *Confronting Consumption* (Cambridge, MA: MIT Press).

Simon, J. (1981) *The Ultimate Resource* (Princeton: Princeton University Press).

White, M. (2002) *Perfectly Japanese*. Los Angeles: University of California Press.

10

Agriculture and Food Systems

Icebreaker: The Global Food Crisis

Between April 2007 and March 2008, average global food prices rose 50%, and up to 100% for some grains such as rice (see Figure 10.01). While this sudden surge in prices was a challenge for all segments of the population, it was especially problematic for the urban

An Introduction to Human–Environment Geography: Local Dynamics and Global Processes,
First Edition. William G. Moseley, Eric Perramond, Holly M. Hapke and Paul Laris.

Published 2014 by John Wiley & Sons, Ltd.

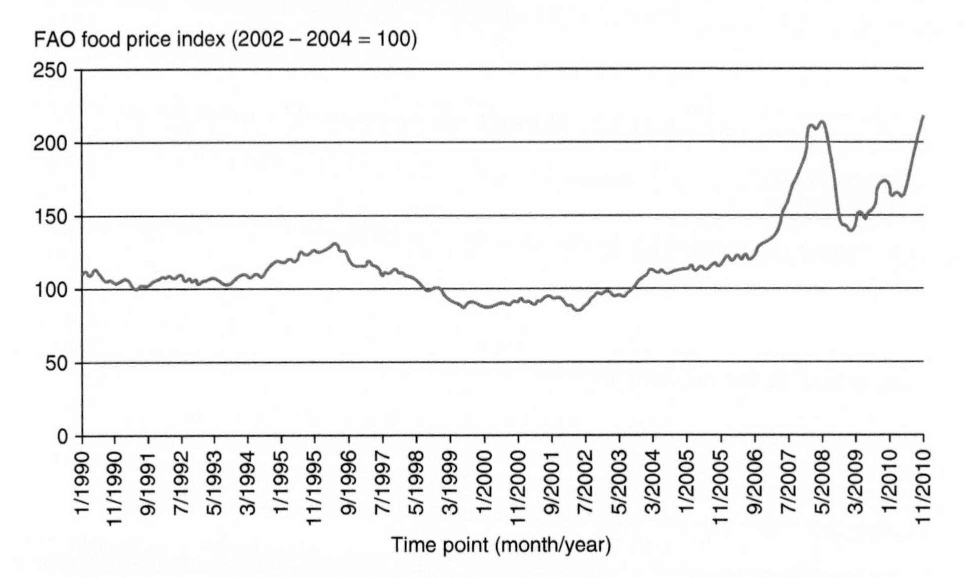

Figure 10.01 Food Price Index. *Source*: Reprinted by permission of the Food and Agriculture Organization of the United Nations.

poor, who spend a disproportionate amount of their incomes on food and have limited abilities to increase their monthly food budgets. This increase in prices caused social unrest around the world, from a protest outside a Safeway store in Los Angeles which was rationing rice, to a pasta riot in Milan, Italy, to demonstrations in cities across the Global South (especially in Africa and Asia).

While we usually associate dramatic rises in food prices with significant harvest failures, global food production was strong in 2007–8. The explanation for this food crisis actually had more to do with a gradual global change in consumer preferences and, most immediately, increasing energy costs. Over the longer term, the world has been becoming more urban (surpassing the 50% mark in 2010). This urbanization has spawned dramatic changes in diets which have put additional pressure on the global food system, including increasing meat consumption (with about 10 kilograms of grain required to produce a kilogram of beef), demand for produce in the off-season, and greater consumption of processed foods. In the short term, the leading causal factor behind higher food prices was increasing energy prices (with crude oil reaching nearly US$150 per barrel in July of 2008, up from about US$60 per barrel at the start of 2007). While the connection between food costs and the price of energy might not be evident, this becomes clear once one understands how energy-intensive the global food system has become. Not only is energy needed to power tractors and ship food around the world, but most pesticides and fertilizers are fossil fuel-based, and large amounts of energy are needed to produce processed and packaged foods. These high-energy prices also put pressure on the global food system by increasing investment in ethanol production, which absorbed 25% of the US maize crop in 2008.

Chapter Objectives

The objectives of this chapter are for the student to be able to:

1. Describe the major agricultural systems around the world.
2. Understand basic principles of agroecology.
3. Identify major environmental problems, and related solutions, associated with different agricultural systems.
4. Understand the influence of agribusiness and government policy on agricultural systems.
5. Critically assess different concepts and terms related to hunger.

Introduction

You could argue that food is all around us, as the earth has perhaps 30,000 plant species with parts that people can eat. However, only about 15 plant species and eight animal species supply 90% of our food. The number of crops that humans depend on for the bulk of their food has steadily declined over time. Today, the world's food supply is based on about 30 major crops. Four of these crops – wheat, rice, corn, and potatoes – make up more of the world's total food production than all other crops combined. Seven animal species provide more than 95% of our meat production. Cattle, poultry, and pigs supply most of our meat, followed by much smaller contributions from sheep, goats, buffalo and horses. How is it that we came to develop such a food system and what are the implications of such a system for the environment and human well-being?

Food consumption and agriculture are two of the most fundamental ways in which humans are connected to their environment. Understanding how we manage the environment for food production, the environmental impacts of different agricultural strategies, and the evolving nature of global food systems, are all themes we will explore in this chapter.

Systems of Agricultural Production

Agricultural systems may be classified along at least three continuums. First, they may be considered in terms of where they fit in on a range of extensive to intensive agriculture. **Extensive agricultural systems** are characterized by limited labor or energy inputs per unit of land. **Intensification** refers to the process of applying more and more labor or energy inputs to obtain higher yields (output per unit area). The end result is **intensive agriculture**, wherein you have land cultivated every year (sometimes with multiple plantings and harvests per year) with very high yields. Intensive agriculture could be achieved via a strategy **of low external**

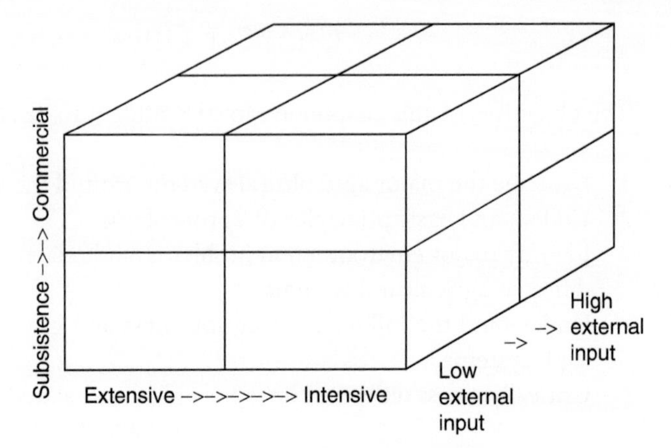

Figure 10.02 Different dimensions of agricultural systems.

input or **high external input agriculture**. A low external input strategy for intensification would entail considerable labor inputs (for transplanting and weeding for example), heavy use of organic inputs produced locally such as compost, manure, and/or night soil; and a diverse planting strategy to control pests and utilize every **niche** (or space) in the farm field. A high external input strategy of intensification would rely more heavily on the use of inorganic fertilizers, pesticides, and mechanization to achieve high yields. Finally, these various systems – either extensive or intensive, or low external input or high external input – may be subsistence or commercially oriented. That said, most intensive systems (especially those relying on a high external input approach to intensification) are more commercially oriented as you typically need cash income in order to purchase such inputs. **Subsistence agriculture** refers to farming which is done primarily for household consumption. In contrast, **commercial agriculture** produces crops destined for the market. Many farming households pursue a hybrid model where they are producing some crops for home consumption and some for the market. All three of these continuums by which we classify or categorize agricultural systems may be seen in Figure 10.02.

Traditional Systems of Agricultural Production

A unifying feature of traditional agricultural systems is that they are low external input approaches. Many of these systems were historically subsistence-oriented, but pure subsistence agriculture is less common, as traditional farmers often sell a part of their crop to obtain cash income. These systems range from extensive to intensive. Traditional agriculture is practiced by about 2.7 billion people, providing roughly one-fifth of the world's food crops and occupying three-fourths of the globe's farmland. Most forms of traditional agriculture that persist today are in developing countries in the tropics.

Many traditional farming systems (both extensive and intensive) are characterized by mixed cropping strategies or **polyculture** (e.g., corn and beans, sorghum

Figure 10.03 A polycrop of sorghum and cowpeas *Source*: W.G. Moseley. Used with permission.

and cowpeas, etc.), as well as **agroforestry** (the mixing of crops and trees in a field). Both of these strategies have sometimes been misunderstood as haphazard and backward, but there are lots of sound ecological and economic reasons why such approaches are undertaken (Richards 1985). One example of a common intercrop or polycrop is featured in Figure 10.03, that of sorghum and cowpeas. This is typical of many mixes which involve a grain crop and a legume (such as most beans). Legumes fix nitrogen, which is used by the grain crop. The grain crop may, in turn, provide structural support to some climbing bean species. The presence of more than one crop in a field also tends to reduce pest damage because many insects have difficulty finding a target plant in a mixed plant field (see more on this in the agroecology section below).

Agroforestry systems are created by either planting trees or selectively leaving trees in a field when an area is cleared for farming. In some ways, agroforestry is a form of polyculture, with a key difference being that the combination involves not just annuals (crops that live one season), but woody-stemmed perennials as well (plants that live multiple seasons). As such, many of the same benefits of polyculture are also found in agroforestry, as well as a number of other benefits related to the more robust structure of trees. Trees are also retained in fields for their economic benefits (often fruit and nut trees) rather than ecological ones. Some trees are leguminous, such as the acacia albida trees in Figure 10.04, which fix nitrogen for the benefit of associated crops. The acacia albida is also prized by farmers in West Africa because it leafs out during the long dry seasons, protecting

Figure 10.04 Nitrogen-fixing acacia albida trees being grown in association with millet.
Source: © Joost Brouwer / Brouwer Environmental & Agricultural Consultancy.

the soil's organic matter from the decomposing heat of the sun, but then loses its leaves during the rainy season so that it does not compete with crops for sunlight. Furthermore, the acacia albida produces a pod which is collected for animal fodder (or animal food). The way the trees were selectively left in the field in Figure 10.04 is known as **parkland agroforestry**, wherein trees are just spread throughout the field. During the long, hot dry season, when Harmattan winds blow across this landscape, these trees will slow down wind speeds across the fields and thereby limit **aeolian** (wind) **erosion**.

The classic extensive, traditional farming strategy is **shifting cultivation**, also known as swidden or slash-and-burn agriculture. While this approach has been demonized in the literature for being wasteful and a cause of deforestation, such claims are often misplaced. Shifting agriculture is a rational strategy if labor resources are limited and one has a relative abundance of land. It is most commonly practiced in the more lightly populated areas of tropical Amazonia, Africa, and Southeast Asia (see Figure 10.05). This process begins by clearing an area of forest, leaving useful trees behind, and then burning the cleared vegetation. This burning returns some nutrients to the soil, which is important in the tropics where many soils (such as oxisols and ultisols) are nutrient-poor. The farmer will cultivate this plot until yields become too low (often after three to four years of farming), and then let it lie **fallow** (or rest) for 20 to 40 years, while he or she moves on to another plot to farm. The surrounding forest will typically recolonize the abandoned plot and the soil fertility will gradually be restored over a number of years.

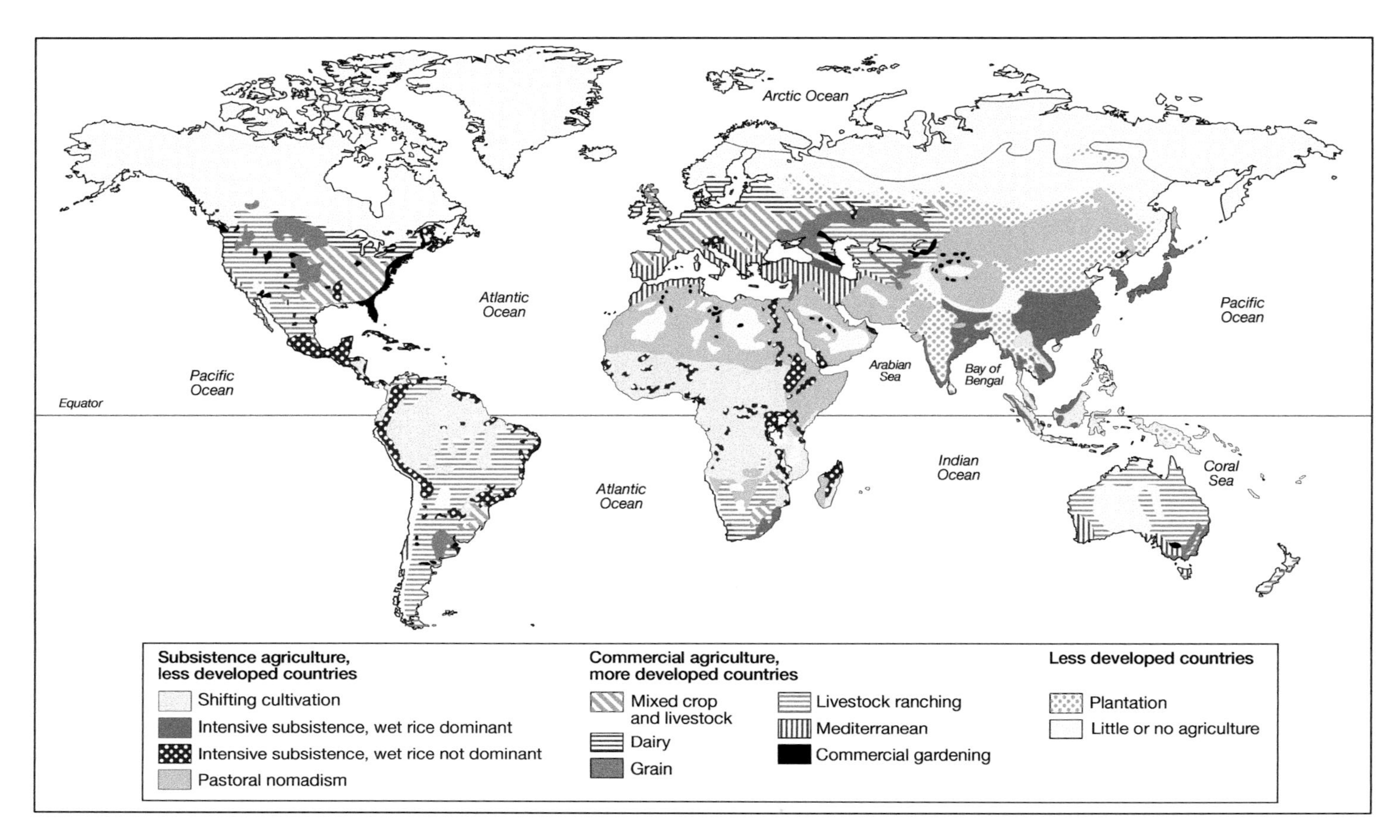

Figure 10.05 World map of food systems.

Traditional farmers often look for certain indicator vegetation to know when the plot is ready to be cleared again and farmed. As mentioned, this is an ecologically sound approach to farming under certain conditions. In many ways, farmers are using the services of nature to restore soil fertility rather than relying on external sources to do this.

In several areas of the world, farmers are gradually transitioning away from shifting cultivation to more intensive agriculture. This tends to occur as population densities increase and farmers start to till a plot of land for longer and longer periods of time (Boserup 1965). Farmers may do this and simply accept lower yields (farming larger areas to compensate for lower yields) or they may begin to add organic inputs (such as manure and compost) to maintain soil fertility. This intermediate stage – between shifting cultivation and more intensive traditional agriculture – is sometimes referred to as **bush fallow rotation**. Under this system, cultivation time may extend to 8–10 years with a similar length for fallows. This relationship is sometimes expressed in terms of the **R Factor** (50 in this case), which = [(years under cultivation) / (years under cultivation + fallow)] × 100. This compares to R factors of around 10 for shifting cultivation or 100 for many forms of intensive traditional agriculture.

Intensive traditional agriculture refers to situations where farmers produce relatively high yields per unit of agricultural land as a result of heavy inputs of labor, draft power (i.e., the use of animals to pull plows), organic fertilizer (compost and manure), and water. They often produce enough food to feed their families as well as to sell for income. As Figure 10.05 shows, this approach is very common in South and East Asia, the highlands and scattered areas of Africa, as well as the highlands of Latin America. A common factor in all of these areas of intensive traditional agriculture is higher population densities. Intensive traditional agriculture is sometimes divided into those systems dominated by wet rice and those dominated by other crops such as wheat, maize, upland rice, and small grains. In Asia, wet or paddy rice is the principal crop in areas with a long, warm, and rainy growing season. Wet rice is also grown in swampy areas of West Africa and under similar circumstances along the northeast coast of South America. Wheat, maize, upland rice, and other small grains are grown in drier climates.

Another form of traditional agriculture is **pastoralism** (see Figure 10.06). This refers to extensive nomadic herding (as opposed to intensive and fenced animal-rearing). This is a way of life for 15–20 million people who tend livestock over vast areas of arid and semi-arid land in the Old World. The core area is from Sahelian and Saharan Africa, across the Arabian Peninsula and Southwest Asia, into central Asia. Nomadic herders have regular migration patterns that are dictated by the seasonal conditions of grasslands. Many of these herders would historically trade milk for grain from farming communities, whereas now most such transactions are market-based. Like other practitioners of traditional agriculture, pastoralists have often been blamed for land degradation, as well as wasteful and inefficient herding practices. For example, many range management scientists used to believe that use of fenced pastures on a rotation schedule with much higher stocking

Figure 10.06 A herder and his goats near a seasonal water hole in Douentza, Mali.
Source: W.G. Moseley. Used with permission.

densities was a superior form of animal husbandry. While such an approach may be appropriate in the temperate zone, more recent scholarship on non-equilibrium ecology has shown that nomadic pastoralism is the most ecologically and economically rational strategy where rainfall patterns and pasture conditions are highly variable (e.g., Behnke et al. 1993). In the past, many governments (colonial and postcolonial) attempted to sedentarize pastoralists based on a belief that settled crop farming was a superior or more modern form of existence. Today, the existence of these groups is threatened by restrictions on their movements across national boundaries.

The Industrialization of Agriculture

Over time, the advances of the industrial revolution have been applied to agriculture. Industrial agriculture involves large amounts of external inputs, and **monocultures** (or single-crop fields), to produce high yields of crops and livestock for sale. These external inputs include fossil fuel energy, water, hybrid and GMO seeds (see Box 10.01), inorganic fertilizers, and pesticides. Industrial agriculture is practiced on about 25% of all cropland, mostly in the Global North. Industrial agriculture is often broken down into a number of subcategories, including that focused on grain production, horticulture (or commercial gardening), livestock ranching, mixed crop and livestock farming, and Mediterranean farming. Climate and topography often explain the spatial pattern of industrial farming activities, with ranching occurring in the driest environments, followed by grain farming, mixed crop and livestock, dairy, and finally horticulture in the areas of higher precipitation (see Figure 10.05). Mediterranean is a special category of agriculture occurring in environments characterized by dry summers and wet winters (and the considerable use of irrigation). The industrial agricultural approach spread to a number of countries in the Global South in the 1960s following the Green Revolution (an approach which will be discussed below).

While industrial techniques and approaches initially came to crop agriculture, this philosophy would eventually transform animal husbandry as well. Rather than free-range chickens, pigs, or cattle which roamed fenced fields and ate a mixture feed, grass, and hay, industrial animal husbandry increasingly emphasizes **concentrated animal feeding operations (CAFOs)** wherein animals are raised in confined situations (often large warehouses for smaller animals) and feed is brought to them. As animals are raised in close proximity to another disease is often a problem; this is controlled via the heavy use of antibiotics. The waste produced by such facilities is enormous and the management of this is a growing problem where CAFOs have proliferated.

Industrialized agriculture has not only meant changes in farming technology, but changes in the organization of farming enterprises. North American farmscapes were originally dominated by **family farms** of various sizes. With the increasingly capital-intensive nature of farming, this region began to see the proliferation of **corporate farms**, farms that are owned and run by a corporate entity rather than a family. While there has been a tendency to romanticize family farms in North America (some of which are quite large), corporate farms tend to be larger and there is a concern that the management and investors of these operations have less of a long-term connection to the land and local communities.

Box 10.01 Seed development and industrial agriculture

An understanding of the basics of plant breeding is useful for comprehending the science behind increasing agricultural yields. While plant selection (the saving of seeds from plants with the most desirable characteristics) has occurred since the dawn of agriculture roughly 10 million years ago, formal plant breeding emerged at the end of the 18th century. Here, crops were systematically cross-fertilized in hopes of obtaining plants with a desirable mix of characteristics.

In the early 20th century, a significant advance was the development of a technique known as hybridization. Normally, a plant must be cross-pollinated with another. A plant that pollinates itself is known as an inbred, and it usually produces offspring that perform less well than the parent. **Hybridization** involves cross-breeding two inbreds from desirable parentage, a technique that produces highly productive plants (Tripp 2001). From the early 20th century moving forward, extremely productive hybridized plants were developed which were highly responsive to fertilizer use. Because these plants were bred for yield, and grown as monocultures, they were often more susceptible to pest damage and thus needed to be used in combination with pesticides. Furthermore, the seeds produced by hybrid plants were not as productive as their parents (with productivity declining successively from generation to generation). What this meant was that new seeds needed to be purchased for each planting (rather than saved). Annually purchased hybrid seeds, and their requisite inputs, gave rise to future agricultural input company giants such as Monsanto and Pioneer. While hybrid seed were good for business and food production, this approach to farming would provide to be very problematic for the environment.

The most recent advancement in seed technology is **genetically modified (GM) seeds**. The key advantage of GM seeds over hybridization is that plant breeders are now able to insert genes from other (unrelated) species to create desirable characteristics in a food crop. While GM seeds are purported to have several advantages (e.g., higher yields, lower need for pesticides), they also evoke a host of other concerns. These include unknown human health effects, the escape of genes from GM plants into other crops and plants, the need to continually invent and introduce new GM seeds in response to insect populations that become resistant to old GM seeds, and concern about undue reliance on the companies who develop such seeds.

A Mixture of Traditional and Industrial Agriculture

While it is tempting to neatly categorize all forms of agriculture as either traditional or industrial, there are a number of hybrid situations that are worthy of discussion, including Green Revolution agriculture, dual-agriculture economies, plantation agriculture, and urban agriculture.

In the 1960s, a concerted effort, known as the **Green Revolution**, was undertaken to boost food crop production in the developing world. The Green Revolution, involving highly productive hybridized crops in conjunction with pesticides and inorganic fertilizers, largely benefited Asia and South America because it devoted most of its attention to food crops prevalent in these regions, mainly rice, wheat, and, to a lesser extent, maize. Africa was bypassed by the Green Revolution for the most part, with a few significant exceptions such as maize in Zimbabwe, or rice in some areas of West Africa. While the Green Revolution of the 1960s and 1970s did boost food production (by about 75%), it has been criticized for not really resolving the hunger issues it was designed to address (see a fuller discussion of this later in the chapter), exacerbating wealth differences by favoring wealthier farmers, and spawning a host of new environmental problems related to chemically intensive agriculture. Given that the first Green Revolution largely bypassed Africa, there have been calls for a new Green Revolution for Africa since about 2007 (Annan 2007). Some are concerned that architects of this new Green Revolution have not learned from the problems encountered during the first Green Revolution (Thompson 2007; Moseley et al. 2010). In either case, the Green Revolution has produced agricultural landscapes throughout the Global South which are a mixture of industrial and traditional agricultural practices. A common scenario is to have cash crop (or crops grown for cash income) fields where hybrid seeds, inorganic fertilizers, and pesticides are used next to a field of food crops where such purchased inputs are absent or limited.

Another common hybrid situation is the **agricultural dual economy**. This was discussed in Chapter 4 as one form of dualism known as rural dualism. This exists when there is an export-oriented commercial agricultural sector which exists alongside smaller-scale, mostly subsistence farming. The classic situation is one wherein small subsistence farmers spend the day laboring on nearby large plantations, returning to work on their own farms in the evenings and during the weekends. The linkages between the two sectors are often pernicious for the small farmer/laborer. Large commercial farms often need access to cheap labor. The advantage of laborers who moonlight as small farmers is that they need not be paid a living wage (as they have their own production to cover a portion of their annual food needs). Many small farmers who end up working on larger farms may initially take on such employment in order to bridge a food production shortfall. In so doing, however, they may embark on a slippery slope of declining production as they spend critical time off the farm which compromises their own ability to produce food. This is a hybrid farming situation because you have large-scale (more industrial) commercial farming which exists alongside smaller-scale, traditional farming.

In many cases, the large-scale farms which exist in an agricultural dual economy have a style of organization known as **plantation agriculture**. Plantation agriculture is a form of industrialized agriculture found primarily in the Global South where cash crops, such as sugar, bananas, coffee, tea, and cacao are largely grown for sale in the Global North. While these farms use a variety of external inputs, they often are not mechanized to the same degree as many farms in the Global North because farm labor is so cheap and/or this farming involves labor processes which are not easily mechanized (see Figure 10.07). Plantations employ large numbers of laborers who may live on the farm or in surrounding communities.

The final type of hybrid agriculture is **urban agriculture**. Urban agriculture has long occurred in some cities of the Global South and it more recently had a resurgence in urban areas of the Global North. Because of generally smaller plot size, mechanization is not usually possible in urban agriculture. Some municipal regulations may also limit the degree to which pesticides can be used. The result is farming that uses traditional, organic practices as well as newer techniques. Urban agriculturalists typically specialize in the production of vegetables for the urban market. A much-studied case of urban agriculture is that in Havana, Cuba (see Figure 10.08a). The context for this case was a US embargo of Cuban trade which left the island nation virtually isolated after the loss of support from the Soviet Union in the early 1990s. In order to avoid starvation, the country turned to low external input agriculture on a massive scale, including the promotion of urban

Figure 10.07 Worker on a tea plantation in Sri Lanka. © Tony French/Alamy.

agriculture (Miller 2007). There has also been a revival of urban agriculture across the United States. This has been particularly pronounced in a number of rust belt cities (such as Detroit and Milwaukee; see Figure 10.08b) where urban agriculture has been promoted as a source of employment and an alternative land use in derelict inner-city neighborhoods. Many also appreciate the environmental benefits of having more food grown locally rather than transported long distances.

Figure 10.08a Urban agriculture in Havana, Cuba. *Source*: © Enrique de la Osa/Reuters/Corbis.

Ecology of Agroecosystems

All ecosystems are composed of a variety of interactions between plants, amongst animals, and between plants and animals. Although agroecosystems are generally less complex than natural ecosystems, such interactions exist and need to be understood. Agroecosystems may either be polycultures (multiple-crop) or monocultures (single-crop). Knowledge of agroecological principles allows a farmer to work with, rather than against, nature. When pest control or fertilization is not achieved naturally, then farmers address these problems artificially through means that are expensive and environmentally problematic.

Figure 10.08b Urban agriculture in Detroit. *Source*: © Santa Fabio Photography.

There are two general types of insects in agroecosystems. These are carnivores (or predators) and herbivores (or pests). Carnivores eat other insects while herbivores eat plants. There are two basic hypotheses concerning interactions between insects, and between insects and plants, which help us understand the benefits of having more complex agroecosystems. The first is the natural enemy hypothesis, which suggests that there will be more predators that prey on insect pests (herbivores) in complex polycultures than in simple monocultures. The larger number of insect predators means that pest populations are less likely to become unmanageable. The second is the resource-concentration hypothesis, which argues that insect pests are more likely to find and remain on hosts that are growing in monocultures. This may be contrasted with complex, diverse systems where pest species with a narrow host range have greater difficulty in locating and remaining on host plants in small, dispersed patches (as compared to large, dense, pure stands). This is because different crops may camouflage each other visually or by smell. Companion plants in a polyculture may also physically block dispersal.

There are two basic types of plant interaction in an agroecosystem. The first is competition, wherein two crops compete for the same resource (such as sunlight, soil nutrients, and water). This is why one doesn't plant two crops too close together or why weeds are removed (in order to limit competition). The other basic type of plant interaction is facilitation, wherein one species has a beneficial effect on another species. We have already discussed one example of this earlier in the chapter, which is when a legume might fix nitrogen that is used by a grain crop. Polycultures that maximize facilitation and minimize competition between plants are likely to be the most productive.

Equations for analyzing polycultures relative to monocultures may be found in Box 10.02.

Box 10.02 Tools for analyzing intercropping systems

A. *Land Equivalent Ratio* (LER). The LER allows you to compare the number of units of land in a monoculture that it would take to produce the same quantities of crops 1 and 2 in a polyculture. If LER > 1, then there is a polyculture advantage. A shortcoming of this calculation is that it does not account for the value of the crops being grown.

$$LER = (P_1 / M_1) + (P_2 / M_2)$$

P_1 and P_2 = yields of two different crops in polyculture
M_1 and M_2 = yields of two different crops in monoculture

B. *Relative Value Total* (RVT). The RVT allows you to compare the intercrop to the most valuable crop in monoculture. The result of the equation is the factor by which polyculture is more or less valuable than monoculture.

$$RVT = (aP_1 + bP_2) / (aM_1)$$

P_1 and P_2 are the yields of crops 1 and 2 in polyculture
a and b are the respective prices for these crops
M_1 is the yield of the primary crop in monoculture

C. *Replacement Value of Intercropping* (RVI). The difference between this measure and RVT is that it takes account of input costs associated with the monoculture yield (otherwise it is the same equation). The result of the equation is the factor by which polyculture is more or less valuable than monoculture.

$$RVI = (aP_1 + bP_2) / (aM_1 - c)$$

P_1 and P_2 are the yields of crops 1 and 2 in polyculture
a and b are the respective prices for these crops
M_1 is the yield of the primary crop in monoculture
C = costs (e.g., insecticides, fertilizers)

While we tend to think of diversity in terms of the mix of different plants and insects in an agroecosystem, another important scale to consider is the diversity within any particular crop species. Even in a monoculture of rice for example, traditional farmers might plant several varieties of rice to take advantage of varying conditions within a field. This diversity may also render the field less susceptible to pests. Sales of hybrid and GM seeds have led to less genetic diversity in global seed stocks. A related concern is the decline of landraces, domesticated plants (which have not been systematically bred by seed companies) that are adapted to the natural and cultural environments in which they originated. The loss of landraces is a problem, as this is the basic material from which future seeds are generated.

Common Environmental Constraints and Remedies Associated with Traditional Agriculture

There are a number of environmental constraints associated with traditional agriculture. Here we review many of these problems from a narrow, technical perspective, as well as common strategies for addressing these problems. In all cases, it is important to keep in mind the broader systematic or ultimate causes of these problems which may limit resolution of them at the local level.

As population densities increase and farming communities begin to fallow land for shorter and shorter periods of time, soil fertility and crop yields will likely decline. If land is still relatively abundant, farmers may deal with this problem for a time by simply farming more land (which is only possible in cases where there is sufficient labor, draft power, or machinery to do this). As discussed previously in this chapter, the farmer will, sooner or later, be called upon to add inputs to the soil in order to maintain its productivity. Alternatively, farm production may simply be allowed to decline if people are able to find more lucrative employment opportunities in the city. In other instances, farmers may use remittance income (or money sent home by relatives) to invest in the intensification of agriculture, such as occurred in Machakos, Kenya (Mortimore and Tiffen 1994).

The most common ways to maintain soil fertility in traditional agriculture are via polycropping (already discussed) or the addition of manure and compost. In some cases traditional farmers may be reluctant to practice polyculture because these approaches were actively discouraged by state agricultural extension agents who viewed such practices as primitive and backward (Richards 1985; Moseley 1996). Changing views about polyculture within state agricultural agencies is key to reviving this practice. Another potential constraint to intensification is that farmers may not keep enough animals to produce sufficient quantities of manure to spread on their fields. This is often the case for poorer farmers, who may have a limited number of livestock. There are also areas where it is very difficult to keep livestock because of disease (such as tsetse fly-infested regions). In such instances, more intensive production of compost is required and/or the use of **green manure**. Green manure is produced when vegetative matter is buried in the soils of a field, which decomposes and improves the organic content of the soil. This traditionally

Figure 10.09 Traditional mounding as a technique for producing green manure. *Source*: William G. Moseley. Used with permission.

was done when a field was scraped clear of grasses (typically at the start of the farming season), and the grasses buried in mounds on top of which crops were planted (see Figure 10.09). More recently, an agroforestry technique known as alley cropping has been employed to produce green manure. Here, rows of a leguminous bush known as leuceanea are planted about every sixth row in a field. These bushes are coppiced (cut back) once or twice per year and the leaves worked into the soil.

Soil erosion is another problem sometimes encountered in traditional agriculture. The causes of this are often a complex mix of local and external factors. Anytime soil is exposed, it becomes susceptible to wind and water erosion. Susceptibility to erosion is also enhanced as the slope of a field increases. The most desirable agricultural areas have historically been flat bottom lands. However, farmers often get pushed up on to more steeply sloped fields for a couple of reasons. First, some areas may just not have that much flat bottom land, so farmers quickly move to steeper slopes as populations grow. Second, it is also not uncommon for the best agricultural lands to be taken over by more powerful members in the community, more export-oriented farmers, or (in the colonial era) Europeans who had a favored status. As such, soil degradation often relates back to power grabs that marginalized poorer farmers to more erosion-susceptible areas. In some instances, soil erosion may result from overgrazing. Traditional pastoral systems kept animals moving from area to area. This starts to become a problem as stocking densities increase and the movement of herds is more restricted.

Figure 10.10a Terraced rice fields in the Philippines; terracing combats erosion in traditional agricultural systems. *Source*: © jonaldm / istockphoto.

Figure 10.10b Senegal: (i) Contour berm; (ii) Farmers using an A-frame device to determine contours for anti-erosion berms. *Source*: © Mary Cadwallender / marycad.wordpress.com.

The way erosion has historically been managed on steeper slopes is via the use of terraces. Terraces have existed for centuries, even millennia, in some traditional agricultural systems (such as those in the Philippines featured in Figure 10.10a), but they do require significant amounts of labor to construct and maintain. On less steep inclines, rock or grass lines are often laid down that follow a contour line (such as contour berms in Senegal: see Figure 10.10b). Many of the agroforestry practices discussed earlier in this chapter are also traditional means of combating soil erosion.

Common Environmental Constraints and Remedies Associated with Industrial and Plantation Agriculture

Supporters of industrial agricultural approaches often point to high yields as the greatest benefit of this form of agriculture. What is less well understood is that these high yields and outputs are literally fueled by significant energy inputs. Some have argued that the energy efficiency of agricultural systems must be taken into account (Pimentel et al. 1973; Moseley and Jordan 2001). In other words, it is not enough to look at the productivity of agricultural systems; one must also consider their efficiency in terms of a ratio of energy output over energy input. Comparative analysis shows that traditional agricultural systems tend to be much more energy-efficient than industrial systems, which depend on significant fossil fuel inputs (Bayliss-Smith 1982). Furthermore, the energy intensity, and inefficiency, of industrial agricultural production has only grown over time (Jordan 1998). Turning this trend around will mean a major reorienting of industrial agriculture. Rather than relying on external inputs to control pests and maintain soil fertility, a trend towards greater efficiency would require greater use of organic inputs and smart use of plant mixes to control insect pest problems.

The use of pesticides in industrial agriculture has increased steadily since World War II, creating a host of environmental challenges. Many of these chemicals were developed during the war years, and were increasingly utilized for civilian purposes after the war. Pesticide is a broad term which includes insecticides (for use with insects), herbicides (for use with weeds), and fungicides (for use with fungi). The first issue is the problem of **broad spectrum pesticides** which, when applied, indiscriminately kill both good and bad insects. From the human perspective, good insects are the predators (or carnivorous insects) which serve as a natural control on herbivorous insects (or pests). Once natural enemies are eliminated, this may actually exacerbate a pest insect problem over the long term as herbivorous insect populations often recover more quickly than those of carnivorous insects.

Another problem is that many pesticides are made of **persistent chemicals** which do not break down quickly in the environment (they have a long half-life). DDT is a classic example of a persistent chemical. This persistence was created by design as initially it was thought to be a positive characteristic as long-lasting chemicals would be effective over greater periods of time. The problem is that these chemicals lived on in ecosystems, creating problems for many other species. In many cases, persistent pesticides have been banned from use in the Global North, but they continue to be used in the Global South. As such, many persistent chemicals find their way back into the Global North when they are imported on fruit or vegetables as pesticide residues. This problem has sometimes been referred to as the **circle of poison**, wherein persistent chemicals continue to be manufactured in the Global North, are exported for use in agriculture abroad, and then return on food imports. Increasingly strict import regulations mean that, as we saw a few decades earlier in the Global North, persistent chemicals are being replaced by organophosphates, which are less persistent but more toxic at the point of contact. This creates health problems for farm workers in the Global North and the Global South. The problem in the case of the latter is that there may be a lower likelihood of appropriate safety gear being worn during the application of such chemicals.

Bioaccumulation and **biomagnification** are examples of two interconnected problems related to the use of persistent chemicals in an ecosystem. With bioaccumulation, some chemicals are selectively accumulated by organisms that consume or are exposed to them, often in the fatty tissues. The concentration of a pollutant may rise to toxic levels in an organism over time. Biomagnification is a process whereby the concentration of a substance in animal tissues is increased step by step through a food chain. Biomagnification of DDT is what led to the demise of many birds of prey. Rachel Carson (author of *Silent Spring*) first brought this issue to national attention in the USA. Of course, it is not only birds of prey which find themselves at the top of the food chain, but humans as well. Problems have been observed, for example, amongst certain populations which consume high levels of lake fish. In this connection, there has been a higher incidence of birth defects among Native Americans around the Great Lakes in the United States because of their greater propensity to consume fish.

Since beginning to use pesticides in earnest in the post-World War II era, farmers havc had to apply increasing quantities of these chemicals in order to achieve the same effect. This is largely due to a problem known as **pesticide resistance**. This is because no pesticide is 100% effective. It may kill 98–99% of an insect population, but there are always a few individuals who survive because of some rare trait that allows them to fare better in the face of such chemicals than the majority of the population. While a pesticide may prove to be highly effective for a number of years, a few individuals will survive and reproduce, passing on a rare trait of resistance which slows builds in the population over time. While this process of natural selection might take thousands of years in slow-reproducing mammals, the whole process is sped up considerably amongst insects, which have comparatively short life spans. Figure 10.11 shows the increasing number of pesticide resistant species between 1908 and 1998. The increasing use of pesticides (which is a cause and effect of pesticide resistance) in order to achieve the same effect is sometimes referred to as the **pesticide treadmill**. One of the main methods for slowing down the development of pesticide resistance is to employ a refuge strategy, i.e., to plant untreated fields alongside treated ones. This delays the development of insect resistance by providing food for susceptible insects which then mate with resistant insects and dilute resistance. This, however, is an expensive strategy which only those in the wealthiest countries can afford to practice.

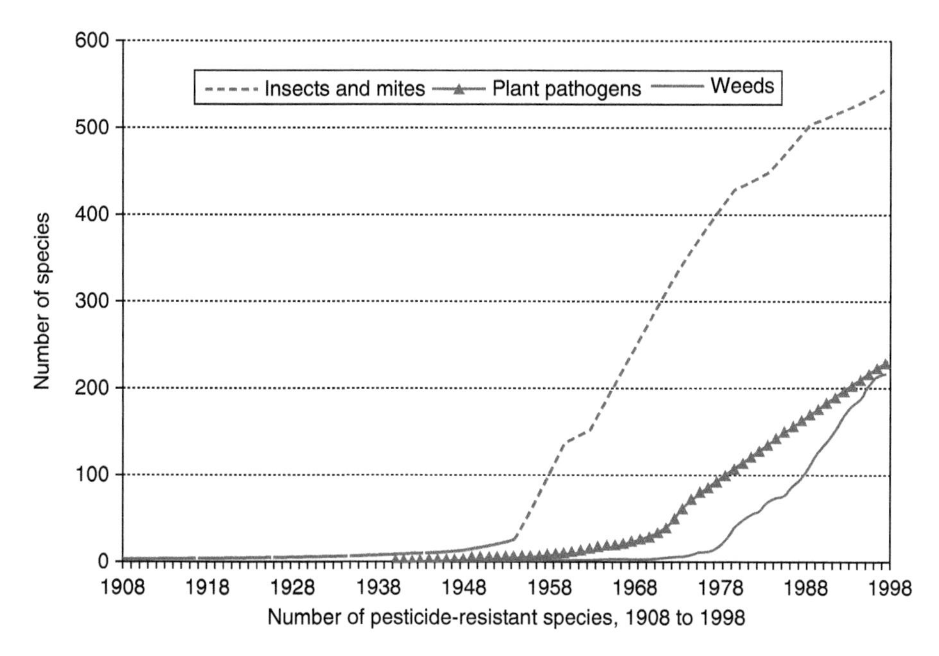

Figure 10.11 Reported numbers of pesticide-resistant species. *Source*: "Number of pesticide resistant species 1908–1998"(Vital Signs, 1999). Reprinted by permission of the Worldwatch Institute.

It is often asserted that many of these pesticide-related problems could be avoided by increased use of GM crops. This is a controversial assertion. It is true that some crops have been genetically modified to include a pesticide. An example of this is Bt cotton. Most pests consuming Bt cotton will die, and thus the problem is dealt with without spraying a chemical that will run off and travel elsewhere via surface and groundwater. But Bt also kills useful insects such as butterflies, so the concern about broad-spectrum pesticides remains. Furthermore, even the pests which are the target of Bt cotton, such as cotton bollworms and beetles, develop resistance to the chemical over time (Liu et al., 1999), which means that new strains of Bt cotton must be developed. One of the greatest concerns with GM crops is the issue of **genetic escape**. This threat was only imagined until 2001, when it was reported in the scientific journal *Nature* that the Bt gene from US-manufactured Bt maize had made its way into local maize varieties in Mexico, where it had been illegally planted (a claim which was later disputed) (McAfee 2003). Crops like maize openly pollinate, which means that genes from one crop may cross over to another crop, or to a wild relative. The most likely scenario is that Bt maize was illegally planted in Mexico where it crossed with local varieties. Also of concern might be genetic escape from a product like Monsanto's Roundup Ready® Corn (maize), which is widely used by North American farmers. This maize is resistant to the common Roundup herbicide (manufactured by the same company), which means that the farmer may apply this herbicide to control weeds without damage to the maize crop. Having such herbicide resistance, or the Bt gene, spread to wild plants would be problematic for many reasons.

Other approaches aimed at reducing pesticide use in industrial agriculture include biological control (introducing insect predators such as ladybugs) or mechanical control (physically removing pests). Sometimes an approach known as integrated pest management (IPM) is undertaken, which involves using a mix of biological control, mechanical control, and chemical control (small doses of pesticides). Rotating crops on fields from year to year also helps minimize the build-up of crop-specific pests in the soil. There is also a modified version of intercropping, known as strip cropping, where about 20 rows of one crop are grown, followed by 20 rows of another, across a field.

Pesticide-related environmental problems are often just as bad or worse in plantation than in industrial agriculture. This may be because environmental regulations are more lax, or less strictly enforced, in the Global South. Interestingly, two of the most common commodity crops grown on plantations, coffee and cocoa, were traditionally grown under an upper story of trees i.e., as an agroforestry system. Plant breeders then developed coffee and cocoa plants which grew in open sunlight, were more productive, and also required use of pesticides. Monocropped, open-sunlight cocoa plantations have been particularly hard hit by black fungal disease over the past 15 years. These disease problems, plus the organic food movement in the markets of the Global North, have meant that growers of these crops are now moving back to shade-grown coffee and cocoa.

Water resources are also often impaired by industrial agriculture, including both ground and surface water. Groundwater contamination is a serious issue in most areas where industrial agriculture is the dominant land use (see Figure 10.12). Persistent pesticides do not break down quickly and plants are not able to absorb all of the fertilizer that is applied. These compounds either seep into the groundwater (which humans often use as a source of drinking water) or travel along the surface to streams. Pesticides and fertilizers may remain in aquifers for long periods of time as the rate of water turnover is often quite slow.

Surface waters are damaged by the run-off of pesticides and inorganic fertilizers. Inorganic fertilizers add extra nitrogen and phosphorous to surface waters, nutrients which cause algae blooms and inadequate oxygen supplies for fish as aquatic plants die and decay. This anthropogenic nutrient enrichment of lakes is referred to as **eutrophication**. Large inputs of organic waste into surface waters are also problematic. While we may think of this as the result of the dumping of unprocessed sewage, run-off of animal waste (which has been applied as fertilizer) from farm fields is also a problem. For example, chicken farms on the Delmarva Peninsula, which forms the eastern side of the Chesapeake Bay (near Washington, DC, in the USA), produce large amounts of waste which is spread over the fields of farms in this area. So much of this organic waste ran off into the bay that it created a haven for pfiesteria bacteria at various times in the 1990s. These bacteria not only led to large fish kills but were also harmful to humans (sometimes leading to dementia for those who swam in pfiesteria-infested waters).

Many forms of industrial agriculture have also required massive investments in dams, canals, and irrigation infrastructure. Water diversion out of lakes and

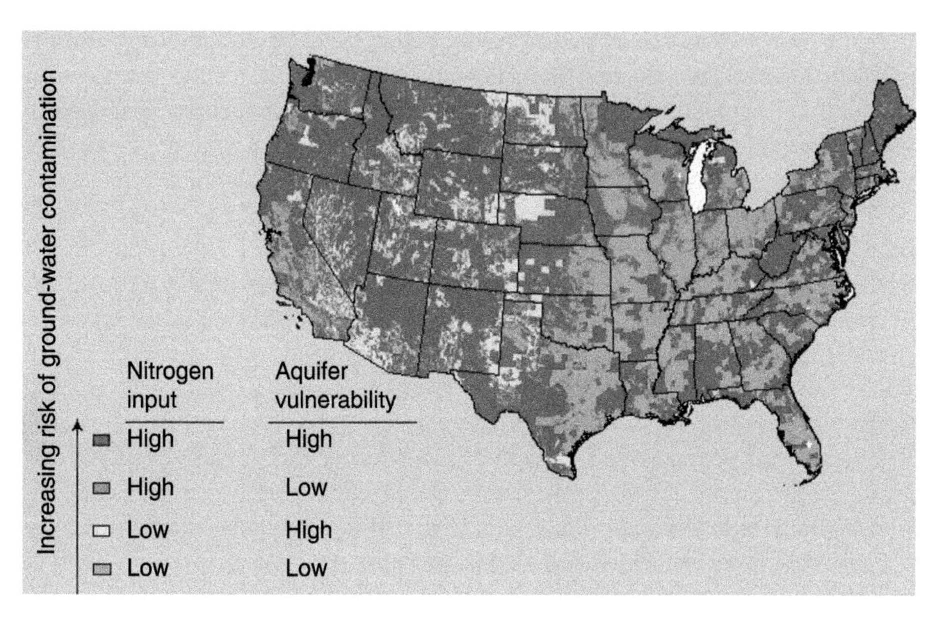

Figure 10.12 Risk of groundwater contamination in the USA. This map represents the risk of nitrate contamination in shallow groundwater. Areas shown in red have the highest risk of nitrate contamination based on three risk factors: the level of nitrogen input in an area (typically from fertilizer, livestock waste, or high human population density); the drainability of soils; land area devoted to cropland. *Source*: USGS National Water Quality Assessment website. http://water.usgs.gov/nawqa/nutrients/pubs/wcp_v39_no12/#FIG1 (accessed March 13, 2012).

rivers for irrigation significantly changes lacustrine and riparian environments (as discussed in Chapter 12, on water resources). Unlike water diverted for other purposes (such as power generation, or municipal or industrial use), which is eventually returned to the lake or river, water for irrigation is never returned to the system. One of the most extreme cases of a river and lake system impacted by water diversion for agricultural purposes is the Aral Sea in contemporary Kazakhstan and Uzbekistan. The two main rivers flowing into the Aral Sea, the Syr Darya and the Amu Darya, were both diverted by the Soviet Union to provide for massive irrigated cotton production in the Kyzyl-Kum Desert. The Aral Sea was once the fourth-largest lake in the world. By 2007, it was one-tenth its original size and had split into three separate lakes. By 2009, the smallest of these three lakes had disappeared and the second-smallest lake was nothing but a thin strip. The region's once prosperous fishing industry has been destroyed and dust storms are now a problem on the dry lake bed. A related problem is that many of the fields for which this water was used employed flood irrigation and were not particularly well drained. This led to the destruction of agricultural soils via a problem known as **salinization**. Salinization occurs when water evaporates and salts are left behind on a repeated basis. This gradually destroys the productivity of soils over time and is very difficult to reverse.

Maintaining water quality in and around industrial agricultural systems means better managing soil erosion, reducing water use, and curtailing the use and dispersal of pesticides. Soil erosion is just as much a problem in industrial agriculture as it sometimes is in traditional agriculture. The use of **contour plowing**, or plowing along the contour rather than up and down a slope, is one technique for reducing erosion. Wind breaks, planted along the windward side of fields, also help reduce wind speeds across fields and thus aeolian erosion. **No-till** and conservation tillage are also practices known to substantially reduce soil erosion, but no-till is a controversial practice which has recently been adopted on many farms in the Global North. The idea is to do away with aggressive tillage which destroys the soil's organic matter and loosens soil so that it may erode. Those practicing no-till use special seeders to plant seed directly and then often manage weeds through the use of herbicides rather than tillage. It is this last aspect, use of herbicides, which makes it a more controversial practice for environmentalists. Farmers will also plant **vegetative buffers** of bushes and trees along the shores of lakes and streams to reduce agricultural run-off into these waters.

Given that agriculture accounts for the largest share of water use in many countries, the more efficient use of this resource is key. Drip irrigation, while expensive to install, is one of the most efficient means of irrigation. Such systems deliver water via a tube to the base of each plant (as opposed to flood or pivot/sprinkler irrigation). The more widespread adoption of such systems might occur if the true cost of water (currently subsidized in many countries) were charged to farmers.

Finally, one could not conclude this section without noting that one of the biggest engines for change in industrial agriculture has been the burgeoning organic food movement in the Global North. While this is still only a very modest proportion of the total agricultural market, it is the fastest-growing segment. This has allowed for the flourishing of a number of smaller, niche organic producers.

Agribusiness and Government Agricultural Policy

The companies that supply inputs to farmers (equipment, seeds, fertilizers, and pesticides) and purchase, process, and market their outputs, also known as **agribusinesses**, are a powerful group that shapes markets and influences agricultural policy worldwide. It is these global entities, rather than farmers, which have tended to benefit from national and international policies which promote production over other factors in farming systems. While the profit margins earned by farmers have declined over time in most circumstances, the opposite has often been true for agribusiness.

The United States has long pursued what is sometimes referred to as a cheap food policy. In fact, relative to other industrialized countries, the percentage of income spent by Americans on food is amongst the lowest. One of the main advantages of a cheap food policy (from the perspective of business leaders) is that it reduces upward pressure on wages. Food prices have been kept low by encouraging overproduction. Subsidies and price support are some of the major government policies which have encouraged farmers to keep producing more in the face of adverse market conditions. Some inputs into the farming process, such as water and public grazing land in the western United States, have also been provided to farmers at below-market prices. The impact of these policies which encourage overproduction has been agriculture-related environmental problems and indebted farmers in the USA, and low global commodity prices which hurt farmers in other parts of the world.

Famine and Hunger

If you read reports and analysis from the news media, then you will know that famine and hunger in the Global South are often presented as environmental problems. These arguments suggest that famine is the result of a catastrophic natural event (such as a prolonged drought or locust infestation) which destroys local food production over a number of years. Hunger then ensues in response to this decline in production. In a similar way, recurring hunger is the result of inefficient food production techniques and the degradation of natural resources (particularly soil resources). Many geographers have long found such assertions to be problematic, arguing that famine is much more of a political than an environmental problem (Dando 1980; Watts 1983; Blaikie 1985). Furthermore, many geographers question whether conventional measures undertaken to reduce hunger, such as the Green Revolution approach to produce more food, will work because they don't address the root causes of hunger (Watts and Bohle 1993; Yapa 1993).

Let us first examine the difference between food security, food self-sufficiency, and food sovereignty, terms that are sometimes errantly used interchangeably. **Food security** has been defined as access to enough food at all times for an active and healthy life (World Bank 1986). The opposite of food security is food insecurity, which can be acute (famine), chronic (at low levels all the time), or seasonal (e.g., happening for a period every year before the new harvest). While food security may be obtained by a mixture of local production and food imports, **food self-sufficiency** more narrowly refers to producing enough food within a country's borders to satisfy the caloric needs of a population. Food self-sufficiency was an explicit national policy of many national governments in the 1960s and 1970s. Many governments sought to produce enough food within their borders for reasons of national security. While such an objective is understandable, it sometimes led to environmentally problematic agricultural endeavors such as irrigated wheat production in Saudi Arabia. By the 1980s, most food policy analysts argued that a food security approach (rather than food self-sufficiency) that involved some trade made more sense. This view held until about 2007, when the global food crisis, discussed at the start of this chapter, began to unfold. Now, because of the price volatility associated with global food supplies, many national governments are beginning to reconsider their decisions to move away from food self-sufficiency policies. **Food sovereignty** is similar to food self-sufficiency in some ways, because of a shared emphasis on local food production. A key difference is the means by which such food is produced. Food self-sufficiency initiatives are mainly aimed at boosting food production through high external input agriculture. In contrast, food sovereignty calls for increasing local food production through locally accessible and controlled technologies.

In order to understand why simply producing more food will not solve the world's hunger problems, it is crucial to be able to distinguish between the notions of food availability and food access. **Food availability** refers to the amount of food available on national markets. In fact, it used to be assumed by the UN Food and Agriculture Organization (FAO) that hunger would be avoided if the total amount of food produced nationally, plus net imports, was equal to or greater than the caloric needs of a population (Moseley and Logan 2001). But when the UN monitored national food stocks in this manner (known as the food balance sheet approach), they were taken aback by instances of mass hunger when their calculations indicated that there should not be a problem. This paradox is resolved once it is understood that food availability is not the same thing as **food access**, or people's ability to acquire the food that is available through purchase, gifts, or their own production. Amartya Sen, Nobel laureate in economics, is often credited with bringing this problem to the attention of those in the food policy community (Sen 1981). He conceptualized food access in terms of entitlements, or people's legal ability to access food. Sen observed that there was plenty of food on the market during most famines, with the real problem being a collapse of entitlements when food simply became too expensive, and therefore out of reach of the poor.

What the conceptual insights described above suggest is that solving the hunger problem is not just a question of producing more food as has been suggested by some (e.g., Borlaug 1995). Technocratic approaches that emphasize improved seeds (or Green Revolution approaches) are in some sense blind to scale. There is an assumption that a shortage of food at one scale (household, class, or community) implies that there are shortages at much broader scales (provincial, national, regional, or global). What technocratic approaches offer is a set of techniques for more production at these broader scales (by empowering wealthy farmers to produce more), but they don't address problems of access (or entitlement failure) at finer scales of resolution, such as poorer households and marginalized groups. Approaches that address hunger at this level often involve appropriate and accessible food production technologies, income-generation projects and cooperative grain banks. It is also worth mentioning that many technocratic approaches to resolving hunger through increased production are highly dependent on the availability of cheap fossil fuels. If we are coming to the end of an era of cheap fossil fuels, then this raises many questions about the long-term viability of industrial agricultural production as currently designed.

Chapter Summary

The study of agriculture and food systems has long been the bread and butter of human–environment geographers. This chapter described agricultural systems around the world as well as the basic principles of agroecology. Environmental problems related to traditional and industrial agriculture were examined as well as strategies for mitigating these issues. The chapter concluded by examining the influence of agribusiness and government policy on farming systems as different ways of thinking about the global hunger issue.

Critical Questions

1. What is the general influence of increasing population density on agricultural practices?
2. Why is it that more complex agroecosystems tend to have fewer pest problems?
3. Describe some strategies for addressing soil erosion in traditional agricultural systems.
4. What are some advantages and disadvantages of the cheap domestic food policy pursued by the United States?
5. Why is it that simply producing more food will not solve the global hunger problem?

Key Vocabulary

aeolian erosion
agribusiness
agricultural dual economy
agroforestry
bioaccumulation
biomagnification
broad spectrum pesticides
bush fallow rotation
circle of poison
concentrated animal feeding operation (cafo)
commercial agriculture
contour plowing
corporate farm
eutrophication
extensive agriculture
fallow
family farm
food access
food availability
food security
food self-sufficiency
food sovereignty
genetic escape
genetically modified (GM) seeds and crops
green manure
Green Revolution
high external input agriculture
hybridization

intensification
intensive agriculture
intensive traditional agriculture
intercropping or polyculture
low external input agriculture
monoculture
niche
no-till agriculture
pastoralism
parkland agroforestry
pesticide resistance
pesticide treadmill
persistent chemicals
plantation agriculture
R factor
salinization
shifting cultivation (aka swidden and slash and burn agriculture)
subsistence agriculture
urban agriculture
vegetative buffers

References

Annan, K. (2007) Remarks on the launch of the Alliance for a Green Revolution in Africa. World Economic Forum, Cape Town, South Africa, June 14.

Bayliss-Smith, T. (1982) *The Ecology of Agricultural Systems* (New York: Cambridge University Press).

Behnke, R.H., Scoones, I., and Kerven, C. (1993) *Range Ecology at Disequilibrium: New Models of Natural Variability and Pastoral Adaptation in African Savannas* (London: Overseas Development Institute).

Blaikie, P.M. (1985) *The Political Economy of Soil Erosion in Developing Countries* (London: Longman).

Borlaug, N. (1995) Mobilising science and technology to get agriculture moving in Africa. *Development Policy Review*, 13(2), pp. 115–29.

Boserup, E. (1965) *The Conditions of Agricultural Growth: The Economics of Agrarian Change under Population Pressure* (Chicago: Aldine).

Dando, W.A. (1980) *The Geography of Famine* (London: Edward Arnold).

Jordan, C.F. (1998) *Working with Nature: Resource Management for Sustainability* (London: Hardwood Academic Publishers).

Liu, Y.-B., Tabashnik, B.E., Dennehy, T.J., Patin, A.L., and Bartlett, A.C. (1999) Development time and resistance to Bt crops. *Nature*, 400, pp. 519 (doi:10.1038/22919).

McAfee, K. (2003) Corn culture and dangerous DNA: real and imagined consequences of maize transgene flow in Oaxaca. *Journal of Latin American Geography*, 2(1), pp. 18–42.

Miller, S. (2007) Epilogue: Cuba's latest revolution. In *An Environmental History of Latin America*, pp. 229–336 (London: Cambridge University Press).

Mortimore, M. and Tiffen, M. (1994) Population growth and a sustainable environment. *Environment*, 36(8), pp. 10–20, 28–30.

Moseley, W.G. (1996) A foundation for coping with environmental change: indigenous agroecological knowledge among the Bambara of Djitoumou, Mali. In W.M. Adams and J. Slikkerveer (eds.), *Indigenous Knowledge and Change in African Agriculture*, pp. 11–130. Studies in Technology and Social Change 26 (Ames: Iowa State University).

Moseley, W.G., Carney, J., and Becker, L. (2010) Neoliberal policy, rural livelihoods and urban food security in West Africa: a comparative study of The Gambia, Côte d'Ivoire and Mali. *Proceedings of the National Academy of Sciences of the United States of America*, 107(13), pp. 5774–9.

Moseley, W.G. and Jordan. C.F. (2001) Measuring agricultural sustainability: energy analysis of conventional till and no-till maize in the Georgia Piedmont. *Southeastern Geographer*, 41(1), pp. 105–16.

Moseley, W.G. and Logan, B.I. (2001) Conceptualizing hunger dynamics: a critical examination of two famine early warning methodologies in Zimbabwe. *Applied Geography*, 21(3), pp. 223–48.

Pimentel, D., Hurd, L.E., Bellotti, A.C., Forster, M.J., Oka, I.N., Sholes, O.D., and Whitman, R.J. (1973)

Food production and the energy crisis. *Science*, 182(4111), pp. 443–9.

Richards, P. (1985) *Indigenous Agricultural Revolution: Food and Ecology in West Africa* (London: Hutchinson).

Sen, A. (1981) *Poverty and Famines* (Oxford: Clarendon Press).

Thompson, C.B. (2007) Africa: Green Revolution or Rainbow Revolution? *Foreign Policy in Focus*, July17.

Tripp, R. (2001) *Seed Provision & Agricultural Development* (London: Overseas Development Institute; Oxford: James Curry; Portsmouth, NH: Heinemann).

Watts, M. (1983) *Silent Violence: Food, Famine and Peasantry in Northern Nigeria* (Berkeley: University of California Press).

Watts, M. and Bohle, H. (1993) The space of vulnerability: the causal structure of hunger and famine. *Progress in Human Geography*, 17(1), pp. 43–67.

World Bank (1986) *Poverty and Hunger: Issues and Options for Food Security in Developing Countries* (Washington, DC: World Bank).

Yapa, L. (1993) What are improved seeds? An epistemology of the Green Revolution. *Economic Geography*, 69(3), pp. 254–73.

11

Biodiversity, Conservation, and Protected Areas

Icebreaker: Jaguar Habitat

The 1996 sighting of a rare jaguar (Panthera onca) in southeastern Arizona created excitement. Local conservationists delighted in the news, given that the more common sight was of old jaguar pelts, found in small ranch cabins in Arizona and on the other side of the international border in Sonora, Mexico. Was this native predator making a comeback? And why was there, relative to the Arizona news, so little interest on the Mexican side of the border?

An Introduction to Human–Environment Geography: Local Dynamics and Global Processes,
First Edition. William G. Moseley, Eric Perramond, Holly M. Hapke and Paul Laris.

Published 2014 by John Wiley & Sons, Ltd.

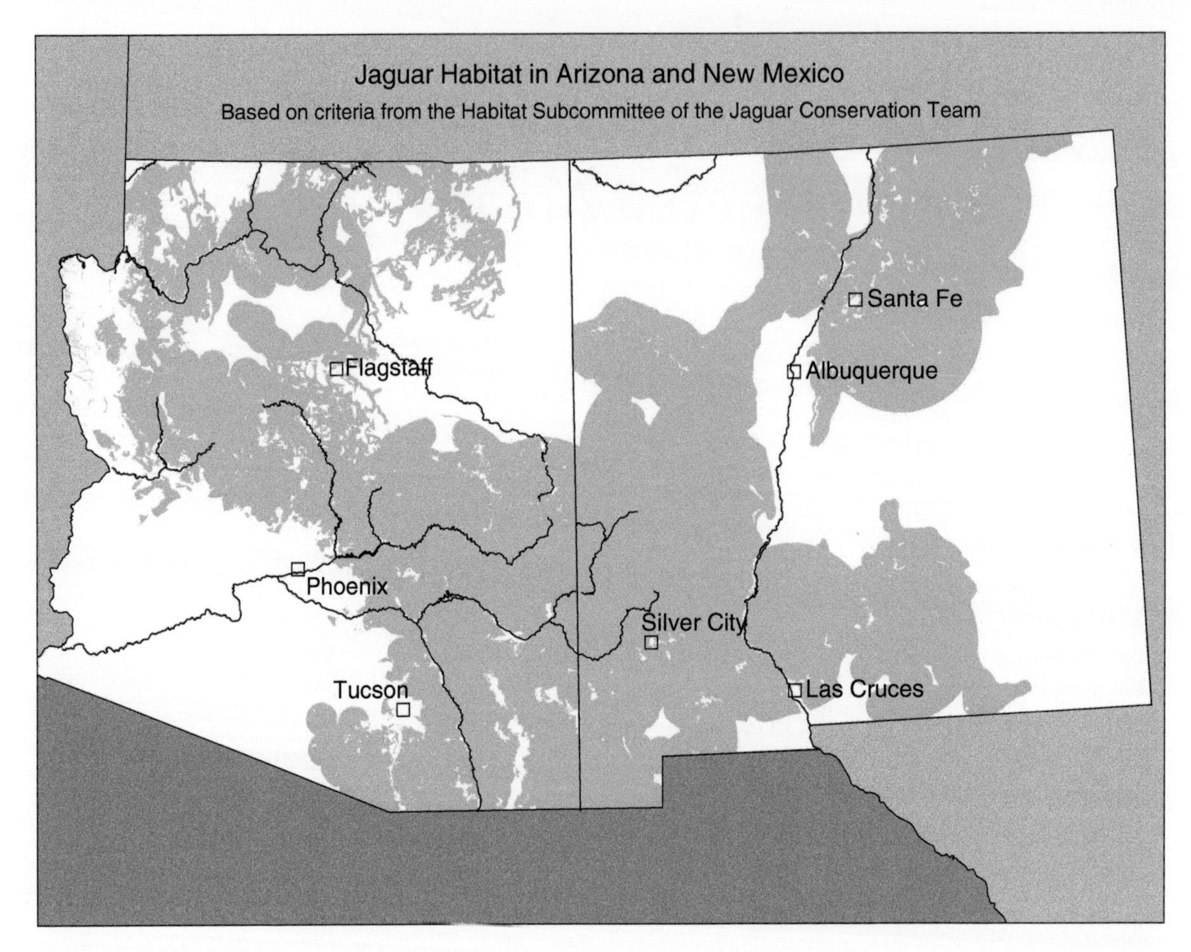

Figure 11.01 The hypothetical "range" of the jaguar in the past, based more on potential habitat range than accurate historical records of jaguar sightings. *Source*: http://jaguarhabitatusa.files.wordpress.com/2008/06/jaguar_az26nm1suitable_pp.jpg.

For conservation organizations in the United States, the jaguar was a sign of hope, that a species recovery was in effect and that jaguars were returning to the northern fringes of their original **distribution range** *and territory (see Figure 11.01). As a tropical species, jaguars were never plentiful in what is today the American Southwest. In rural Mexico, however, there is little enthusiasm for jaguars because of the practical realities of keeping sheep, goats, and cattle alive. The jaguar is not warmly received as a biological partner in the Mexican landscape.*

This transnational narrative of the jaguar is only one example of how complicated biodiversity conservation can be in real life. We value certain species over others, conservation is a cultural value that varies, and there are no simple solutions to trying to balance biological needs with human livelihoods. What underlies this story, however, is the question of who controls nature. These are matters of **environmental governance**, *or who decides conservation outcomes, who manages and for what purpose, and how do we do this without trampling on local peoples.*

Chapter Objectives

The objectives of this chapter are for the student to be able to:

1. Appreciate the importance, distinctions, and terminology associated with biodiversity.
2. Learn key aspects, along with illustrative case studies, of conservation and preservation efforts worldwide.
3. Be able to evaluate the environmental and policy differences of establishing protected areas in a number of national and regional settings.
4. Understand how human landscapes and environmental settings are strongly intertwined.
5. Contextualize conservation that stretches across multiple jurisdictional and political boundaries.
6. Apply critical thinking and new concepts to conservation areas, examples, and tools in their own home region.

Introduction

The main goal of this chapter is to explore different conceptions of biodiversity, the reasons we care about it, and how humans seek to protect it. After reviewing the concept and import of biodiversity, and key ecological ideas related to this term, we examine the complicated history of biodiversity conservation. We also, at a more concrete level, review examples or case studies of national and international policies and programs designed to protect biodiversity.

Biodiversity and Habitat Conservation

What do we mean by **biodiversity** and what are the concerns over biological declines? Scientists and conservation organizations are concerned about the worldwide decline in biodiversity, the world's biological diversity. Outright disappearance of particular species, or **extinction**, has also been common. Indeed, most scientists refer to our current era as the "sixth wave of extinction," in that the rate of species disappearance is hundreds of times faster than natural, or background, extinction. There have been five prior **mass extinctions**, where a major percentage of all life on Earth disappeared because of a natural event. These past episodes are typically attributed to a variety of factors, such as climate change, or asteroid strikes on the Earth. This current sixth wave, however, has been principally attributed to humanity. So not only are humans engaging in a giant biochemistry experiment with the atmosphere, but we may be fundamentally altering genetic and evolutionary pathways for whole sets of organisms on Earth. But the concern for biodiversity reaches well beyond single species. Typically, scholars of biological diversity think of three levels of diversity that are of concern: **genetic**, **species**, and **habitat**. These are also scaled to the level of concern, so that genetic diversity is important at the micro-scale, when planning for the conservation of a single group or sub-population of a single species. Is there enough genetic diversity, for example, within a species, to sustain the longevity of the particular population in question?

What conservation biologists call **genetic bottlenecks** occur when a lack of genetic diversity is apparent in a sub-population, such as the lions of the Ngorongoro Crater in East Africa. Through long-term isolation and disconnection from other groups of lions, this small sub-population finds itself in a genetic bottleneck – they may not have the **minimum viable population** (MVP) to sustain genetic diversity in their group alone for long-term survival. While this need not imply doom for the entire species in question, this varied genetic vulnerability, if repeated in multiple places, can place genetic stress on the populations of lions in the larger region. This process of isolation can occur naturally, for a variety of reasons. But it can also be caused by humans: for example, deforestation can segment large areas of forest into smaller patches, a process known as **habitat fragmentation**. This example also illustrates that genetic and spatial isolation are often related: **conservation biologists** are scientists interested in connecting populations in larger habitats to ensure that genes and species are conserved over the long term.

At the global scale of diversity, conservationists and scientists are concerned with the diversity of habitats, landscapes, and ecosystems available for all species in any given region. There is also an ongoing debate about whether biodiversity of ecosystems creates more ecosystem stability or, alternatively, if this is simply an accident of environmental history in certain places. Biogeographers and conservation biologists, in particular, are concerned about this scale of diversity as a whole, even if individual scholars devote attention to smaller-level studies that involve a particular species or sets of species in a landscape. And this attention to the macro-environment is rightly reflected in the changing priorities of conservation groups worldwide (see more below). No longer are conservation groups only interested in the preservation of a single species. Rather, they have realized the importance of the whole landscape, of conserving the entire set of landscapes. Saving a tree, in other words, helps little if the entire forest is burning.

Biologist Norman Myers first coined, and made widespread, the term **biodiversity hotspot** (Myers et al. 2000), referring to a set of biological environments that were home to a remarkable diversity of organisms that were also predominantly endemic (Figure 11.02). The species and the ecosystems that hosted them could not be found elsewhere, and for this reason Myers wanted particular attention drawn to these sets of ecosystems that are under direct human threat and difficult to reproduce elsewhere. This is a clear example of where the human–environment perspective is useful. It is clearly a human construct to designate one area as a "hotspot" even if there are clear criteria and reasons for doing so. It is also a perfect

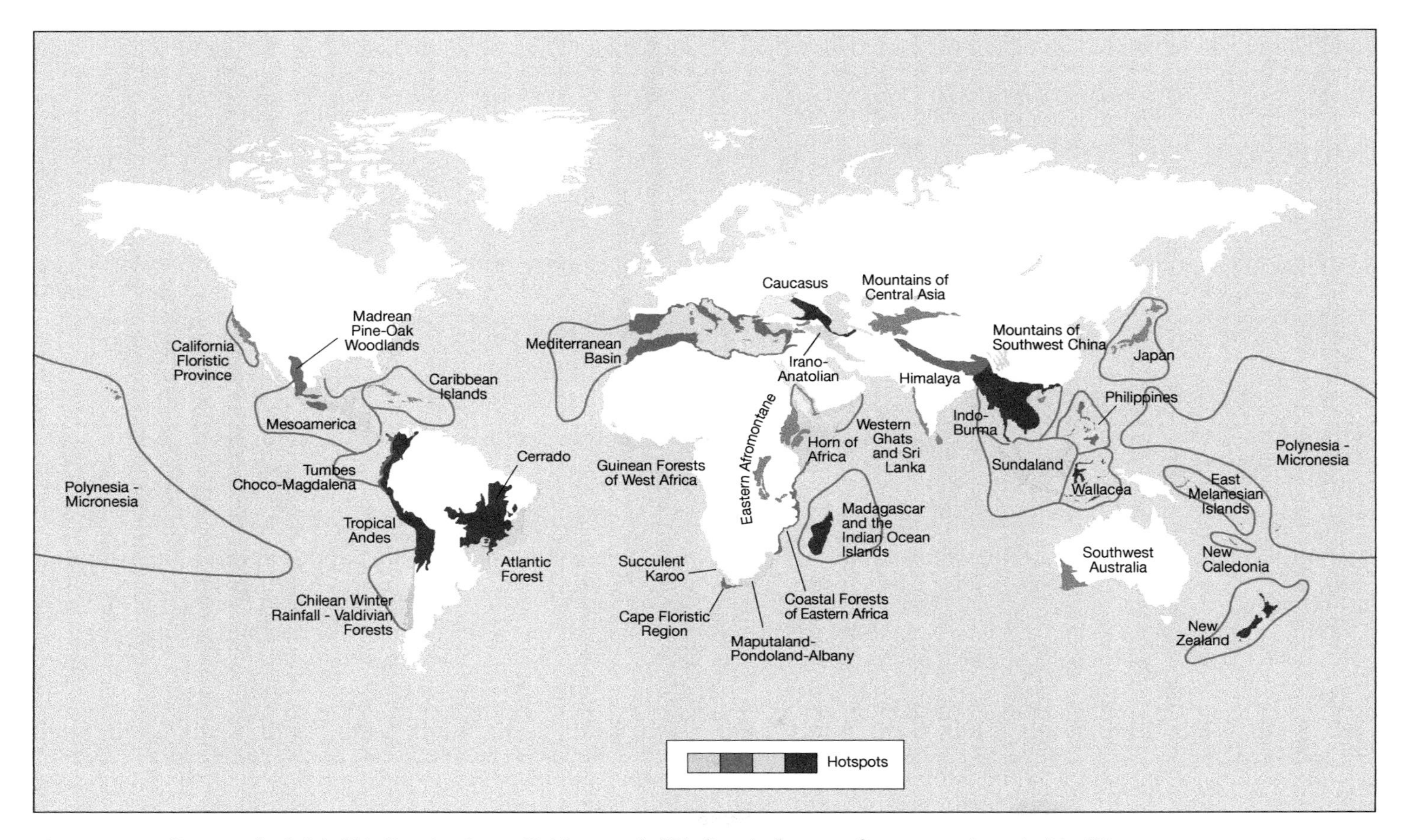

Figure 11.02 "Hotspots" of global biodiversity. *Source*: N. Myers et al., "Biodiversity hotspots for conservation priorities," *Nature*, 403, pp. 853–8. Nature Publishing.

fusion of matters that seem biophysical with the concerns of conservationists, a direct confrontation of why humans in particular environments face challenges.

How does biodiversity align or clash with our own quite human efforts to conserve, use, or set aside natural areas for protection? If species exist and function on a local scale, the efforts to preserve both rare endemic species and important groups of **keystone species** (see Chapter 3), biological individuals that are vital to entire habitats, are increasingly global in nature.

Conservation and more strict **preservation** efforts are still focused on creating **protected areas** for regions and species of interest. Biodiversity is important for several reasons, some fundamental to human livelihoods, but also because the concept gives other species consideration for conservation efforts and protected area designs. In terms of setting aside areas for conservation, conservation planners like to talk about centrally important ecosystems that can serve as **cores** to overall plans, and multiple areas of overlapping **buffer zones** where some human use or intervention is allowed. This language survives in most park designs or conservation discussions. One of the central goals is conservation area design is to minimize the so-called **edge effect**, decreasing the exposure of native species and habitats to outside influences such as natural hazards or even non-native species.

Species diversity relates to the larger association of mixed species, and how diverse or concentrated species arrangements tend to be within any given ecosystem, such as a savanna or a wetland environment. As you may recall from Chapter 3, species diversity correlates nicely with latitude, and there is higher species diversity in the tropics than in the higher latitudes. The same is also true with elevation changes, as the number of overall species generally declines as one goes up in elevation, although there are a few exceptions. For example, in tropical climates, the addition of topography and elevation can locally increase the level of diversity. So-called **cloud forests** display remarkable levels of species diversity, much of it based on the variety of niches and microenvironments available for different plants, animals, and insects. Biodiversity and complex ecosystem also produce what have been termed **ecosystem benefits** (or ecosystem value). In contrast to human-centered notions of benefit, ecosystem benefits are the collective ecological values we derive from clean water, clean air, and complex and intact ecologies. Wetlands, for example, filter toxins carried in by flooding rivers. Good stream quality supports healthy fish populations and the wildlife that depends on surface water. Trees take in atmospheric carbon dioxide and produce oxygen, a benefit for the entire biosphere. These benefits are hard to value quantitatively, but many ecological economists are trying to estimate "how much" these ecosystem benefits are worth to humans.

The current, or potential, **economic value** of biodiversity to humans is a more common way of valuing natural ecosystems. For example, roughly half of our current pharmaceuticals and medicinal products are based on natural compounds derived from plants. So in this case, a particular species loss may directly impact a future cure, an unknown resource use value which humans could therefore not enjoy. Humans potentially lose out on all of the possible future uses of species if they are allowed to become extinct. That possibility is increased if we continue with habitat fragmentation, the spatial dismantling or division of natural habitats, in areas of high biodiversity. But there are other possible consequences to habitat and biodiversity decline, some of which directly impact humans. The recent growth in **ecotourism**, tourism based on natural environments, could be affected if those habitats or landscapes were suddenly or even incrementally altered. If a company promoting forest canopy tours no longer has a set of forest canopies to show because of localized or regional deforestation, its efforts may have limited economic benefits for local communities. If you were on a guided tour of, say, the sub-tropical Chaco region of South America, and you found large clearings of new agricultural fields of soybeans, this would obviously impact your perception of the value of this area as a natural habitat.

But what about the **intrinsic value**, those non-economic "values" of nature and other species? Even if you do not have a national park next door, surely you have strolled through a green area, even a city park, and felt some sense of ease and peace. This is "intrinsic" since it is difficult to put an outright economic "price" on the value of this experience. But the argument about intrinsic value can be taken even further, away from the notion that humans are the most important species. Doesn't every species deserve the right to persist? While this form of **biocentric** and philosophical thinking is now more common in the more spiritual aspects of conservation as a form of cultural, even religious, stewardship, the emphasis on why humans "need" nature to be complete cannot be dismissed. Literate and oral-historical traditions exist across most societies and nations to support the importance of what biologist E.O. Wilson has called our propensity for **biophilia**, a natural draw and attraction to life itself, a fundamental feature of human character and human nature. So although biodiversity and conservation have been constructed as or understood to be scientific pursuits, the humanists, poets, and eco-critics have much to say. Or to be blunter about it, nature is too important to be left only to natural scientists; we need writers, critics, and poets to describe much of the inherent and apparent beauty in terms that do not neutralize or objectify nature to a simple subject of science.

Our ideas about nature, about protecting sets of environments and landscapes, have greatly changed within our own cultures (see Chapter 2). These are not fixed philosophies of saving the Earth, however that earnest and well-intentioned notion might be translated into your own way of thinking. So the changing definition of "conservation" or "preservation," over time is critical to understanding these changing stances about protecting particular areas, species, and landscapes. Even if it's clear that biodiversity matters, how we maintain a set of ecosystems and biomes that have ecological integrity is not always a product of perfect knowledge. Robert Wilson's (2010) *Seeking Refuge* illustrates the complicated notions of how some areas become valued, over time, for different purposes (see Box 11.01).

Box 11.01 Conservation is for the birds

Seeking Refuge examines the challenges of conserving migratory animals in the working landscapes of western North America. Migratory waterfowl – ducks, geese, and swans – have lengthy migration routes stretching from the Arctic all the way to southern California (and even Mexico). Yet, over the past two centuries, people have destroyed the wetland habitats they need and divided the landscape into a grid of private and public lands. Robert Wilson's book tells the story of how Americans and Canadians carved out refuges in this grid for the birds to survive. It also examines the challenges of managing animals in working landscapes – places intensively used by people for non-conservation purposes such as irrigated industrial farming. Although the book is an environmental history, it has relevance to contemporary conservation issues today. If migratory animals and other species are to survive, we must manage them in these working landscapes. This book looks at the century-long history of how refuge managers, bird lovers, and environmentalists have sought to sustain these magnificent birds in some of North America's most intensively used landscapes.

A fundamental dimension to biodiversity is that not all biodiversity is equal. Biologists and ecologists argue that biodiversity should be reflective of the site or location in question; in other words, we want a high diversity of the species that are from that area. Species **native** to a particular region and that don't occur elsewhere are called **endemic**. Why is this a challenge or sometimes problematic? For one, species can colonize and spread rapidly or slowly. While plants may need decades to colonize an island, for example, some forms of life like spiders and birds have more rapid rates of spread. Perhaps the most famous historical experience we have on record is the eruption of Krakatau in the Indonesian Islands. The explosion in 1883 wiped clean an entire ecosystem, as if a "reset" button had been pressed, and the recovery of those volcanic remnants gave observers the chance to record what returned to the islands and in what order. Small size and the ability to quickly disperse were clear advantages for the early colonizers of what was left after the eruption. But an imperfect inventory of what used to be on the island, versus colonizers to the clean-slate island, made it challenging to distinguish previous inhabitants from new ones.

This means that it is difficult to ultimately conclude what species are native to the area in question versus those that are considered **exotic** (those from elsewhere, see Figure 11.03). Complicating the picture further is that humans are quite good at introducing or accidentally bringing over species in boats, on planes, even on the soles of our shoes. This biotic baggage follows us and is one of the principal reasons why many early explorations of islands led so quickly to native species declines. New biological arrivals, like rats, made quick work of ground-birds who did not fly, and eliminated smaller mammals, even if their introduction was accidental.

Figure 11.03 A cultivated buffelgrass pasture in northern Mexico. This exotic and invasive grass has been widely planted in the US–Mexico borderlands. It is at once loathed in the US conservation community and appreciated by ranchers in Mexico. *Source*: Eric Perramond.

What makes an exotic species an **invasive** one? Scientists attempt to specify the effects of detrimental exotic species by distinguishing organisms that lack direct competitors in new settings that multiply aggressively and quickly, or are remarkably general in their demands. If we think of the typical "weed" in a grass lawn, such as dandelion (*Taraxacum officinale*), the organism is remarkably capable of quickly dispersing and can survive even repeated chemical treatments, having few natural predators. Without competition, or similar species sharing the same resources, a new species can thrive in a new ecosystem setting. Species that have high reproduction rates, high dispersal rates, and with limited predators can simply overwhelm local ecosystems in a short time-span. And new arrivals that have a wide range of tolerance for temperature, diet, and water needs are often termed **generalists** and can survive and succeed in a variety of settings. This largely explains the success of the common dandelion in most of our lawns. **Specialists** to a particular environment, then, have less tolerance for being moved to other settings.

Not all exotic species rapidly and aggressively colonize. There are some plants, animals, and insects that gradually can fill in unoccupied or less competitive ecological niches. If they are not problematic for existing (native) species, they can be termed **naturalized** in many of these new ecosystem settings. It is the aggressive and displacing exotics that are, and are viewed as, truly problematic.

Buffelgrass (*Pennisetum ciliare*) is an exotic, invasive species that was first introduced into the United States in the early 20th century as a potential soil stabilizer and a potential rangeland grass (seen in Figure 11.03). Quickly adopted by many ranching outfits, buffelgrass was perceived to be a boon for restoring overgrazed grasslands in semi-arid environments. In Sonora, Mexico, ranchers continue to plant buffelgrass widely in pastures that are less than 1200 meters in elevation as a complementary feed resource during the dry season (February through June). Just across the international border in Arizona, however, local officials, residents, and conservation groups are frantically trying to remove buffelgrass invasions from roadsides and slopes where the dry grass can burn out native Sonoran Desert species.

The differences in how each species is valued are plain to see in news stories. On the US side of the border, buffelgrass is an enemy, a spreading scourge that invades native pastures and cactus habitat principally through fire. It is viewed as a threat to regional (Sonoran Desert) biodiversity because of its aggressive characteristics, its ability to range far and wide from its original point of origin. In northern Mexico, however, buffelgrass found a ready and welcoming home on private and communal grazing lands (Brenner 2011). There was no dominant concern about this species spreading; it is encouraged by ranchers. Poor, landless, Mexicans in fact collect the seed from buffelgrass right along the main highway between Nogales and Hermosillo, Sonora.

The portrayal of the villainous buffelgrass in Arizona was in contrast to local Mexican use and acceptance of this exotic invasive for use in pasturing. A Tucson-based non-governmental organization went so far as to put out a "Wanted dead" poster with a drawing of buffelgrass on it (Perramond 2010). So even if conservation efforts are now global, national, cultural, and economic borders still reveal stark differences between peoples and their priorities. Next, we turn to how early nation-state efforts at conservation shaped later global concerns about biodiversity, and then use several concrete examples to explain the continued relevance of national efforts.

Nation-State Efforts at Conservation

Early nation-state efforts at conservation were commonly focused on **sovereign resources**, such as forested areas. Even in the late 19th century, efforts to "conserve" natural resources were for future use, even if there was also some concern with preservation, or setting aside areas for "nature itself" without much human interference. But the early period of conservation clearly focused on the human uses and values for natural resources. Today, this worldview is reflected when agencies and parks argue for **multiple uses** in conservation areas. We have discussed this earlier (see Chapter 5), in the example where a colonial power invented the idea that there must have been a forest in the past but where evidence of such past forests was scant or non-existent. But whether couched in conservation or preservation language, these early efforts largely viewed locals as having eliminated forest in the past or recently, and therefore management was necessary to restore forests to some previous state (see Davis 2007). The dominant concern was for standing timber, preserving watersheds (with forest), and setting aside landscapes for their aesthetic value. Only later in the 20th century was there any momentum for couching conservation in strictly biological or ecological terms. Almost every country with environmental legislation now has some form of designation for protecting species of interest.

Created in the midst of late 20th-century environmentalism, the US **Endangered Species Act** (ESA) of 1973 was ground-breaking legislation for the protection of species vulnerable to extinction or in severe decline. Currently, more than 1,200 animal species and some 750 plant species are listed under the ESA as either endangered or threatened (a lesser classification). The difficulties of achieving meaningful protection under the ESA, heightened by sometimes local opposition to new species listings, illustrate how thoroughly human (political in this case) the criteria and chances for listing can be in the long run. Yet these species and habitats fall under different agency jurisdictions and missions. The terrestrial species in the United States, for example, are administered and protected by the US Fish and Wildlife Service (FWS), while the National Oceanic and Atmospheric Administration (NOAA) handles oceanic fisheries and aquatic species. The larger conception of species "protection" is also inherently geographic: the Fish and Wildlife Service is under the Department of the Interior, while the NOAA is in the Department of Commerce. Most countries, of course, have their own versions of protecting valued species, and many of these policies share similar language. Since 1973, the Convention on International Trade in Endangered Species of Wild Fauna and Flora (**CITES**) has been in force for signatory countries. And since 1993 the larger framework of the **Convention on Biological Diversity (CBD)** has operated as even larger, more inclusive umbrella for countries addressing protected areas.

Within the context of the United States, the turn from species protection to habitat conservation plans has been notable, even if the ESA has remained a powerful guide and constraint for setting specific conservation goals. Variations of **habitat conservation plans** (HCPs) exist, and California created an interesting framework in 1991 with the passage of the Natural Community Conservation Planning Act. Currently, areas and species habitats viewed as critical can be protected through a concerted public–private partnership in which the California Department of Fish and Game works to provide guidelines for private landowners and developers. These efforts in many ways parallel international schemes for "conservation with development," as it is often referred to in the literature. While critics have noted the potential oxymoron of combining residential or commercial developer schemes with conservation habitats, and that ultimately conservation has not been effective enough in these designated areas, some 7 million acres are under NCCPs in California. It is laudable that these forms of locally settled conservation plans involved a wide range of people who have an interest in these plans, often termed **stakeholders**. Critics note, however, that these HCPs may ultimately put at risk the very species that are of concern if the plans are too flexible for locals.

Most conservation organizations and efforts, however, continue to focus on the terrestrial ecosystems we depend on as a species. Still, the bias towards forest conservation as opposed to, say, grassland, wetland, or peri-urban conservation is notable for its longevity. Even less attention has been focused on marine resources, and the tricky dimension of conserving biodiversity in aquatic environments (see Chapter 12 for more).

Although Chapter 3 made clear that most environments share a variety of habitats, many can be grossly characterized by their dominant life-forms, such as vegetation, and by their climate. Until recently, ecosystems such as large tracts of wild grasses and areas of extensive wetlands have rarely been designated as protected areas. While most countries in the developed world now have some form of protective designation for wetlands or grasslands, with biome- or ecosystem-specific labels, the dimensions and resources for set-asides are limited. Nevertheless, there are both "national grasslands," though these are minor in extent, and more commonly land trusts used to protect grassland environments.

To use one well-known example, the United States does have a National Wetland Inventory (NWI), a set of large-scale, fine-detail, maps that provide a good spatial resolution. Yet even the resolution on these wetlands, as they appear on the maps, belies the difficulty in designating an area as a wetland. And many areas that were seasonally, or even permanently, inundated have been drained. This is as true in high-altitude environments such as Colorado as it is for the more extensive lowland wetlands of the Gulf Coast region of the US. At this regional state scale, Colorado law distinguishes between wetlands that are connected to water bodies and those that are seemingly disconnected from water tables and aquifers. Just as challenging and important are the increasing number of protected areas and parks close to cities. How can we protect or maintain areas that are highly used by the public for social recreation? The extensive trail system in the city of Colorado Springs is a challenge to maintain, to name but one example of the peri-urban conservation challenge, and the soils are highly prone to wind and water erosion (see Figure 11.04).

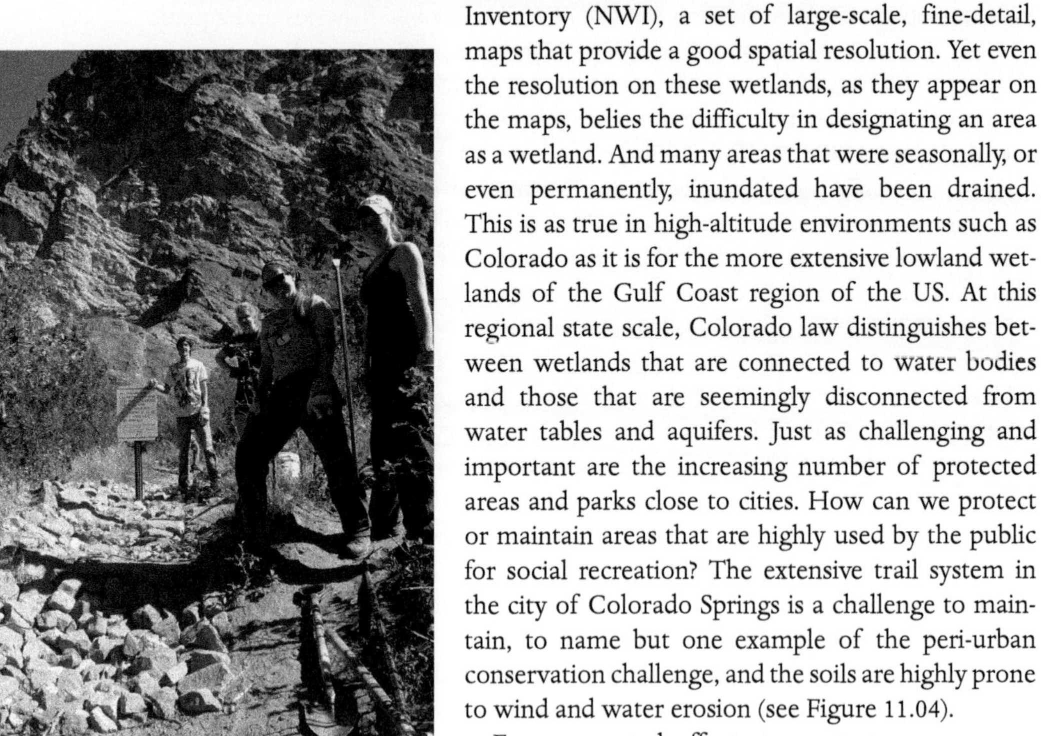

Figure 11.04 The long-term degradation of intensively used municipal parks can leave soils prone to entrenchment and severe erosion. Here, a restoration crew from Colorado College's environmental management class works to armor a soil gully formed by heavy trail use and highly erodible soils. *Source*: Eric Perramond.

Few concerted efforts to protect open oceans were made prior to the late 20th-century environmental movement. Oceanographers and marine biologists have focused attention, during the last decades, on the worldwide vulnerability and decline of near-shore habitats such as coral reefs. These are of particular concern because of their remarkable marine biodiversity, and the role they play for hundreds of organisms dependent on the particular symbiosis of algae and polyp for not only habitat but reproduction.

National Protected Areas in the Pyrenees

In the Pyrenees, Andorra, France, and Spain all manage protected areas for similar purposes, yet designate different areas with categories that involve varying levels of biological protection and human presence. There are no (American-style) wilderness areas in the Pyrenees, a notable difference in both material terms of landscape and in symbolic notions of whether humans belong in or are foreign to a particular environment. Human use is assumed, and is usually part of the official park "charters," rather than shunned (see Figure 11.05). In the central Pyrenees, in France, the Parc national des Pyrénées (created in 1967) has some 40,000 inhabitants in the area. These locals are also compensated in the language of **sustainable development** for direct aid to livestock support.

Figure 11.05 View of the town of Vernet-les-Bains, France, with the eastern boundary of the Catalan Pyrenees Natural Regional Park in the background. In contrast to a national park, the designation of a natural regional park entails preservation of the cultural and natural features of a region. The rules and regulations on biological protections are less strict. *Source*: Eric Perramond.

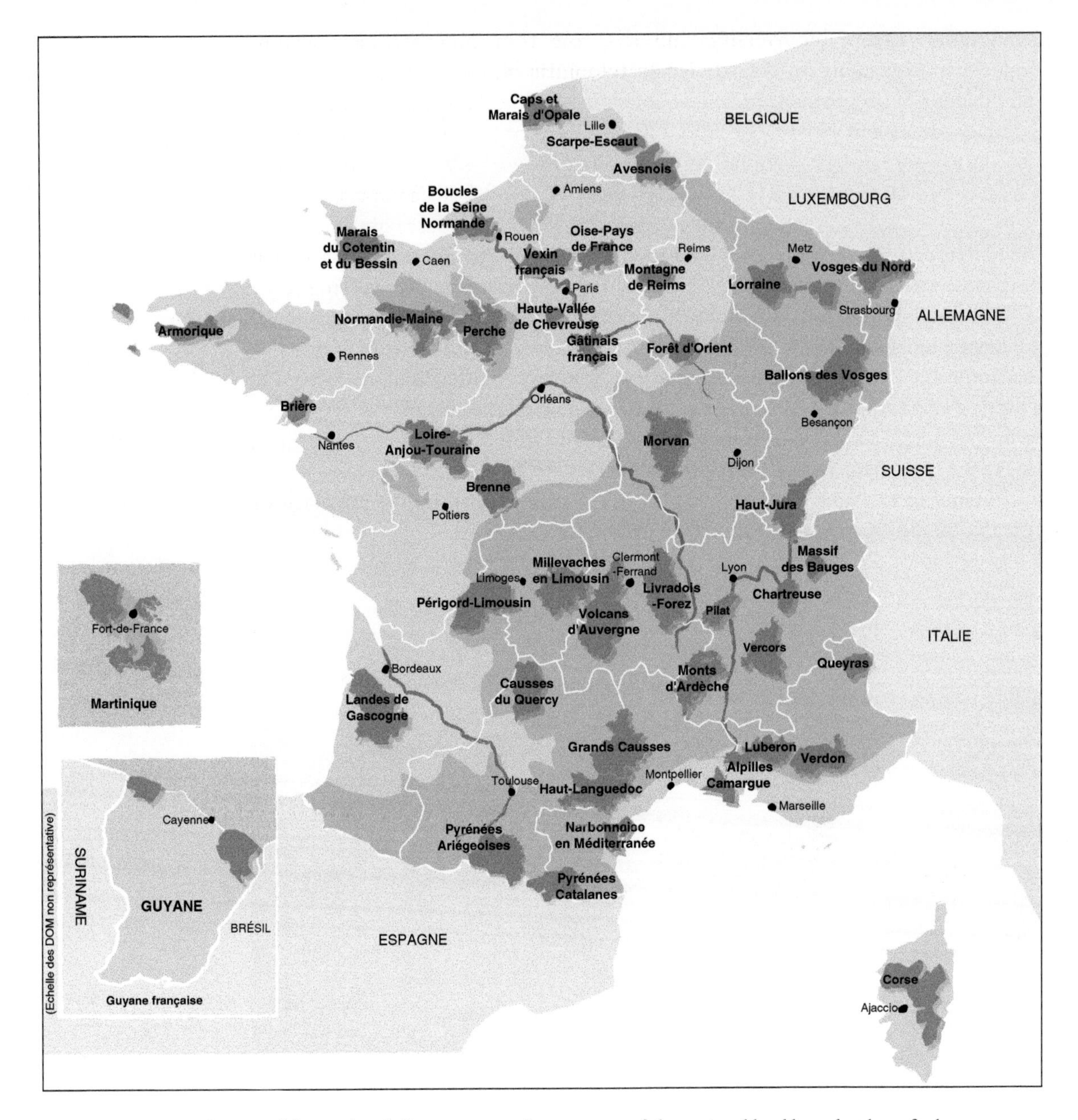

Figure 11.06 French regional "natural parks" occupy a good percentage of the national land base, but have far looser conservation rules than the national parks, which are far fewer. These regional parks reflect both culture and nature since they are designated for both scenic and productive consumption. *Source*: http://www.parcs-naturels-regionaux.tm.fr/fr/decouvrir/parcs.asp. Reprinted by permission of the Fédération des parcs naturels régionaux de France.

France and Spain have national parks, nine and 12 respectively, that enforce a higher level of protection to natural or biodiversity features. While Andorra has what it terms **natural parks**, as a principality it has no national parks. France also has what are termed **regional natural parks**, now numbering around 46 in total, and these explicitly recognize the human presence in scenic areas and acknowledge the importance of the local cultural landscape (see Figure 11.06). These latter natural parks also serve as conduits for local sustainable development initiatives, many based on local agriculture and heritage tourism. But are these biologically protected landscapes? In the eastern Pyrenees, locals were ambivalent about the new Parc naturel régional des Pyrénées Catalanes (Regional Natural Park of the Catalan Pyrenees) established in 2004. Since the park's establishment coincided with recent immigration of northern Europeans to the area, and newcomers have different notions of park protection and access rules, locals view conservation rhetoric with a cautious eye. They have embraced the notions that these are inhabited landscapes, that they express local heritage, foods, and commerce. Yet locals bristle at the notion of taking agricultural landscapes out of production, even if more recent northern European migrants interpret some Pyrenean landscapes as "wild" when they may be highly used and grazed by nearby villages.

In southern Spain, the Parque nacional de Sierra Nevada (National Park of the Sierra Nevada) also has a "parque natural" ("natural park") designation for the cultural component. The overall entity is also part of Spain's participation in the **United Nations "Man and the Biosphere" program** as part of its "Alpine" biogeography protection plan. Here again, we have visible fusion of a physical biome (Alpine setting) and a quite political construct (Man and the Biosphere) driven at an international level. And even with multiple designations and roles recognizing its natural international importance, the area does not exclude local peoples. If there is indeed a national park core dedicated to preserving the unusual endemics within the protected area, it is still an inhabited and human-used area where the "natural park" attempts to preserve the distinctive cultural landscape harking back to the Moorish period of rule in Spain. Thus, if the core of the park is for biodiversity protection, the periphery is for culture, the inhabitants who still make use of the high-altitude pastures and forests for plant collection (Wright and Campbell 2008).

Globalizing Conservation and Preservation Efforts

While globalization has been a term typically associated with the ever-increasing networks of human economies and cultural media, it is useful for discussing conservation organizations. After all, how would the reach of a fictitious "Local Wildlife Fund" ever improve if it was only local? Even conservation organizations have adapted to this new globalization of nature protection, although their names may not (yet) indicate their now far more systemic efforts (Sierra Club, Audubon, and World Wildlife Fund, for example).

As you read in Chapter 2, the terms conservation and preservation have changed over the course of the last 150 years. And the terms mean different levels of species and area protection in different countries and regional settings.

While the old debates may seem quaint and only "historical" in nature to you today, these environmental non-governmental organizations continue to compete for donors and dollars. Distinguishing between the organizations can be daunting, even if you are knowledgeable about ecological issues. When I compare the literature by conservation organizations placed in my mailbox, I ask: Are the messages from the Sierra Club the same as those of Audubon or the World Wildlife Fund? How is the message of the Nature Conservancy aligned with those groups? What are the major differences and similarities in their missions, pitches, and the imagery they use? If we compare their mission statements, you can see real similarities and differences. For example, the Sierra Club's mission statement reads:

> To explore, enjoy, and protect the wild places of the earth;
> To practice and promote the responsible use of the earth's ecosystems and resources;
> To educate and enlist humanity to protect and restore the quality of the natural and human environment; and to use all lawful means to carry out these objectives. (http://www.sierraclub.org/policy/)

If I compare this to Audubon's mission, which is "to conserve and restore natural ecosystems, focusing on birds, other wildlife, and their habitats for the benefit of humanity and the earth's biological diversity" (http://www.audubon.org/nas/), how am I to distinguish between these two groups?

Finally, the Nature Conservancy brochure claims that they seek to "preserve the plants, animals and natural communities that represent the diversity of life on Earth by protecting the lands and waters they need to survive" (http://www.nature.org/aboutus/howwework/). All of these environmental groups are seemingly earnest and engaged in meaningful conservation work. Note the similarity in their messages. Who would you give to in this case?

One of the central challenges all of these environmental NGOS (or ENGOs) face, however, is to work more effectively with local populations living near parks and protected areas. In regions that have been host to political and economic conflict, efforts such as **peace parks**, conservation areas that stretch across national borders, have been popular.

Transboundary Conservation Approaches

A number of **transboundary conservation** efforts, which involve managing areas that span political boundaries, can be found on the African continent. Several countries have attempted to implement international parks regardless of existing administrative boundaries, even if the politics of park management and the economics of associated ecotourism still remain a challenge for management authorities and the nation-states involved. A few of these have also been considered "peace parks," adopting language that has been in use at least since the 1920s.

The first, and commonly cited, example of a peace park was along the Canada–United States border in 1926 (Duffy 2001). So even if these peace parks are commonly interpreted as yet one more example of the globalization of conservation schemes, they exist and have to be treated with some serious consideration and analysis. Do they fulfill, or can they fulfill, human and biodiversity goals at the same time as the title of peace park implies? Critics of incorporating humans argue that efforts to protect biodiversity must be in areas largely free of humanity (Terborgh 1999). The impacts that established parks can have on local livelihoods of nearby residents become clear, say, when an elephant leaves a park boundary and crushes your household garden (Naughton et al. 2005).

A recent analysis by King and Wilcox (2008) has illustrated the problems of transboundary conservation in Africa, and contrasted those very mixed results with the challenges of conservation along the US–Mexico border for jaguar protection, our first example in this chapter. The establishment of transnational conservation schemes, such as the African examples, minimizes the political context of such efforts and may lead to localized domination by non-governmental conservation groups and idealists (see Figure 11.07). In the case of the Limpopo Transfrontier Park (LTP), a larger park shared between, and crossing the boundaries of, South Africa, Mozambique, and Zimbabwe, separate conservation areas were established by forcing out local populations only to, in theory, serve as some sort of solution to resettling refugees of nearby countries. And the new efforts in the LTP may, in fact, contribute to the continuing legacy of national parks of forcing locals out of management decisions, and entrusting conservation to the larger NGOs or creating the opening for such a framework.

In the US–Mexico boundary case, there is no transnational protected area, and increasing human immigration with subsequent militarization of the border seem to preclude a park. Efforts to conserve the jaguar are riddled with ambivalence and different cultural receptions for the animal, as we have seen. The net result of all this has been the resurgence of **public–private partnerships**, in this case taking the form of environmental NGOs buying up land or partnering with private landowners in the region. These NGOs are doing so to conserve the jaguar, but may not have the same concerns about what this privatization of conservation may mean to nearby residents on both sides of the border. Even native species can occasionally require "exotic" answers to conservation problems.

A more promising combination of protected areas, local peoples, and conjoining conservation with development is illustrated by Childs (2004, 2006). In this case, conservation efforts with park development are combined with strict attention to actual revenue distribution and possible democratization of the conservation practices in place. An elaboration of the earlier vision to combine conservation with poverty alleviation (Logan and Moseley 2002), this more recent set of ideas has clear merit for those critical of conservation without benefits, much less access, to protected areas. While the record of programs that combine conservation with local redistribution of benefits, such as CAMPFIRE in Zimbabwe, has been mixed depending on era and manager efforts, they have transferred as templates to

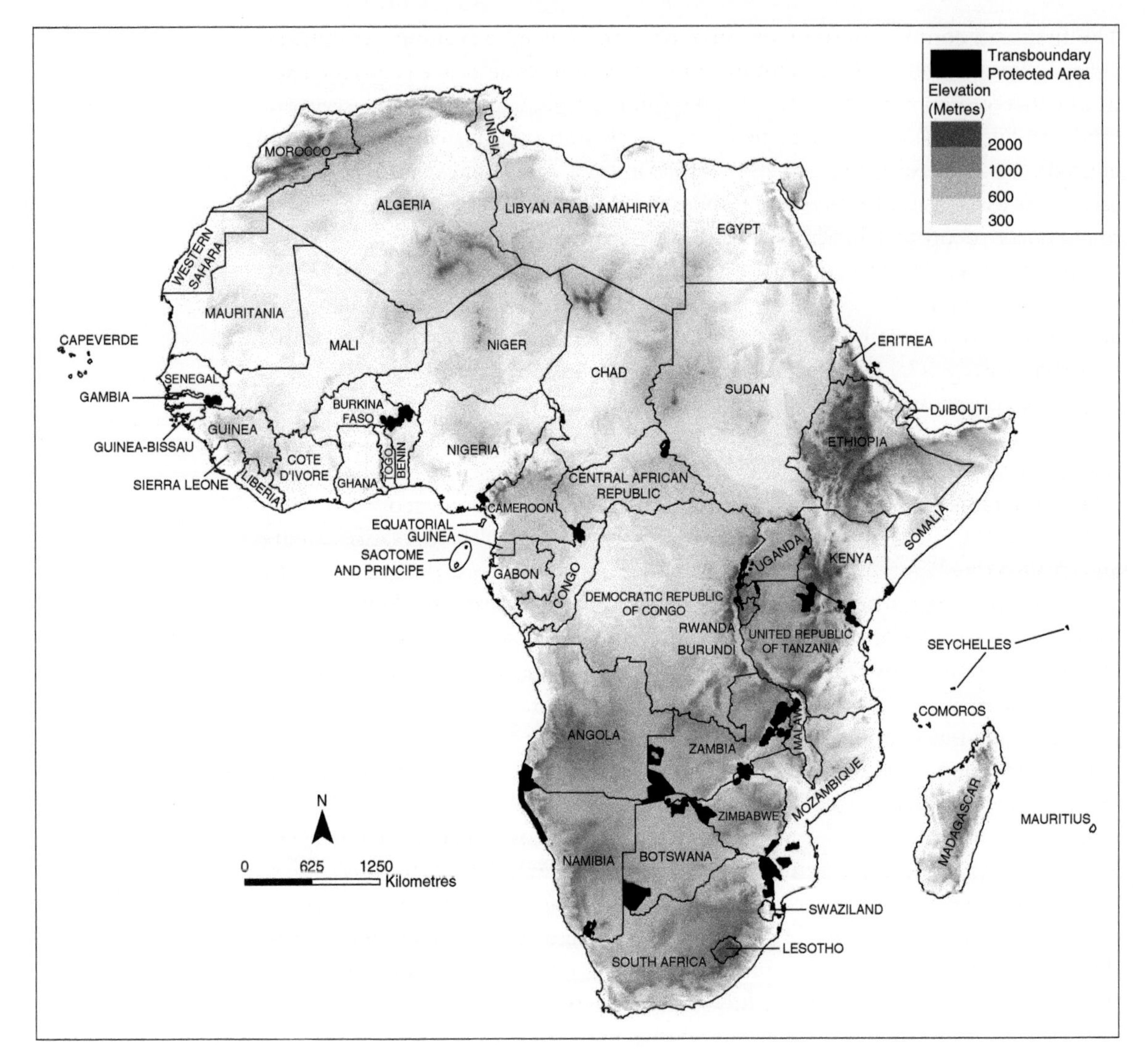

Figure 11.07 Transboundary protected areas, including so-called peace parks, on the African continent. *Source*: http://psugeo.org/Africa/African%20UNEP%20Atlas/transboundary-protected-areas.jpg. Reprinted by permission of UNEP.

other countries such as Zambia (Childs 2006). Protected areas, in other words, are being asked to perform multiple functions and not just preserve biological diversity. If local human needs are ignored, the results are predictably disastrous.

Cultural and Natural Biodiversity

Early work on conservation issues involving indigenous peoples stressed the spatial overlap of biodiversity with indigenous presence. We have gone well beyond this seeming coincidence. Local, if not indigenous, control over the shaping of biodiversity and the use of the nearby natural resources, continues to be a conservation issue. Clearly, most organizations dedicated to conservation do not want to create what Dowie (2009) has termed **conservation refugees**, people excluded from areas because of conservation efforts. But the dream of environmental conservation continues to play out in both developed and less developed areas of the world, even as local peoples contest park development as a form of exclusive practice by NGOs (Sundberg 1998; Heatherington 2010). This is not to say that the preservationist dream is dead. But it is clear that conservation organizations, and their overseers and critics, now pay far more attention to the people living in areas proposed or targeted for some form of conservation or protected area. And taking seriously the inhabitants and their ability to control their own destiny and identity is crucial to any real and sustainable notion of conservation in the long run, one that does not run afoul of ethical breaches with indigenous peoples.

One of the most notable areas for indigenous diversity is, of course, the Amazon Basin. A long record of human habitation, and increasing levels of public awareness, have paralleled the continued existence of large swaths of humid tropical rainforest. Even with continued incursions by logging companies, cattle ranches, and illegal gold mines, Amazonian tribes have adapted with the latest technologies and political tactics to survive into the 21st century. The Kayapo tribe of central Amazonia is one example of a native group that has struggled to maintain the cultural and physical integrity of their territory (see Posey and Balick 2006).

The Kayapo and many of the Amazonian tribes illustrate the struggles of indigenous peoples, to retain some degree of **territoriality**, or the control of original areas of settlement. Their access to resources is typically a priority, but in some cases, indigenous peoples may be more concerned about the impacts of their own popularity. In the high Himalaya, the overlooked impact of ecotourism is of particular concern to local residents, and while early efforts by locals may have had some limited success, their ability to shape tourism policy and logistics within their own settings has led to greater influence on the practices of visitors and government policies that foster ecotourism (see Stevens 1997 for more). In addition to these tourist pressures related to economic globalization, more recent concerns involve the constantly changing climate.

Discussions of anthropogenic climate change, or more specifically global warming, have been ongoing for decades at the highest levels of international politics. We can already see and read that protected areas are being recycled as one partial solution to the dilemma of our carbon legacy in the atmosphere. There is much debate, (as was clear in Chapter 8, on climate change), as to the role of forests as **carbon sinks**. Even less clear is how much carbon is fully sequestered in biological settings that originally were designed for biodiversity protection. While deforestation and land cover changes (biomass removal) are still important factors in driving carbon loads into the atmosphere, an emphasis on reforestation or emphasizing protected areas only for carbon sequestration may have its own problems. The **carbon cycle** is a well known if not fully quantified cycle. Therefore, adding a temporary carbon sequestration tool, such as a tree to a park, does "something" for atmospheric **carbon uptake** in the short term (30–300 years, depending on the species) but does little to change the overall balance of carbon in the long term. You still end up with more carbon in the atmosphere once the tree dies and decomposes, releasing stored carbon. Minute portions are stored in the deep soils as **inorganic carbon**, a permanent removal of carbon, but these are minuscule compared to the ongoing release of anthropogenic carbon. So efforts like the **REDD** (Reducing Emissions from Deforestation and Forest Degradation) and REDD + (which adds financial incentives for conservation) programs should be viewed as a promising start, but critically analyzed for what they do not include or address (see Chapter 8).

The human–environment relationship has changed over time, a point we have repeatedly made in this book (especially Chapters 2 and 5). And our concepts about the role of humans in nature have likewise changed over time, and by region. It should come as no surprise that the same is true with parks or protected areas. Protected areas have served multiple and changing roles over the last centuries. Early ideas reflected the priority of natural resource use conservation (future use). Protected areas and parks then developed a focus on preservation, as areas for "avoiding" deforestation, to "biodiversity" purposes. Today, the trend is to set aside areas as "carbon sinks" and for **carbon sequestration**.

The protected area concept serves multiple purposes, but has changed over time and by regional context. As dominant stories of human–environment relationships, or **environmental orthodoxies**, the concepts of conservation, preservation, or even parks have evolved but are still present and can suggest some promising future pathways for conservation efforts. Human–environment research that pays attention to the promise of conservation, without neglecting human needs nearby, can better inform environmental policy and local environmental management.

Chapter Summary

Biodiversity of all stripes is important not only for ecological integrity, but also for the material human benefits and the cultural and intrinsic dimensions of nature's presence. The act of conservation is challenging, however, because of the politics of environmental governance. Nation-states, NGOs, and individuals take multiple pathways to setting aside or integrating conservation areas that are deemed important for biodiversity. Protected areas, natural parks, national parks, biosphere reserves, and wildlife refuges and preserves have all been used in these efforts, to name only a few strategies. These are almost always well-intentioned efforts to conserve, if not preserve, parts of the planet's flora and fauna deemed valuable and irreplaceable. Not all of these efforts, however, have been beneficial to local residents. Important lessons from earlier national efforts aimed at conservation, even at the regional or local level, must be considered.

The record of national park development must be remembered especially as nation-states, non-governmental organizations, and locals attempt to work together in increasingly trans-border efforts. Displacement of locals, for one, is not a satisfactory or equitable start for a conservation area and may result in limited-access or environmental justice issues (Chapter 7). Exclusion of nearby locals and indigenous peoples from protected areas rarely works without local buy-in or at least incentives for participation in those conservation efforts (Dowie 2009). Increasingly, as population growth and density infringe or pressure local natural resources, more models and solutions for a humanized but conserved landscape will be needed. Transboundary efforts offer intriguing models for possible conservation that crosses political boundaries, but do have certain challenges. Most landscapes in the world are at least partly humanized, and if continuing work can be informed by the human–environment perspective, this may go a long way towards creating better long-term solutions for conservation and for human livelihoods (Zimmerer 2006).

Box 11.02 Protected areas in your region

1. What protected areas, if any, are in your region or town?
2. Are there any notable species of concern, threatened or endangered, that the area protects?
3. How do the conservation rules in these protected areas affect you, your family, or community members?

Critical Questions

1. How can we protect biodiversity without curtailing local human uses of the environment?
2. How can conservation and development be balanced in societies where basic needs have not yet been met?
3. Whose nature is it? Who gets to decide what is nature, or what is natural and what is unnatural, in any given setting?
4. What responsibility do conservation organizations have to local peoples in areas without national parks? What are the ethics of doing so in areas with a national park presence?
5. Conservation can clearly produce ecosystem and economic benefits, but for whom, by whom, and for what purposes?
6. Can so-called "peace parks" be made to work across political boundaries?

Key Vocabulary

biocentric
biodiversity
biodiversity hotspot
biophilia
buffer zones
carbon cycle
carbon sequestration
carbon sink
carbon uptake
CITES (Convention on International Trade in Endangered Species of Wild Fauna and Flora)
cloud forest
conservation biology
conservation refugees
Convention on Biological Diversity (CBD)
cores
distribution range
economic value
ecosystem benefits
ecotourism
edge effect
Endangered Species Act
endemic species
environmental governance
environmental orthodoxies
exotic species
extinction
generalists
genetic bottleneck
genetic diversity
habitat diversity
habitat conservation plans
habitat fragmentation
inorganic carbon
intrinsic value
invasive species
keystone species
mass extinction
minimum viable population (MVP)
multiple uses
native species
natural park
naturalized species
peace parks
preservation
protected areas
public–private partnerships
REDD
regional natural park
sovereign resources
specialists
species diversity
stakeholders
sustainable development
territoriality
transboundary conservation
United Nations "Man and the Biosphere" program

References

Brenner, J. (2011) Pasture conversion, private ranchers, and the invasive exotic buffelgrass (*Pennisetum ciliare*) in Mexico's Sonoran Desert. *Annals of the Association of American Geographers*, 101(1), pp. 84–106.

Childs, B. (ed.) (2004) *Parks in Transition: Biodiversity, Rural Development and the Bottom Line* (London: Earthscan).

Childs, B. (2006) Revenue distribution for empowerment and democratization. *Participatory Learning and Action*, 55, pp. 20–9.

Davis, D.K. (2007) *Resurrecting the Granary of Rome: Environmental History and French Colonial Expansion in North Africa* (Athens: Ohio University Press).

Dowie, M. (2009) *Conservation Refugees: The Hundred Year Conflict between Global Conservation and Native Peoples* (Cambridge, MA: MIT Press).

Duffy, R. (2001) Peace parks: the paradox of globalization. *Geopolitics*, 6(2), pp. 1–26.

Heatherington, T. (2010) *Wild Sardinia: Indigeneity and the Global Dreamtime of Environmentalism* (Seattle: University of Washington Press).

King, B. and Wilcox, S. (2008) Peace parks and jaguar trails: transboundary conservation in a globalizing world. *GeoJournal*, 71, pp. 221–31.

Logan, B.I. and Moseley, W.G. (2002) The political ecology of poverty alleviation in Zimbabwe's Communal Areas Management Programme for Indigenous Resources (CAMPFIRE). *Geoforum*, 33, pp. 1–14.

Myers N., Mittermeier, R.A., Mittermeier, C.G,. da Fonseca G.A.B., and Kent, J. (2000) Biodiversity hotspots for conservation priorities. *Nature*, 403, pp. 853–8.

Naughton-Treves, L., Holland, M.B., and Brandon, K. (2005) The role of protected areas in conserving biodiversity and sustaining local livelihoods. *Annual Reviews in Environmental Resources*, 30, pp. 219–52.

Perramond, E.P. (2010) *Political Ecologies of Cattle Ranching in Northern Mexico: Private Revolutions* (Tucson: University of Arizona Press).

Posey, D. and Balick, M. (eds.) (2006) *Human Impacts on Amazonia: The Role of Traditional Ecological Knowledge in Conservation and Development* (New York: Columbia University Press).

Stevens, S. (ed.) (1997) *Conservation through Cultural Survival: Indigenous Peoples and Protected Areas* (Washington: Island Press).

Sundberg, J. (1998) NGO landscapes in the Maya Biosphere Reserve, Guatemala. *Geographical Review*, 88(3), pp. 388–412.

Terborgh, J. (1999) *Requiem for Nature* (Washington: Island Press).

Wilson, R. (2010) *Seeking Refuge: Birds and Landscapes of the Pacific Flyway* (Seattle: University of Washington Press).

Wright, J.B. and Campbell, C.L. (2008). Moorish cultural landscapes of Las Alpujarras, Spain. *FOCUS on Geography*, 51(1), pp. 25–30.

Zimmerer, K. (ed.) (2006) *Globalization and the New Geographies of Conservation* (Chicago: University of Chicago Press).

12

Water Resources and Fishing Livelihoods

Icebreaker: Privatized Water in Cochabamba, Bolivia

In January 2000 a "war" broke out in the Andean nation of Bolivia that grew into one of the country's most serious social crises in the recent past. The conflict erupted when the price of drinking water was tripled in the central Bolivian city of Cochabamba, and peasants in the surrounding arid region suddenly found that the water they had been drawing freely for generations no longer belonged to them. City-dwellers accustomed to subsidized water supplies were confronted by the true market price, while the peasants – mostly

An Introduction to Human–Environment Geography: Local Dynamics and Global Processes, First Edition. William G. Moseley, Eric Perramond, Holly M. Hapke and Paul Laris.

Published 2014 by John Wiley & Sons, Ltd.

Quechua Indians who had owned the water for centuries – involuntarily found themselves customers of Aguas del Tunari, a subsidiary of the British firm International Water.

These changes were the result of Law 2029, a bill passed at the end of 1999 that privatized state drinking water and sewage disposal. The law was pushed through without any kind of public hearing and under pressure from the French firm Lyonnaise des Eaux, which is in charge of the water supply in the capital city of La Paz. Groups of local residents and civil society organizations came together to oppose the privatization. Once begun, the water conflict escalated into a rebellion that lasted 10 months and came close to toppling the government. An 11-day state of siege ensued and a dozen people died.

The resistance was successful. After the protests, the Aguas del Tunari contract was canceled, and the company filed a claim seeking US$25 million in compensation, but in the settlement that was reached no compensation was paid. The citizen's movement (the Coordinadora), which got Aguas del Tunari thrown out of Cochabamba, became a worldwide symbol of the struggle against free-market reforms and structural adjustment policies, and its techniques began to be emulated in other cities in the region, such as the sprawling working-class city of El Alto, Bolivia, and Buenos Aires, the capital of Argentina. In spite of the citizens' victory in the "water war," Cochabamba's basic problems have not been solved. Although water services have improved, the city has no more than five hours of water a day, and only 40% of farmers in the surrounding area have access to clean water. The dam remains unconstructed, and plans to tap into underground rivers have so far made little headway. Efforts to form a cooperative or limited company with the involvement of local people have also not materialized.

Sources: Chavez 2006; Cuba 2000

Chapter Objectives

The objectives of this chapter are for the student to be able to:

1 Describe the hydrologic cycle and human interventions in it.
2 Discuss geographic patterns of global water resources and the current status of the world's water resources.
3 Understand the nature and extent of the global water crisis from various perspectives: physical; technological; management/planning; political.
4 Understand the politics surrounding water and fishery resources management.
5 Critically assess debates about common property resource management and resource privatization.

Introduction

Water is the essence of life and center of all things on Earth. All living beings need water to survive. Water shapes landscapes and creates weather patterns (see Chapter 3). It is the major component of plants and animals – comprising 60–70%

by weight of all living organisms – and is necessary for photosynthesis. Water plays an important role in nearly every environmental system. Water resources support critical habitats and provide the base on which many human livelihoods depend. Agriculture could not exist without water, and water availability is crucial to economic development and the reduction of poverty. Culturally, water has been central in the beliefs and practices of every human civilization that has existed. The ancient Greeks, for example, believed that everything in the universe originates and ends as water (Cech 2010), and indigenous peoples in many parts of the world define their cultural identity in terms of water.

Indeed, it is difficult to overstate the importance of water as it permeates all dimensions of life and landscape on Earth. With over 70% of its surface area covered by water, Earth is "the water planet" of our solar system. Yet Earth's water resources are under stress, and water scarcity threatens the well-being of hundreds of millions of people worldwide. It is anticipated that access to water will be the major political issue of the 21st century.

In this chapter you will learn about water resources and one human livelihood that is rooted in water resources – fisheries. We begin with a discussion of the circulation of water between land, bodies of water, living beings, and the atmosphere and then consider the geography of water – where water resources are located, patterns of water use, and human interventions in the **hydrologic cycle**. Next we describe the current global water crisis and the factors driving unsustainable water-use practices before turning to a discussion of two political issues surrounding water: privatization and transboundary water issues. Finally we turn to a consideration of ocean resources and global fisheries and conclude with a discussion of common property resources and possible solutions to the governance of water and fishery resources.

The Global Water (Hydrologic) Cycle

The global water (hydrologic) cycle is the natural process by which water circulates between land, oceans-lakes-streams, living beings, and the atmosphere and by which freshwater is produced. The cycle is driven by solar energy and is depicted in Figure 12.01. Heat from the sun causes water on the surface of Earth to evaporate. **Evaporation** is the transformation of water from liquid to gas and is one means by which water is transported from Earth's surface to the atmosphere. The other way this occurs is through **transpiration**, which is the release of water vapor into the atmosphere by plants through their leaves. Plants soak up water from the soil and then transpire it back to the atmosphere. Evaporation and transpiration (collectively referred to as **evapotranspiration**) act as natural processes of distillation, effectively creating pure water by filtering out dissolved minerals.

Water vapor is stored in the atmosphere, but as warm air rises it expands and cools, leading to **condensation**, the transition of water vapor from a gas to liquid form. As water droplets form and coalesce they fall to the ground as **precipitation**

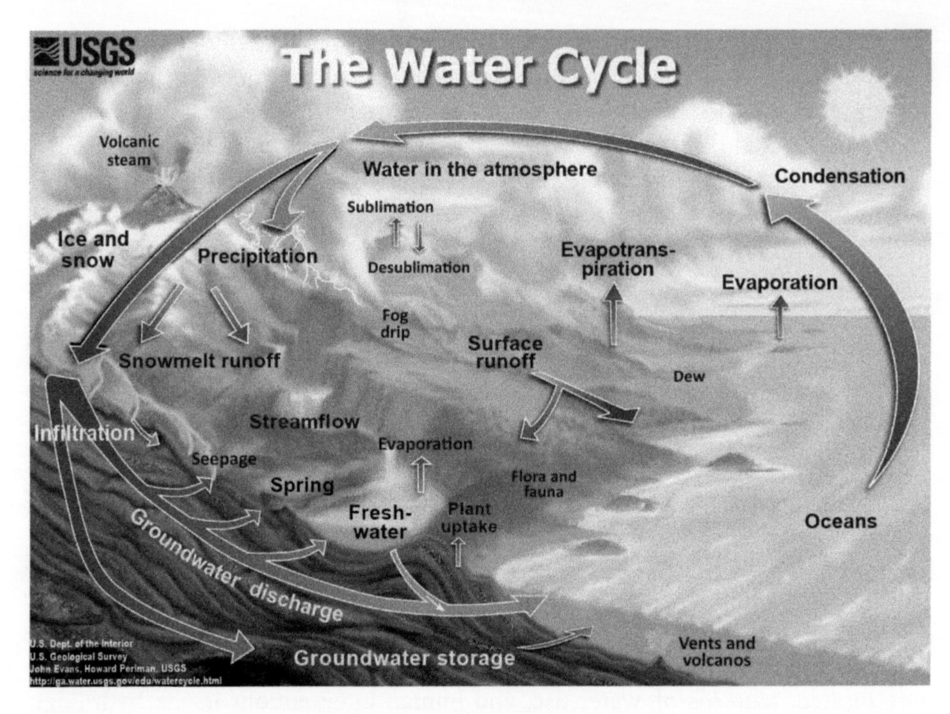

Figure 12.01 The hydrologic cycle. *Source*: US Geological Survey http://ga.water.usgs.gov/edu/watercycle.html (accessed March 13, 2012). Courtesy of the US Geological Survey.

in the form of rain, snow, sleet, or hail. Much of this precipitation flows as **runoff** into streams, rivers, lakes, and oceans, where it is stored for some time and then again evaporates. Some precipitation is absorbed by soil and taken up by plants through their roots. Some precipitation freezes and becomes glacial ice. Some precipitation and surface water soaks down through soil and rock to recharge underground reservoirs known as **aquifers**. Over time, this **groundwater** also eventually flows into lakes and oceans, although it takes a significantly longer time to do so than surface runoff.

Warm temperatures and strong winds speed up evaporation, and runoff is influenced by factors such as climate, terrain, soil type, vegetative cover, the extent of developed rooftops and concrete surfaces, and the intensity and volume of precipitation. Steep slopes and land areas that have been deforested or converted to agriculture lose water more rapidly than areas that are heavily vegetated. Paved areas in residential or commercial districts accelerate runoff, and a large volume of precipitation over a very short period flows more rapidly across land surfaces than light rainfall. As water flows across land it conveys organic matter and bioactive chemicals (including pollutants) from their source areas on land to lakes, streams, and oceans. Thus, human activity on land can have a significant influence on the quality and integrity of water resources. Of course, we also remove and replace water from its natural cycle in several ways. We extract or divert water for drinking and for use in agriculture, industry, sanitation, transportation, and the generation of energy.

Animals, including humans, have their own hydrologic cycles that connect them to the global water cycle. We take in water through consumption of food and liquids, which, after providing nourishment, leave the body in the form of urine and solid waste. This waste is then cycled back to the environment via water. Water dilutes urine and breaks down solid waste, thereby allowing it to be absorbed in soil or washed into waterways by runoff from precipitation. In the case of humans, most of our bodily wastes are removed by sewage systems either to septic tanks or municipal wastewater treatment facilities where waste is treated and then returned to rivers as effluent water. A growing concern is the way in which our use and ingestion of various chemicals is impacting Earth's water resources. In addition to traditional sources of pollution from human activity such as industry and agriculture, which are discussed later in the chapter, evidence is emerging that we are causing serious damage to water resources and aquatic organisms such as fish through our increasing use and bodily consumption of chemicals – such as certain pharmaceuticals – that do not biodegrade easily – see Box 12.01.

Box 12.01 Why the contents of our medicine cabinets are turning up in our waterways

A recent scientific investigation by the US Geological Survey found that the active ingredients of hundreds of pharmaceuticals and personal care product chemicals were detectable in streams and rivers of the United States. These chemicals include food additives such as caffeine, antibacterial compounds such as Triclosan, and synthetic hormones such as ethinyl-estradiol, the active ingredient in many birth control medications. The consumer products from which these active ingredients originate are typically found in our households, specifically in medicine cabinets and bathrooms. While these chemicals are thought to be beneficial to humans, they are stable enough such that much of the chemical passes through humans without being metabolized. When we take a tablet for a headache for example, a large portion of that pain reliever goes through our body and is excreted by us in its original form.

Once these chemicals are excreted or come off our bodies during showering, they enter the wastewater stream. In rural areas and many coastal areas, wastewater typically ends up in a septic tank for treatment. In larger towns and cities, wastewater is treated in a municipal wastewater treatment plant. Unfortunately, septic tanks and municipal wastewater treatment plants were not designed to "treat" and destroy the chemicals found in pharmaceuticals and personal care products. As a result, many of these chemicals are still present in the treated effluent water that leaves standard septic tanks and wastewater treatment plants. As an additional concern, in many locations, underground septic tanks can corrode and leak into groundwater, which may eventually serve as drinking water for some people. The result is that

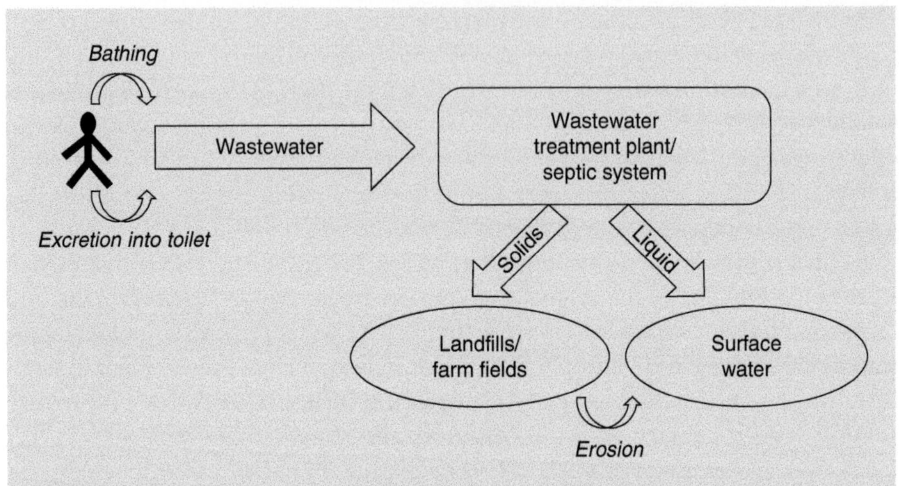

Figure 12.02 Chemical structures. *Source*: Siddhartha Mitra. Used with permission.

many of these drugs and consumer products often end up in surface waters such as rivers or streams, which are the ultimate end-point for treated effluent, as described in Figure 12.02.

What if those natural water bodies that receive treated (i.e. "clean") effluent are in fact lakes in which people swim? Or what if the streams and rivers that receive treated effluent contain fish that are consumed by humans? Even more basic than that, what if these water bodies receiving effluent serve as the raw water supply for a city's drinking water? Scientists have already shown that some pharmaceutical and personal care product chemicals persist in water as it is being treated and made potable via a drinking water treatment plant. Are we at risk of exposing ourselves to some of these chemicals through our drinking water?

We should be concerned about exposure to pharmaceutical and personal care products in water because some of the chemicals have been demonstrated to affect the ecosystem in a negative way. For example, over the course of a few years, scientists conducted an interesting experiment and added a few micrograms per liter of ethinyl-estradiol to see what would happen to fish in an experimental lake in Canada. Within three years, so many individuals of one species of male fish were feminized that there weren't enough productive males available to inseminate females. As a result, over the course of the experiment, the entire species of fish collapsed! Could similar exposure scenarios exist in other water bodies throughout the world receiving these chemicals?

So why are we only now hearing about these chemicals and why isn't something being done to control for the presence of these chemicals in our waters? The problem is that detectable levels of pharmaceutical and personal product chemicals in natural waters are considered to be "emerging" threats to water quality, meaning that there are no laws in place to control for their abundance in the natural environment. In fact, many knowledgeable scientists and government officials who deal with environmental issues have not even heard of this emerging threat to water quality. Moreover, analyzing and detecting these chemicals in water is challenging. Only recently have technological advancements been made that allow scientists to isolate and detect these chemicals at low levels in water. Finally, to really control for a source of water pollution, we must know the point of origin of that source. Unfortunately, as we are finding out, there may be other sources of pharmaceutical chemicals to natural waters that people have not yet accounted for, such as agriculture. For example, much of the agricultural industry in the United States supplements the diets of farm animals with excessive levels of hormones and antibiotics. The waste excreted by these farm animals often contains these same hormones and antibiotics, in their unmetabolized form. Furthermore, solids from wastewater treatment plants are often applied as fertilizer to some farm fields. Inevitably, the pharmaceuticals and personal care products in these solids can also end up in groundwater or in nearby rivers and streams as a result of erosion during rainstorms. As of 2010, the US government is finally considering legislation to regulate the presence of these "emerging" chemicals in water. But as consumers, if we are to make any headway in controlling for the presence of pharmaceuticals and personal care product chemicals in natural waters, we have to consider their ultimate fate beyond their presence in our medicine cabinets and our bodies.

Source: Contributed by Siddhartha Mitra, PhD, Department of Geological Sciences, East Carolina University.

The Geography of Water

As stated in the introduction, water covers 70% of Earth's surface. Ninety-seven percent of this water exists as salt water in oceans and is not directly available for human use. Of the remaining 3% that is fresh water, two-thirds is frozen in glaciers, snowfields, and ice caps or exists in soil or as water vapor in the atmosphere. Thus, less than 1% of the water on Earth is readily available for human use.

Water is distributed unevenly across Earth's surface and over time. Different regions possess different amounts of water, and precipitation varies by season throughout the year as well as from year to year. While some regions experience regular precipitation more or less consistently, in other regions rainfall is intermittent. Regions that have monsoon seasons, for example, receive a large portion of their total annual rainfall in highly concentrated storms lasting just a few hours. Often precipitation does not occur when it is needed most. In other places too much precipitation causes flooding. Thus, striking global inequities in water resources and supply exist between regions and countries.

Fresh water resides on the surface in lakes, reservoirs, rivers, streams, and ponds (**surface water**) or as groundwater beneath the surface. The greatest single body of fresh water by volume is Lake Baikal in Siberia, Russia. Lake Tanganyika, which crosses four countries in east Africa, is the second-largest lake in volume in the world, and Lake Superior, one of the Great Lakes of North America follows third, although it is largest freshwater lake in surface area. Overall, about 70% of lake water is located in North America, Africa, and Asia (Christopherson 2006).

Rivers and streams are fed by runoff. Runoff is highest in the tropics along the equator and in the coastal mountains of the northern hemisphere. The most water-rich world regions in terms of runoff are Asia and Latin America, each accounting for one-quarter to one-third of total global runoff. North America (including Mexico) accounts for about 15%. Sub-Saharan Africa's share of global runoff is about 10% – although some countries in Africa, such as the Congo, are among the most water-rich in the world. Europe accounts for about 9% of global runoff, and Australia-Oceania 5%. The Middle East and North Africa is the most water-limited region with only 1% of global runoff (WWAP 2009).

Groundwater is found beneath the land surface in rocks, fine clay material, sand, and gravel. It is the largest source of freshwater on Earth. While about 126,000 cubic km (30,300 cubic miles) of fresh water exists in lakes and streams, it is estimated that 4 million cubic km (1 million cubic miles) of groundwater may be found within 800 meters (half a mile) of Earth's surface. Despite this volume, the spatial distribution of groundwater is quite uneven and variable. In some regions inaccessible or insufficient amounts of groundwater make the installation of wells economically unviable (Cech 2010). Groundwater resources are also threatened by pollution from hazardous wastes and chemical fertilizers and pesticides and from over-pumping, which depletes groundwater reserves faster than they can be replenished. Groundwater supplies are closely tied to surface runoff, and the interaction of surface and groundwater is increasingly a concern in regions where excessive groundwater extraction is adversely affecting surface water. Similarly, where climate change is reducing precipitation and surface runoff, groundwater recharge may be negatively impacted.

One of the challenges facing society is that the distribution of the world's human population does not correspond to the distribution of the world's water resources. Some of the most densely populated regions of the world are water-poor while other sparsely populated regions are water-rich. Although Asia accounts for a large share of global runoff, its population is nearly 4 billion – more than half of the world's total. Thus with the exception of Southeast Asia, Asia as a whole is relatively water-poor when water supply is adjusted for population size. In contrast the Amazon River is the greatest river in terms of runoff, but it is almost completely remote from centers of human population. In fact, about 20% of global runoff is remote and not readily available for human use. Timing – when runoff occurs – also determines actual usability of water. Roughly 50% of total global runoff occurs as uncaptured floodwater. Hence, a useful way of thinking about water resources from a human perspective is in terms of *accessibility* and *per capita availability* (see Figure 12.03).

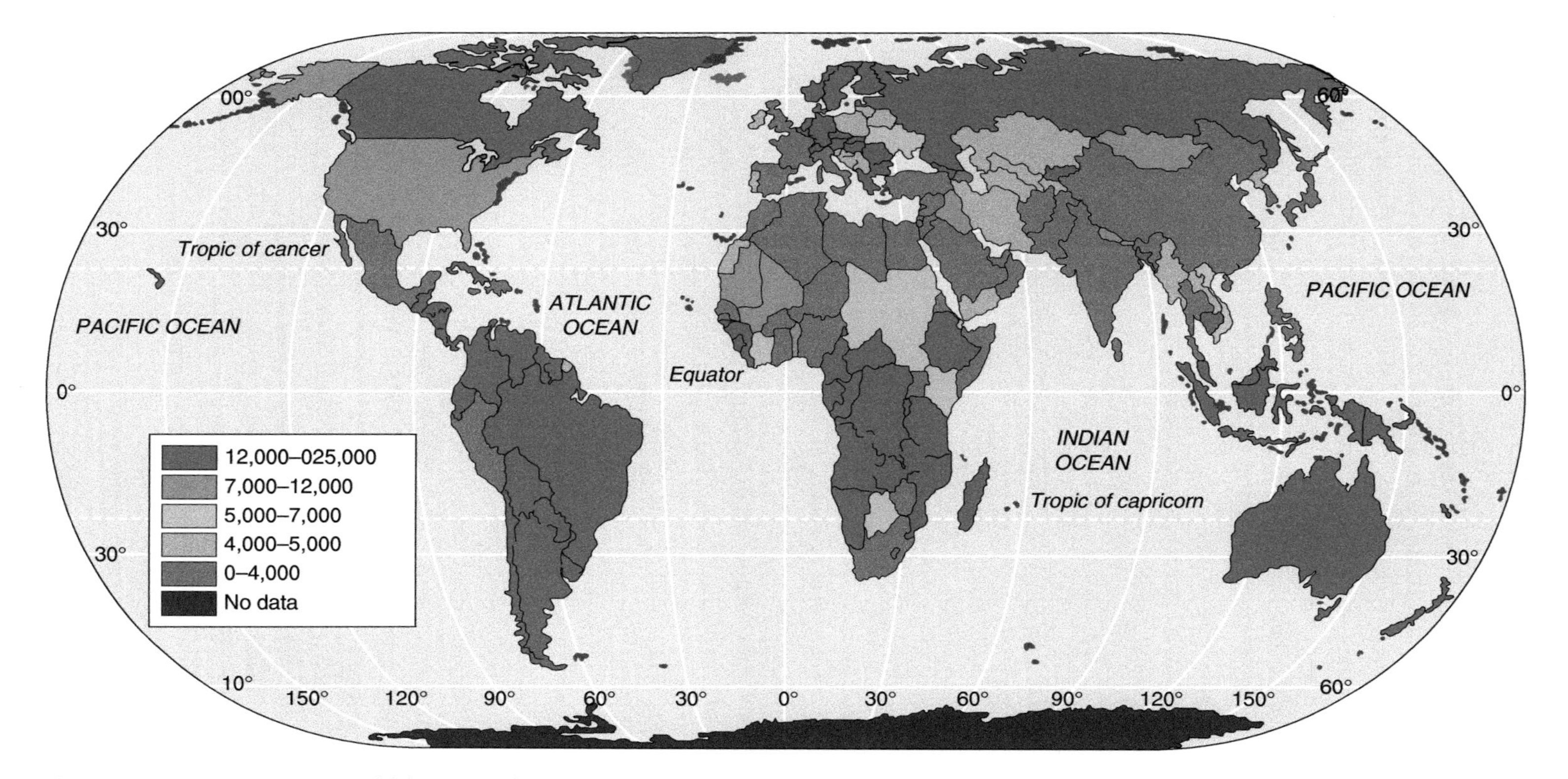

Figure 12.03 Per capita water availability. *Source*: http://www.wrsc.org/attach_image/global-freshwater-availability-capita-2007

Factors that determine water availability include physical factors, such as precipitation (when and where it occurs) and climate conditions (how quickly evapotranspiration takes place), and social factors such as economic structures and political relations. In addition to physical variations in water supply distribution, water resources within particular world regions are often unevenly and unequally distributed between urban and rural populations and between the rich and poor. In many countries of the world, glaring disparities exist in the access that different groups of people have to clean water. The story about water conflicts in Bolivia that opened this chapter is illustrative. Bolivia is relatively water-rich in terms of per capita availability, and yet, as the story above suggests, many Bolivians experience significant deprivation from an inadequate water supply.

Human Interventions in the Hydrologic Cycle

Humans have intervened in the global water cycle in several important ways that have significantly impacted both water resources and the environments in which they are located. Most societies have developed complex technologies for the management of water resources. In addition to devising systems for drinking water, sewage removal, irrigation, flood control, and transportation, human civilizations have built elaborate systems for bathing, recreation, and aesthetic expression (e.g., fountains, water sculpture) (see Figure 12.04).

Canals, Irrigation, and Flood Control

The first developments in hydraulic technology took place nearly 6000 years ago in Mesopotamia and Egypt and were driven by agriculture (Mays 2008). Frequent flooding of the Tigris and Euphrates rivers and the lack of good drainage led ancient Mesopotamia to develop canals and at least one diversion dam for the irrigation of desert areas. The Egyptians began building dams on the Nile River to store water around 2650 BC and also developed artificial basin irrigation, deliberately flooding and draining land with the use of sluice gates and transverse dikes. Similar technologies were in use in South Asia around the same time, and beginning in 2280 BC the Chinese undertook major construction of dams, dikes, and other waterworks along the Huang He River to control its flooding. Chronic food shortages subsequently prompted the development of extensive irrigation works between the 4th and 6th centuries, and Chinese empires were deemed "good" or "bad" on the basis of their ability to maintain these irrigation systems (see Cech 2010).

In the Americas, large-scale irrigation systems first appeared about 3000 years ago. Of note are the Purrón Dam, canal networks, and terrace irrigation built in

Figure 12.04 The Vaillancourt Fountain, San Francisco © Gregory Olsen / iStockphoto.

and around the Tehuacan Valley in southern Mexico between 750 and 500 BC (Christopher and Neely 2006). The Aztecs built an impressive system of aqueducts (above-ground systems for conveying water) in their capital, Tenochtitlan (Mexico City) in the 15th century AD, and stonewalled terraces with elaborate ditch systems were constructed by the Incas on the steep hillsides of the Andes Mountains (Cech 2010) (see Figure 12.05). Further north, in the present-day state of Arizona in the United States, the Hohokam constructed almost 300 miles of canals that irrigated 250,000 acres of desert land in the 9th century AD. Approximately 150 years later the Anasazi developed community irrigation projects in the desert lands of southwest Colorado.

A more recent example of hydraulic engineering is found in the Netherlands, one of the most densely populated countries in the world. Approximately 27% of the Netherlands lies below sea level, and two-thirds of this small country are vulnerable to flooding from the North Sea and the Rhine and Meuse rivers. For 2000 years the Dutch and their ancestors have been working to protect their land from flooding and to reclaim land from the sea through an interconnected system of dikes, dams, floodgates, drainage ditches, canals, and pumping stations. The contemporary system, the North Sea Protective Works, constructed between 1950 and 1997, has assisted the Dutch in controlling floods and reclaiming land from the Zuiderzee (South Sea). Recently land reclaimed by the project was of sufficient size to create a whole new province.

Figure 12.05 Terraces for irrigation, Colca Canyon, Peru. *Source*: © Paolo Aguilar/epa/Corbis.

Groundwater Tapping

The tapping of groundwater has an equally long history. The early Somalians of eastern Africa dug groundwater wells across the desert to provide drinking water for caravans, nomads, and cattle. These wells served as the foundation for small desert communities that later grew into large cities (Cech 2010). More elaborate groundwater systems developed in several parts of the world in the form of **qanats** (underground water collection and conveyance systems; see Figure 12.06). Qanats consist of an underground tunnel that uses gravity to transport water from springs or the water table at higher elevations to irrigation canals on the surface of lowlands. Along the path of the tunnel vertical shafts are installed for circulation, light, and excavation. Qanats first appeared in Persia as early as 3000 years ago. Between the 6th and 4th centuries BC the technology spread to other parts of the Persian Empire as far east as the Indus River valley in present-day Pakistan. Qanats were also constructed across the Mediterranean, in North Africa as far west as Morocco, and along the Silk Road in central Asia and Chinese Turkistan (Mays 2008). This ancient technology continues to function in many places today. In Iran alone over 30,000 qanats are in operation. However, modern groundwater pumping technologies driven by diesel engines are increasingly replacing the qanat in other regions where they were traditionally used (see Box 12.02). This is raising concern about the sustainability of groundwater as the depletion of groundwater resources threatens environmental stability (see Lightfoot 1996).

Box 12.02 Moroccan *khettara*: traditional hydraulic technology under modern technological pressure

In southern Morocco, on the margins of the Sahara Desert, lies the Tafilalt oasis, a historically important caravan crossroads and trading center. Two great rivers – the Oued Ziz and the Oued Rheris – fed by seasonal runoff from the Atlas Mountains cross the alluvial basin of the Tafilalt. However, crops can only be dependably produced in the region with irrigation. Fortunately, though scrub vegetation and stony, barren plains surround the Tafilalt basin, it sits atop a rich reserve of groundwater shallow enough to support a large oasis. Since the 14th century AD, residents of the northern part of the Tafilalt have tapped this reserve through an extensive network of earthen canals and *khettara* (qanats). Approximately 80 *khettara* provided perennial water for 28 dispersed villages, each with its own ruling lineages and mechanisms for organizing labor at the local level. *Khettara* are village-operated and collectively maintained, and intricate relationships have evolved to manage them and distribute their benefits according to each shareholder's inputs of land, labor, tools, and money.

Khettara and the basin's two rivers provided a reliable source of irrigation water for north Tafilalt villages until the early 1970s, when new technologies and government policies forced changes in traditional water management. Availability and distribution of water in the Tafilalt changed dramatically after the 1971 opening of the Hassan Addakhil Dam, which impounded the Oued Ziz. Water from the Oued Ziz had flowed unimpeded into the Tafilalt basin. Now water from the reservoir is released through government canals only three to four times per year for a period of 20–23 days. This water contributes to the irrigation of over 75% of all arable land currently in use in the Tafilalt, but the water is so thinly spread that no longer can fields be irrigated solely with water from the Oued Ziz. Privately owned diesel-pumped wells have now become most important to Tafilalt irrigation and have replaced *khettara*. Almost 750 private diesel-pumped wells are operating. In contrast to the way *khettara* traditionally operated, there is no governing authority – national, regional, or local – that regulates diesel pumps once they have been installed. Consequently they have been overused, which, combined with insufficient water from the dam, has resulted in a dramatic lowering of the water table underlying the oasis. As the water table drops, *khettara* are abandoned.

Since diesel-pumped wells are privately owned, the traditional ties that bind village society are breaking down. Non-farm sources of income continue to draw young men away from villages and out of the oasis, disrupting the social organization of *khettara* systems. Of the original 80 *khettara*, only 19 remain in use. Furthermore, the traditional source of wealth in the oasis,

Figure 12.06a Moroccan *khettara* (qanat). *Source*: © Yoshihiro Takada / amanaimages / Corbis.

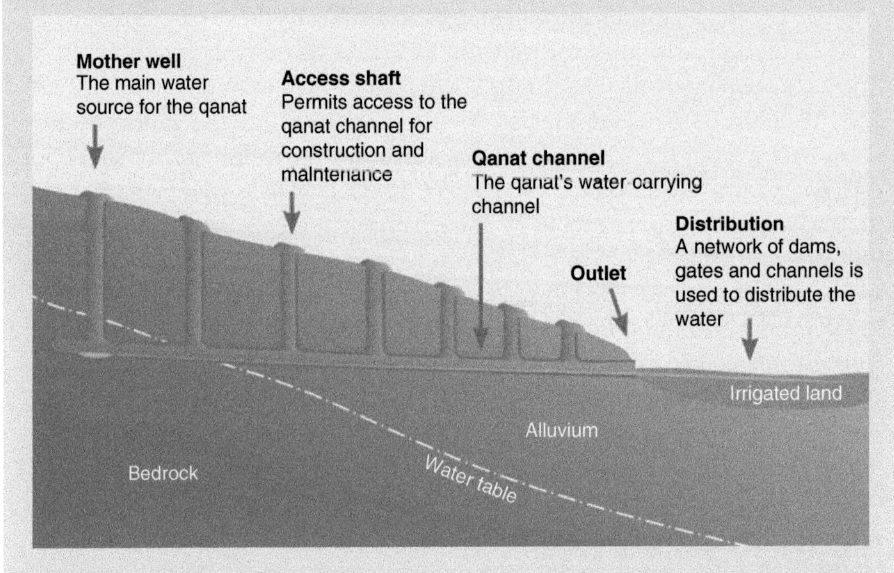

Figure 12.06b Diagram of a qanat.

trade in dates, has been irreparably altered. Only 60% of the palm trees in the Tafilalt still produce a date crop today. The others no longer produce dates, or have died from periodic date blight and / or sustained **desiccation** (extreme drying). Retail and tourist trade and foreign remittances have replaced agriculture as the principal source of wealth. The last of the *khettara* may dry up in the near future, as the water table becomes so low that "following the water" by extending the depth of mother wells will prove prohibitively labor-intensive and expensive.

The competition between traditional and modern water systems is both environmental and cultural. Environmentally, diesel-pumped wells and government canals have led to the abandonment of a sustainable technology in favor of systems that are capable of providing greater quantities of water but are not sustainable. Culturally, the adoption of newer technologies has altered land use patterns that had evolved through the historic reliance of villages on *khettara*. There has been some loss of local control over water resources, because much of the water villages need comes only from the reservoir and drinking water pipes, both regulated by the national government. Injudicious attempts to expand the use of surface and groundwater without regard to the sustainability of withdrawal are depleting groundwater resources and feeding social and environmental instabilities. There is some hope, however. Recently a few *khettara* have been refurbished with government and external assistance and are providing water to farmers. And the national agricultural office has recently been experimenting with bore wells to recharge groundwater. These efforts need to be expanded if the *khettara* system is to survive.

Source: Adapted from Lightfoot 1996.

Dams

The first prime minister of modern independent India, Jawaharlal Nehru, once stated, "Dams are the new temples of India." In doing so, he was affirming India's commitment to modernization. At the time (mid-20th century), big dams signified rationality, progress, and modernity. A **dam** is any obstruction placed in a river or stream to block the flow of water so that it may be stored in a reservoir. Dams are built to control flooding, promote irrigation, provide drinking water, and generate energy. Currently there are 45,000 large dams (higher than 15 m or 49 ft) in the world, and tens of thousands of smaller dams. Almost every major river in the world has a dam erected over it, and some of these represent remarkable engineering feats. Large dam construction was enthusiastically pursued throughout the 20th century. In the United States, President Franklin Roosevelt was a strong proponent of large public works projects, and during his administration (1933–45) the construction of many large dams such as the Hoover and Parker dams on the Colorado River took place.

Dams and reservoirs have enhanced the health and economic prosperity of humans around the world, but their construction has extracted a heavy price in the

Figure 12.07 An aerial view of the Three Gorges Dam on the Yangtze River in Yichang, Hubei province November 28, 2008. *Source*: © Reuters/Corbis.

form of serious environmental problems. Such problems include reduced stream flows, degraded water quality, negative impacts on migratory fish such as salmon, and disastrous flooding when they fail. The construction of large dams furthermore inundates large tracts of land, which destroys forests and human settlements. China's Three Gorges Dam (see Figure 12.07), the largest dam in the world, was completed in 2003. It cost US$25 billion to construct, flooded 22 cities, and forced the relocation of over 1 million people. Its reservoir submerged ancient archeological sites, productive farmland, and wildlife habitat. The slow movement of the river's flow is causing sediment to accumulate behind the dam, depriving downstream tidal marshes of sediment needed for rejuvenation. The Narmada Dam Project in India involves the construction of 30 large dams on the Narmada River. Although the largest structure planned, the Sardar Sarovar Project, is expected to irrigate more than 18,000 square kilometers of drought-prone land and provide drinking water to 30 million people, many argue its environmental costs far outweigh its benefits. The project has spawned a huge citizens' resistance movement (the Narmada Bachao Andolan movement) and created discord between the government and the communities to be displaced.

Humans have for millennia achieved impressive engineering accomplishments in their attempts to harness freshwater sources and cope with the environmental challenges of flooding and drought. However, in meeting our needs for water resources, we have greatly modified the movement, distribution, and quality of water, altering many environmental systems in the process. Several early urban civilizations are believed to have collapsed as a result of the mismanagement of water resources. Today, many of the world's major rivers are running dry before reaching their outlets to oceans. Large quantities of water from the Colorado River, for example, are diverted to serve the cities of Los Angeles and San Diego and agriculture in California and Arizona. The ensuing reduction in flow of the Colorado River has dramatically altered the ecology of the river delta and devastated the fisheries of the Sea of Cortez into which the Colorado River flows. One of the most striking examples of surface water depletion is the plight of the Aral Sea that sits on the border of Uzbekistan and Kazakhstan. Once the fourth-largest lake in the world, the sea is now dying. Irrigation for large-scale cotton farming during the Soviet era depleted over 80% of its total volume. It is estimated that 60,000 fishing jobs have been lost, and even cotton farming is no longer economically viable due to soil blight. As these examples demonstrate, our current water-use practices threaten the long-term sustainability of the planet's water resources.

Contemporary Water Practices: How Do We Use (and Abuse) Water?

Human water use may be categorized as either **consumptive** – water is removed from a source and not returned, or **nonconsumptive** – water is either not removed from a source or is temporarily removed and then returned. For example, to generate power water may be taken from a river, used temporarily to operate turbines, and then returned to the river downstream. Humans use fresh water for three general purposes: *domestic* (household or municipal drinking, cooking, and cleaning), *agriculture* (crop irrigation and watering livestock), and *industry* (manufacturing, industrial processing, energy). Globally, agriculture uses the highest proportion of water (70%). Industry uses 22%, and domestic use accounts for 8% (WWAP 2009). However, within and across world regions, the proportion of water directed to each of these three uses varies significantly. In arid regions, a greater share of fresh water is used for agriculture. Industrialized nations use a significant portion for industry. In the United States, for example, industry accounts for 46% of water use, agriculture for 41%, and domestic use is 13%. In India, in contrast, agriculture accounts for 87% of total water use, industry 6%, and domestic use is 8%. But in the Congo, one of the water-wealthiest countries, agriculture accounts for only 9% of total use; industry uses 22%, but domestic use is 69% (FAO Water).

In addition to using water for different purposes, different countries consume water at different rates, as do different individuals within countries. A person needs a minimum of 50 liters of water per day for drinking, bathing, and sanitation. Yet, while individuals in developed countries in Europe and the United States consume on average between 200 and 600 liters per person per day, people in developing countries in sub-Saharan Africa and parts of Asia barely use 10 liters per person per day, and the poorest of the water-poor consume 5 liters or less a day.

A concept that has been developed to illustrate the consumption of water in relation to population is the **water footprint**. The water footprint indicates the amount of water needed to sustain a population and includes: volume of consumption; consumption pattern (e.g., high versus low meat consumption); climate conditions; and agricultural practice (water use efficiency) (Hoekstra and Chapagain 2007). High consumption of meat and using a lot of irrigated land to produce food result in a larger water footprint. Globally, the average water footprint is 1240 cubic meters per capita/year. The country with the largest water footprint is the United States (2480 cubic meters per capita/year), followed by Greece, Italy, and Spain (2300–2400 cubic meters per capita/year). At the low end is China, with a water footprint of 700 cubic meters per capita/year, and many countries in sub-Saharan Africa have water footprints of less than 100 cubic meters per capita/year (Hoekstra and Chapagain 2007). Underlying the water footprint concept is the idea of **virtual water**, the amount of water embedded in food and other products. For example, it takes 140 liters of water to produce one cup of coffee. One glass of milk needs 200 liters of water while one hamburger requires 2400 liters of water. A cotton T-shirt contains 2000 liters of water, and a pair of leather shoes 8000 liters.

One of the consequences of economic globalization is that transnational forces increasingly drive local decisions on water use in agriculture and industry. Water footprints of core regions of the world economy (North America, Europe) that import a lot of goods from other countries have been externalized to other parts of the world. Cotton, for example, is produced in many water-scarce areas although it is one of the "thirstiest" crops grown. Europe and the US import a lot of cotton and cotton textiles. Thus, through the global market, European and US consumers rely on water resources in other places and thereby influence agricultural and industrial strategies (and water-use decisions) of many other countries (WWAP 2009). The concepts of water footprints and virtual water have become a useful way to think about global connections in our water use and how we might utilize water resources more efficiently.

Global consumption of water is increasing. Water withdrawals have tripled over the last 50 years. Currently we are appropriating 54% of all the accessible freshwater contained in rivers, lakes, and underground aquifers. Although population growth is one factor in the expanding use of water, economic development, urbanization, and related consumption habits play a much more significant role. Over the past century water use has grown at more than double the rate of population growth (FAO Water). The rapid increase in irrigation development since the 1970s and the continued growth of agriculture-based economies largely explain trends in water withdrawals. Also, emerging market economies (e.g., China, India, and Turkey) are experiencing rapid growth in domestic and industrial demands that are linked to urbanization and changes in lifestyle (WWAP 2009).

As economies and populations continue to grow, as demand for food and other goods increases, and as highly consumptive lifestyles become a global norm, water resources are increasingly stressed. A number of questions emerge about the long-term sustainability of our water resources. Will water resources be adequate? How will economic development and urbanization affect global water supply? How will societies address the competition among growing demands for water? Already evidence of a global water crisis is present.

The Global Water Crisis

We all need water to survive. Yet today one of the most fundamental conditions of human life and development remains unmet: universal access to water. Of the world's 7 billion people, at least 1.1 billion lack access to clean drinking water. Another 2.2 billion people lack access to adequate sanitation. Half of the world's hospital beds are occupied by people with an easily preventable water-borne disease, and contaminated water is implicated in 80% of all sickness and disease worldwide. Every 8 seconds a child dies from drinking dirty water.

Lack of access to clean water and sanitation means extreme deprivation. It means that people live more than 1 kilometer from the nearest safe water source and often must collect water from drains, ditches, or streams that might be infected with disease-causing pathogens and bacteria. It means that people are forced to defecate in ditches, plastic bags, or on roadsides. For women and girls it means long hours spent collecting and carrying water for household use (UNDP 2006).

Two billion people now live in water-stressed regions of the planet, and it is estimated that 15 years from now two-thirds of the world's population will face water scarcity. Already water use exceeds natural renewability in many regions of the world (UNDP 2006). But, just how scarce is the world's water? And, what are the underlying reasons for water scarcity?

Water scarcity exists in two ways. **Physical scarcity** occurs when there is not enough water to meet all demands, including environmental flows. Regions most often associated with physical water scarcity are arid regions, but water scarcity also arises when water resources are over-exploited – most often for irrigation when excessive diversion of surface water and over-pumping of groundwater for agriculture and industry deplete these sources of water faster than they can renew themselves. Inefficient irrigation practices also decrease water supply, and consumer misuse of water results in a lot of water wastage. For example, dripping taps in Europe and the US lose more water than is available each day to more than 1 billion people (UNDP 2006). The United States Environmental Protection Agency has predicted that 36 of its 50 states could face water shortages if present use habits continue.

Pollution is another contributing cause of physical water scarcity. Industries dump toxic waste into the world's waterways. Chemical pesticides, fertilizers, and herbicides run off from agricultural fields to pollute rivers, streams, lakes, and oceans. The disruption of sediment renewal in tidal marshes and deltas from the damming of rivers furthermore leads to saltwater intrusion and contamination of groundwater.

Box 12.03 Living with drought

A **drought** is a prolonged, abnormally dry period when there is not enough water for users' normal needs. Droughts occur frequently in Australia. It is the driest inhabited continent in the world, and its rainfall patterns among the most variable over time. Approximately 89% of Australia's total rainfall evaporates or is transpired into the atmosphere. Only 9% runs off into streams and rivers; and only 2% drains into groundwater aquifers. Furthermore, only about 7% of Australia's total runoff occurs where water is most used – in the Murray–Darling Basin in the southeast. More than 50% of Australia's water is used in this region (see Figure 12.08), primarily for irrigated agriculture.

The problem is that while demand for water is on the rise, water availability is decreasing. Population growth, intensive agricultural development, urbanization, industrial growth, and environmental requirements are all increasing demand for water. Between 1983/4 and 1996/7, water use increased 65%. Most of this increase was due to an expansion of irrigated agriculture. If present use trends continue, experts predict that water use in Australia will increase an additional 66% by 2020. Given Australia's climate conditions and current water use trends, declining water availability is one of the most critical resource issues Australia faces.

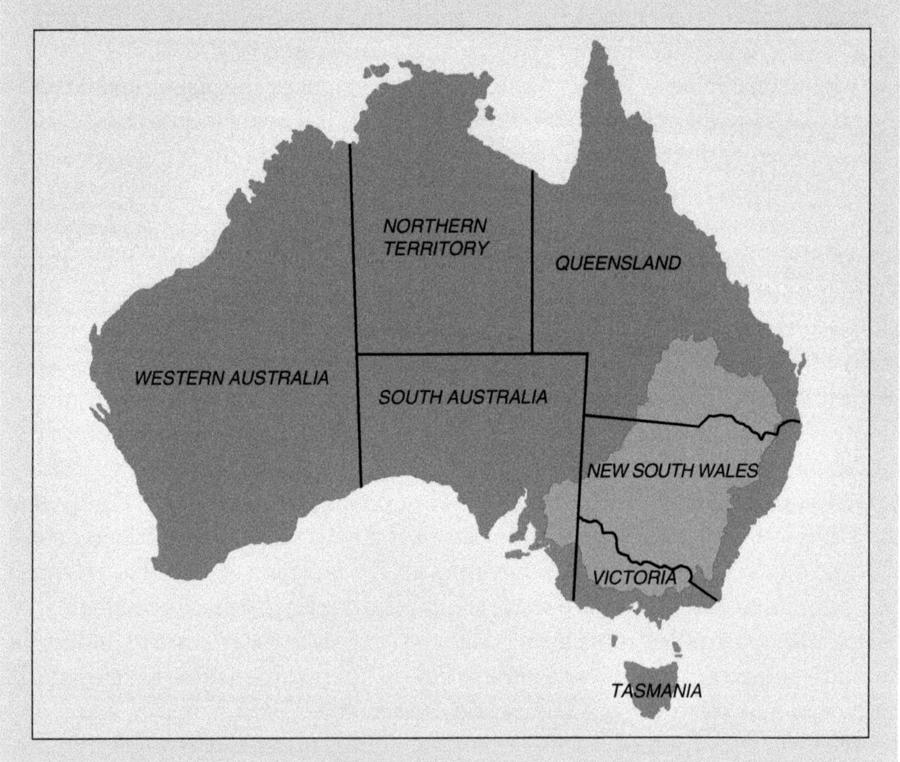

Figure 12.08 Australia: water scarcity.

Sources: Australian government: National Land and Water Resources Audit (NLWRA); *Connected Water: Water Scarcity and Demand*; *Living with Drought*.

The other form of water scarcity is **economic scarcity**. Economic scarcity exists when financial or institutional factors limit people's access to water even though enough water exists in nature to meet everyone's needs. Insufficient services rather than water shortages define economic scarcity. Water scarcity in most of sub-Saharan Africa, Central America, and parts of Asia and South America derives from economic scarcity (see Figure 12.09).

Economic scarcity is about an unequal distribution of resources and is caused by a lack of investment in water infrastructure, mismanagement, corruption, bureaucratic inertia, or by the ways economic and political institutions privilege one group over another. People live too far away from a source of clean water, their communities are not adequately served by water-delivery services, or when water infrastructure and services do exist, people cannot afford to pay for water to meet their needs. Overwhelmingly, the current water crisis is a crisis of the poor. Within the world, those most adversely affected by lack of clean water and sanitation

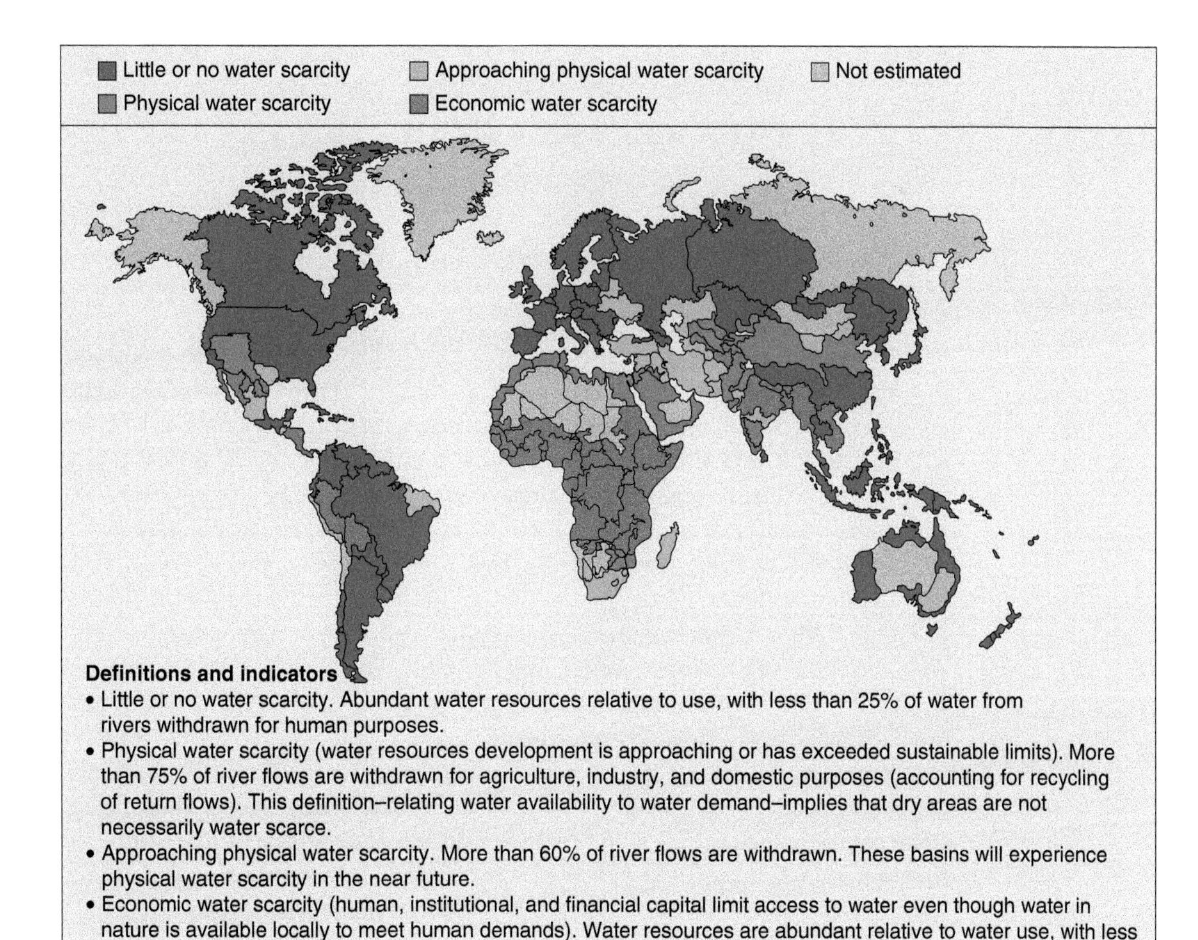

Figure 12.09 Physical and economic water scarcity. *Source*: Molden 2007.

reside in the global south and earn less than US$1 per day. Even in regions where physical scarcity prevails, the poor are the most likely to lose access to water first.

A number of factors underlie water scarcity, be it physical or economic. Population growth and climate change are important factors, but experts increasingly agree that water shortages and pollution are primarily induced by economic, social, and political forces. Global economic expansion impacts water resources through changes in consumption habits and in the way goods and services are produced. Changes in lifestyles reflect human needs, desires, and attitudes, which are influenced by social factors such as culture and education. Poverty constrains people's choices for survival, and what they must do to survive can further stress environmental resources and increase vulnerability to environmental hazards (see Chapter 6) (WWAP 2009). Ultimately, however, the water crisis is about how we, as individuals and in societies, govern access to and control over water resources – questions that are profoundly political. Who gets to control Earth's water, and who decides how water is used, are among the most pressing issues facing human societies in the 21st century.

The Politics of Water: Governance in a Globalized World

> Water promises to be to the 21st century what oil was to the 20th century: the precious commodity that determines the wealth of nations (*Fortune* magazine, May 2000).

Because of its essentialness to life, water has for millennia been viewed as a good to which everyone should have free access. In 2002 the United Nations Committee on Economic, Social, and Cultural Rights recognized water not only as a limited natural resource and public good, but also as a human right. Drawing on a range of international treaties and declarations, it stated: "the right to water clearly falls within the category of guarantees essential for securing an adequate standard of living, particularly since it is one of the most fundamental conditions for survival" (OHCHR et. al. 2003).

Yet the notion of water as a public good and basic human right contrasts starkly with current water policies pushed by other global institutions such as the World Bank, the International Monetary Fund, and the World Trade Organization, which advocate the **privatization** of water resources – that is, handing them over to the control of private transnational corporations. Supplying water today is big business – worth over US$400 billion per year – and evidence suggests that a powerful global water cartel is increasingly gaining control of the world's water resources (see Barlow and Clarke 2002). Since the late 1980s, the private sector has quietly extended its reach into all aspects of water resources. For-profit water companies now provide municipal water services; bottle water for sale; control water used by industry, mining, farming, and energy generation; own and operate dams, pipelines, and water purification systems; and buy up groundwater rights and entire watersheds.

The shift in thinking about water as a public good to water as a private commodity coincided with the rise of neoliberal economic thought in the 1980s, which argues that free-market economics is the best economic policy option for the world. In the 1980s, advocates of neoliberal economics began challenging the effectiveness and legitimacy of publicly owned water supply systems, and by the 1990s several states around the world (Argentina, Bolivia, China, Chile, Great Britain, Indonesia, Morocco, the Philippines, Poland, South Africa, Spain, Thailand, and Turkey) began relinquishing management and/or ownership of water resources to private capital (Barlow 2007; Barlow and Clarke 2002).

In industrial countries such as the United States and the United Kingdom, privatization of water services has grown as municipalities search for seemingly affordable ways to upgrade utilities infrastructure and provide services to growing urban populations. In developing countries, privatization has received impetus from the economic development policies of global lending and financial institutions. Beginning in the 1980s, the World Bank agreed to renegotiate poor countries' loans on condition of undergoing structural adjustment, which included privatizing essential public services. By the 1990s global lending agencies were encouraging developing countries to let big European water corporations such as Suez, RWE-Thames, and Vivendi run their water systems for profit. By 2002 the vast majority of loans for water and sanitation systems were conditional on privatization.

Experiences with the privatization of water resources and services in both industrialized and developing countries have been at best mixed. In some cases, water networks have been revamped and water production has expanded. In other respects, however, considerable evidence suggests that corporate takeover of water has deepened the global water crisis. In numerous examples from both developed and developing countries, privatization has increased water rates and decreased water quality. Water has become prohibitively expensive – especially for the poor. The construction of sanitation facilities has lagged in urban areas, and neither adequate water nor sanitation systems have been installed in poor and rural communities. When they have been installed, the quality of water provided is often substandard (Barlow and Clarke 2002; Barlow 2007).

But the story does not end here because a fierce resistance to corporate takeover of water has grown all over the world, which has become increasingly organized and is emerging as a coordinated global water justice movement. For example, Latin America was the first region in the developing world to experiment with water privatization. Bolivia was the first "water war" to gain international attention. Similar struggles have unfolded in Argentina, Mexico, and Chile since the Cochabamba struggle described in the opening story of this chapter.

Almost every country in Asia has introduced private water management, but local opposition around the region has intensified. In India there are battles against corporate-controlled municipal water systems and multi-national soft drink companies who are accused of depleting groundwater resources and threatening agriculture and poorer people's access to drinking water. In the Global North, too, there are similar struggles over water. The US cities of Atlanta and New Orleans have severed contracts with corporations such as United Water, Suez, and Veolia, citing problems such as faulty infrastructure, dirty water, and corporate resistance to voter rights to approve contracts. And in the USA citizens in the state of Michigan launched a legal challenge against Nestlé Corporation when sinkholes and other environmental problems arose after it starting pumping groundwater to sell as bottled water.

At the national level, the politics of water center on balancing the competing needs and demands of different users: agriculture, industry, corporations, and citizens. At the international level, political conflict emerges from the fact that most of the world's water resources traverse international boundaries. Half of the Earth's land surface is covered by **international water basins**, catchments, or watersheds shared by more than one country. Currently there are 263 international water basins, in which two-fifths of the world's people live and 60% of global river flows occur (UNDP 2006). Fourteen countries share the Danube River, 11 countries share the Nile, and 9 share the Amazon. In Africa, 90% of all surface water in the region is found in transboundary river basins. All of the major rivers in South Asia except the Ganges originate outside the region, as does the Mekong River, on which five countries in Southeast Asia depend (see Box 12.04).

The flow of water through streams, lakes, and aquifers that cross international borders creates **hydrological interdependence** between countries. This simply means that how water is used in one place impacts its use in another place. For example, retaining water upstream for irrigation or power in one country restricts downstream flows for farmers in another country. Similarly, industrial or human pollution released upstream is transported to people in other countries downstream (UNDP 2006). As water scarcity increases, and as countries strive to meet the demands of growing economies and populations, competition between countries for water has the potential to generate international conflict as well as international cooperation. The Convention on the Protection and Management of the [Danube] River, signed in 1994 by all the Danube River basin states, is an example of how institutional cooperation can create a wide range of mutually reinforcing benefits across international borders (see Barraqué and Mostert 2006). Similar international agreements exist in Europe for the Rhine River, and since the end of apartheid, South Africa has initiated a series of agreements to manage shared river basins in the region. In fact, between AD 805 and 1984 more than 3600 water-related treaties were signed throughout the world (Postel and Wolf 2001). These examples of international cooperation indicate optimism. Still, although it has been 4500 years since an outright war was explicitly fought over water (Postel and Wolf 2001), managing hydrological interdependence may be one of the biggest challenges facing the international community in the future.

Box 12.04 Water wars in Asia?

Water has emerged as a key issue that could determine whether Asia is headed toward mutually beneficial cooperation or deleterious interstate competition. No country could influence that direction more than China, which controls the Tibetan plateau – the source of most major rivers in Asia, including the Indus, the Mekong, the Yangtze, the Yellow, the Salween, the Brahmaputra, the Karnali, and the Sutlej. Collectively, these rivers are a lifeline not only to the world's two most populous states – China and India – but also Bangladesh, Pakistan, Nepal, Bhutan, Myanmar, Cambodia, Laos, Thailand, and Vietnam. These countries make up 47% of the world's population.

Concern about potential interstate conflict over river-water resources has arisen from Chinese attempts to dam or redirect the southward flow of rivers from the Tibetan plateau. Unsustainable intensive farming in northern China has aggravated water scarcity within its border, and unbridled industrialization has contaminated its major rivers. In response, China has increasingly turned its attention to the rivers originating in Tibet. It has dammed rivers, not just to produce hydropower but also to channel waters for irrigation and other purposes, and it is currently toying with massive interbasin and inter-river water-transfer projects, which threaten the ecological viability of river systems tied to South and Southeast Asia. China's damming of the Mekong River in particular has inflamed passions in Vietnam, Laos, Cambodia, and Thailand.

The Mekong River has long played a vital role in the livelihoods of people who live on its banks. It extends over 4800 km (2983 miles) and has a drainage basin of 810,000 sq. km (312,743 sq. miles). The Mekong basin has long been an irreplaceable source of water and fish for the villages along its banks. In fact, it supports the largest inland fishery in the world, on which millions of people depend. The Mekong feeds the largest freshwater lake in Southeast Asia, the Tonle Sap in Cambodia, which alone is home to more than 400 species of fish as well as many species of mammals and reptiles. But the Tonle Sap, as well as the livelihoods of all those who live within the Mekong basin, are under threat from China's extensive dam-building activities.

The problem is that dams affect the flow of rivers, and the water level of the Mekong dropped dramatically following China's completion of the Manwan Dam in 1993 and the Dachaoshan Dam in 2003. This adversely impacted the river's ecology downstream, reducing fish stocks and water supply in southwest China, Laos, Thailand, Cambodia, and Vietnam. China's plan to construct at least four more dams has residents and governments in the other Mekong countries worried.

Compounding the concern of these countries is the fact that China has refused to join the Mekong River Commission, which seeks to ensure reasonable and equitable use of the Mekong's water among the nations in its basin. China has also been reluctant to establish interstate agreements on the management of river waters flowing into South Asia (India, Pakistan, Bangladesh), citing in both cases its sovereign right to do what it wants on its portion of these rivers.

Clearly the way to forestall or manage water disputes in Asia is to build cooperative river-basin arrangements involving all riparian neighbors. In the absence of institutionalized cooperation over shared resources, peace would be the casualty in Asia if water became the new battleground.

Source: Adapted from Chellamy (2007, 2008) and Salidjanova (2007).

Ocean Resources and Fishing Livelihoods

Oceans are one of Earth's most important natural resources. Oceans play an important role in converting carbon dioxide in the atmosphere into oxygen and regulating climate. They are mined for minerals and drilled for oil. Humans use oceans for transportation and recreation. Increasingly, oceans are a source of biomedical organisms that may be able to help fight disease. Oceans also provide habitat for at least 250,000 species of plants, fish, and other aquatic organisms, and thus provide a substantial source of food for the world. For thousands of years oceans and the fisheries they support have provided for human needs, but human activities are placing enormous stress on marine resources. Pollution and unsustainable fishing practices threaten marine ecosystems and the livelihoods of tens of millions of people around the world. We have already witnessed the collapse of fisheries in the North Atlantic, and 50–75% of the world's other fisheries are dangerously imperiled. Some scientists predict that if present trends continue, the populations of all ocean species we currently fish will collapse by 2048 (Lotze et al. 2006).

Seafood (fish and shellfish harvested from marine and freshwater environments) contributes at least 15% of average animal protein consumption to 2.9 billion people, and as much as 50% in some parts of Asia and Africa (FAO 2009). For the poor in particular, seafood is a principal source of important micronutrients. Economically, the world's fisheries constitute a significant source of revenue for many developing nations, and they provide livelihoods for hundreds of millions of people. In 2006 fisheries employed 43.5 million people and supported an additional 520 million people with the income they generated.

Seafood is both nutritionally and economically important, and yet the manner in which we are utilizing this resource today threatens the long-term availability of fish and other seafood products. **Overfishing** refers to the extraction of fish and other marine species from the sea faster than they can naturally reproduce. Many marine scientists believe that overfishing poses the greatest threat to ocean environments and marine biodiversity.

Although humans have always harvested seafood and other animals from oceans, sometimes depleting marine species, the problem of overfishing today is the consequence of modern economic forces and technological innovations that increased the production capacity of individual fishing vessels. Throughout the period of European colonialism, world demand for and trade in marine products expanded. As early as the 17th century European explorers viewed the marine species they encountered in the Americas with excitement and quickly set about capturing and killing seals, whales, turtles, fish, and other seafood for profit. Within a very short period of time the population of several species noticeably declined, and a couple of species, such as the sea cow, became extinct (Roberts 2007). Throughout the 19th and early 20th centuries commercial exploitation of marine life expanded around the world as Europe extended its colonial enterprises to Asia and Africa. In addition to expanding world markets, several technological innovations during this period facilitated the intensification and increased the output of fishing activities. For example, trawling vessels, which pull a net through the water, appeared in the early 19th century and were powered by sails. By the 1870s the steam trawl was born, which allowed fishing vessels to trawl back and forth relentlessly over rougher ground and tow larger nets (Roberts 2007). The steam winch reduced the time required to lift nets from the sea, thereby speeding up fishing time. Combined, these two innovations made it possible for individual vessels to capture more fish. Technological innovations in the 20th century, such as factory fishing vessels that can stay at sea for weeks and sonar and high-precision global positioning systems, have led to an unprecedented escalation in fishing intensity that leave fish nowhere to hide (Roberts 2007). Between 1950 and 2008 world fish harvests increased fourfold.

In addition to increased production the second development that has taken place is that the world's fisheries have "globalized." In fact, seafood is the most "globalized" food in that it is harvested by commercial fleets operating in international waters, and it is the most internationally traded food on the world market. World trade in fish products has expanded fivefold since the 1950s. This increase stems directly from the emergence of markets for marine products in industrialized countries of the Global North and a particular form of fisheries development in the Global South to service these markets. Beginning in the 1950s, newly independent countries in Asia and Africa undertook major state-led, planned economic development initiatives. Within the fisheries sector, development programs centered on the transfer of industrial-style fishing technology from the North to the South to target high-value species of seafood such as prawn and tuna for export. More recently, foreign fishing fleets from industrialized countries have obtained access to fishing grounds in developing countries through joint venture and access rights arrangements. The effect of these initiatives has been a dramatic economic, cultural, and environmental transformation of fishing communities and marine ecosystems throughout the world.

Global markets for particular types of seafood (shrimp and shellfish) have changed production patterns of local fisheries in developing countries. This has decreased the availability of traditionally consumed species of fish in domestic markets and reduced fish consumption among the poor in these places. In industrialized countries, overfishing by commercial fleets has led to the collapse of some fisheries – the most famous case being the cod fishery of the North Atlantic – thereby destroying the livelihoods of local fishermen. And overfishing has threatened the long-term viability of other fisheries around the world. Large-scale poaching of fish in tropical waters – most notably off the coast of West Africa – by foreign fishing fleets has depleted stocks available for local fishers from developing countries such as Ghana. Small-scale fishing communities in Africa and Asia relying on traditional craft and gear have been the hardest hit by declining fish stocks as they are the least able to adjust to changing conditions, so their livelihoods suffer.

Finally, declining fish stocks damage marine ecosystems and apparently imperil terrestrial ecosystems as well. Recently overfishing has been linked with the disappearance of large game in West Africa. Research by biologist Justin Brashares and his colleagues (Brashares et al. 2004) suggests that when fish supply declines, people turn to hunting and selling wildlife (bushmeat) on land to meet their food and economic needs. Overexploitation of these resources has seriously decreased terrestrial wildlife numbers causing other ecosystem disruptions, such as an explosion in baboon population growth. Brashares suspects that bushmeat and large game hunting has removed the natural predators of baboons, allowing their numbers to grow into unruly mobs that are now raiding food from and wrecking havoc in human settlements in the region (see National Geographic 2008).

Management of Common Property Resources

Neither water governance nor fisheries management is by any means a new issue facing human societies. But the imperatives of modern economic and political structures combined with rapid technological change and historically unprecedented population growth present a number of challenges for the management and maintenance of Earth's resource systems. Is water a public good to which everyone has a basic right, or is it a commodity to be bought and sold? How should water and fishery resources be divided? If the oceans belong to no one nation or group of people, how do we limit or control the exploitation of their fishery resources? When most of the world's rivers cross international borders, how do we ensure equal access to their waters? And who gets to decide? What is the best institutional framework for governing our use of common resources? Are we doomed to face what has been called the "tragedy of the commons"?

The **tragedy of the commons** refers to a situation in which multiple individuals, acting solely and rationally in their own self-interest, will ultimately deplete a shared limited resource even when it is clear that it is not in anyone's long-term interest to do so (see Chapter 8 for a discussion of this problem in the context of air pollution). The idea dates back to ancient Greece and Aristotle's observation:

> that all persons call the same thing mine in the sense in which each does so may be a fine thing, but it is impracticable ... For that which is common to the greatest number has the least care bestowed upon it. Everyone thinks chiefly of his own, hardly at all of the common interest; and only when he is himself concerned as an individual.

This idea of the self-serving individual re-emerged in the writings of William Forster Lloyd in 1833 and again more recently in 1968 in an essay by Garrett Hardin published in *Science*. Using the example of overgrazing of 19th-century community pastures, Hardin attempted to demonstrate how the depletion of a vital resource was brought about by individuals who continually increased the size of their herds out of self-interest. He argued that human society is inherently destructive toward nature and naturally overexploits common resources. Hardin was primarily interested in human population growth and his essay also focused on larger resources such as oceans. His proposed solutions to the "tragedy" of common resources were to either privatize resources or enact government regulation thereof.

Hardin's essay received considerable criticism for historical inaccuracy and erroneous analysis, for failing to differentiate between **common property** and **open access resources**, and for overlooking the numerous examples of societies successfully managing common property resources. The historical record in fact suggests that market solutions often fail, and government interventions can exacerbate problems rather than solve them. Hardin's thesis also failed to take account of the influence particular economic systems have on resource use. Commercial systems in which profit is the main motivation of economic activity tend to encourage individualized self-interest and accelerate resource extraction by individuals since there is little incentive to not do so.

In contrast to the uninformed pessimism of Hardin, 2009 Nobel laureate and leading scholar in common pool resources Elinor Ostrom has studied how humans interact with ecosystems to maintain long-term sustainable resource yields. Her research has detailed how societies have developed diverse institutional arrangements for managing natural resources, in many cases avoiding ecosystem collapse, though she also recognizes that some institutional arrangements have failed to prevent resource depletion. Ostrom identifies eight principles of stable local common property management, among which are collective decision-making, appropriate adaptation to local conditions, mechanisms for conflict resolution, and recognition of community self-determination by higher-level authorities (Ostrom 1990). The principal lesson that comes out of her research is that there is no one solution to social-ecological problems, but what is remarkable is that throughout history humans have proved highly adaptive and capable of forging solutions to numerous environmental problems through informal collective action. This suggests hope for the future.

Chapter Summary

As the foregoing discussion illustrates, significant challenges surrounding the management of the world's water and fishery resources confront the global community. The production practices of modern society have endangered our most critical resources through pollution and over-extraction. Although population growth and global climate change stress these resource systems, more significant are the social, economic, and political forces that drive unsustainable resource use practices. Given the increasingly globalized scale at which water and fishery resources manifest themselves, the critical issues we confront seem to beg international cooperation. Yet, the question remains, do we have to will to take the action necessary?

Critical Questions

1. Why is it important to differentiate between physical and economic scarcity when thinking about water needs and water management?
2. Should water resources be viewed as a public good or as a commodity? How does the way we view water resources shape water management policies?
3. Discuss the ways water resources and the ecosystems they support are connected to other environmental processes (e.g., climate change, fisheries, wild game).
4. In what ways has economic globalization impacted the development of water and fisheries resources in the world?
5. What measures can you as an individual take to reduce your water footprint?

Key Vocabulary

aquifer
common property resources
condensation
consumptive water use
dam
desiccation
drought
economic scarcity
evaporation
evapotranspiration
groundwater
hydrologic cycle
hydrological interdependence
international water basins
nonconsumptive water use
open access resources

overfishing
physical scarcity
precipitation
privatization
qanat
runoff
surface water
tragedy of the commons
transpiration
virtual water
water footprint

References

Australian Government. *Connected Water: Water Scarcity and Demand*. http://www.connectedwater.gov.au/water_policy/water_security.html (accessed June 5, 2010).

Australian Government (2008) National Land and Water Resources Audit (NLWRA). Available at: http://www.nlwra.gov.au/ (accessed June 11, 2010).

Australian Government Bureau of Meteorology. *Living with Drought*. Available at: http://www.bom.gov.au/climate/drought/livedrought.shtml (accessed June, 5, 2010).

Barlow, M. (2007) *Blue Covenant: The Global Water Crisis and the Coming Battle for the Right to Water* (New York: The New Press).

Barlow, M. and Clarke, T. (2002) *Blue Gold: The Fight to Stop the Corporate Theft of the World's Water* (New York: The New Press).

Barraqué, B. and Mostert, E. (2006) *Transboundary River Management in Europe*. Human Development Report Occasional Paper. United Nations Development Programme.

Brashares, J.P., Arcese, M.K., Sam, P.B., Coppolillo, A.R.E. and Sinclair, A. (2004) Balmford bushmeat hunting, wildlife declines, and fish supply in West Africa. *Science*, 306, pp. 1180–2.

Cech, T.V. (2010) *Principles of Water Resources*, 3rd edn. (Malden, MA: John Wiley & Sons).

Chavez, F. (2006) Cochabamba's "Water War", six years on. *Inter Press Service (IPS)*, November 8. http://ipsnews.net/news.asp?idnews=35418.

Chellaney, B. (2007) China-India clash over Chinese claims to Tibetan water. *Japan Times*, June 26.

Chellaney, B. (2008) Averting Asian water wars. *Japan Times*, October 2.

Christopher, C.S. and Neely, J. (2006) Hydraulic engineering in prehistoric Mexico. *Scientific American*, 295(4), pp. 78–85.

Christopherson, R.W. (2006) *Geosystems: An Introduction to Physical Geography*, 6th edn. (New York: Pearson/Prentice Hall).

Cuba, J. (2000) Free or foreign: the water battle in Bolivia. *The Courier*, December. http://www.unesco.org/courier/2000_12/uk/planet2.htm#top.

FAO (2009) *State of the World's Fisheries 2008* (Rome).

FAO Water. AQUASTAT Database. http://www.fao.org/nr/water/ (accessed June 5, 2010).

Hardin, G. (1968) Tragedy of the commons. *Science*, new series, 162 (3859), pp. 1243–1248.

Hoekstra, A.Y. and Chapagain, A.K. (2001) Water footprints of nations: water use by people as a function of their consumption pattern. *Water Resources Management*, 21(1), pp. 35–48.

Lightfoot, D.R. (1996) Moroccan khettara: traditional irrigation and progressive desiccation. *Geoforum*, 27(2), pp. 261–73.

Lotze, H.K., Lenihan, H.S., Bourque, B.J., Bradbury, R.H., Cooke, R.G., Kay, M.C., Kidwell, S.M., Kirby, M.X., Peterson, C.H., Jackson, J.B.C. (2006) Depletion, degradation, and recovery potential of estuaries and coastal seas. *Science*, 312(5781), pp. 1806–1809.

Mays, L.W. (2008) A very brief history of hydraulic technology during antiquity. *Environmental Fluid Mechanics*, 8, pp. 471–84.

Molden, D. (ed.) (2007) *Water for Food, Water for Life: A Comprehensive Assessment of Water Management in Agriculture*. International Water Management Institute (London: Earthscan).

National Geographic (2008) *Strange Days on Planet Earth: Dangerous Catch*. National Geographic Television and Film. Aired in April. Episode notes accessed at: http://www.pbs.org/strangedays/episodes/dangerouscatch/experts/index.html.

OHCHR, WHO, COHRE, CESR, and Water Aid (2003) *The Right to Water*. Health and Human

Rights Publications 3 (Geneva: World Health Organization).

Ostrom, E. (1990) *Governing the Commons: The Evolution of Institutions for Collective Action* (Cambridge: Cambridge University Press).

Postel, S. and Wolf, A.T. (2001) Dehydrating conflict. *Foreign Policy*, (September/October), pp. 3–9.

Roberts, C. (2007) *The Unnatural History of the Sea* (Washington: Island Press).

Salidjanova, N. (2007) *Chinese Damming of Mekong and Negative Repercussions for Tonle Sap*. ICE Case Studies 218 (May).

Stockholm International Water Institute (SIWI) (2010) Prizes & Awards: John Allan. http://www.siwi.org/sa/node.asp?node=282 (accessed June 5, 2010).

UNDP (United Nations Development Program) (2006) *Human Development Report* (New York: Palgrave Macmillan).

UN Water website http://www.unwater.org/flashindex.html (accessed multiple dates, May–June 2010).

World Water Assessment Programme (2006) *Water: A Shared Responsibility*. United Nations World Water Development Report 2 (Paris: UNESCO Publishing; London: Earthscan).

World Water Assessment Programme (2009) *Water in a Changing World*. United Nations World Water Development Report 3 (Paris: UNESCO Publishing; London: Earthscan).

Part IV

Bridging Theory and Practice

13

Geographic Research

Icebreaker: Declining Fish Catches in Trivandrum, India

One Saturday morning in the summer of 1999, about 1000 men and women angrily marched down Mahatma Gandhi Road, the main thoroughfare of the city of Trivandrum in India. The men waved huge oars and the women carried banners with slogans such as "Save our Seas!" "No Fish, No Life," "No More Foreign Vessels," and "National Fish Workers Unite." They gathered at the central junction and rallied for some time while policemen stood by ready to intervene should violence erupt. After several impassioned speeches, the crowd then quietly dispersed, returning to their villages along the nearby shore or lingering in the city to shop and see the sights.

An Introduction to Human–Environment Geography: Local Dynamics and Global Processes,
First Edition. William G. Moseley, Eric Perramond, Holly M. Hapke and Paul Laris.

Published 2014 by John Wiley & Sons, Ltd.

Who were these people, what grievances did they come to voice, and how did they come to be on this street on this day? What was all the ruckus about?

Later that week, in the early hours of the morning, approximately 35 fishermen stood on the shore in two lines arduously pulling on the ends of a beach seine net. Their voices rang out in a rhythmic chant that kept their movements coordinated with one another. As the net neared the shore the two lines moved closer together until they were almost touching. Amidst angry shouts to "pull harder" the men made one last heave to get the net over the breakpoint of the waves and up on shore. Exhausted the crew gathered around the net as it was opened to assess the catch of their labors. A small heap – 100 at most – of small fish lay in the center of the net – hardly enough to give each man a share to feed his family for the day. Dejectedly some men walked away cursing at one another in frustration. Yet another harvest hardly worth the effort. How would they be able to survive?

Why are harvests declining on the local shores? What factors are causing overfishing? And how will this apparent trend impact the survival of fishing communities in the region? How do geographers approach answering these questions?

Chapter Objectives

The objectives of this chapter are to:

1. Understand the relationship between philosophy and knowledge construction and how different philosophies of science influence the research process.
2. Describe various methods geographers use to collect and analyze data about human–environment interactions.
3. Discuss ways geographers apply research findings to real world situations.

Introduction: What Is Geographic Research?

As the preceding chapters of this textbook indicate, the study of human–environment geography encompasses a wide range of issues, and human–environment geographers engage in research on all kinds of topics. Regardless of the particular topics different geographers investigate, the common objective of geographic research is to gain a better understanding of the relationship between humans, the environment, space, and place. In short, geographic research is about trying to explain and understand the physical, social, and environmental interactions of our world. For most introductory students, however, how geographers actually conceptualize research projects and collect and analyze data to arrive at certain conclusions is something of a mystery.

This chapter introduces students to the various methods geographers adopt to conduct research. We begin with a brief discussion of the theoretical frameworks and scientific philosophies that have influenced geographic research. We then outline different approaches for collecting and analyzing data, including different approaches to fieldwork and the use of remotely sensed data in geographic research. To illustrate how geographic research is used to understand real-world situations, we focus on commodity chain analysis – a particular method for understanding the links between production in one place and consumption in another – and livelihoods analysis, which seeks to understand how people's survival is linked to economic and environmental transformation. Finally, we conclude the chapter with a brief discussion of the relationship between geographic research and social change.

How Geographers Theorize the World

Like all research, geographic research is a process of inquiry and discovery that seeks to expand knowledge. That said, different types of research are undertaken for different reasons. *Exploratory* research investigates little-understood phenomena with the intent to identify important variables and to generate questions for further research. *Descriptive* research documents and characterizes the phenomena of interest. *Explanatory* research seeks to explain what particular forces caused an event and why. This differs somewhat from *understanding*, in which the objective is to interpret meaning and its social and geographic differences. For example, "What does a particular monument *mean* to different groups of people?" or "How do people with ambulatory disabilities perceive urban settings?" Finally, some research seeks to *predict* future outcomes, to forecast likely events or behaviors resulting from different phenomena.

Theory is important in research because, whether we are aware of it or not, it shapes our view of the world. The reason we undertake research and the particular problems we seek to understand are issues that are fundamentally theoretical in nature, and as such are shaped by two philosophical questions. "What exists, or what is the nature of the world?" and "How can we know it?" The first question, the nature of the world, preoccupies a branch of philosophy known as **ontology**. Ontological questions are important to geographers for several reasons, not the least of which is the fact that "geography" literally means "writing about the world." It then follows that "we need to think through the characteristics the world possesses before we can begin writing about it" (Shaw et al. 2010: 10). Ontology thus can be defined as a set of assumptions and theories about "what the world is like." The second question, how we can know the world, belongs to the branch of philosophy called **epistemology**, which is concerned with the nature and scope (or limitations) of knowledge. It ponders questions such as: What is knowledge? How is knowledge acquired? What do people know? and How do we know what we know?

The purpose of research is to contribute to knowledge. But in order to claim to know something one must be able to justify that claim. This inevitably raises philosophical questions about whether and how different knowledge claims are warranted (Graham 2005). How do we justify our claims? On what set of assumptions about the world and our ability to know it do we base our claims? These are important questions because different philosophies of science are premised on different ontological and epistemological assumptions. For example, the natural sciences have been heavily influenced by **positivism**, a philosophy of science that argues that the scientific method is the best approach for investigating the processes by which both physical and human events occur. Positivism's epistemology holds that the only authentic knowledge is that which is based on immediate, direct observation that is then verified through repeated observation or experimentation. Its ontological assumption is that reality exists independently of people and can thus be distantly and objectively observed by scientists. According to this perspective, the objective of scientific inquiry is to identify universal principles that explain the workings of the world. By carefully and objectively collecting data we can determine laws to explain and / or predict natural phenomena and human behavior.

This contrasts with the philosophy of **humanism**, which focuses on human values and concerns and on understanding, experience, and interpretation. Humanism's ontology states that the world that exists is that which people perceive to exist, and knowledge, therefore, is obtained subjectively in a world of meanings created by individuals. There is no such thing as absolute truth; rather, all knowledge is subjective. Furthermore, "we may know something intuitively and create interpretations of the world based on that knowledge, even if we cannot necessarily observe and test it" (Herod and Parker 2010: 62). For example, we know what it feels like to be "happy," but how do we test this? The aim of humanist research is to interpret and understand individual worlds as they are subjectively constructed, in contrast to explanation of universal principles obtained through replicable experimentation.

As these two examples suggest, our epistemological and ontological assumptions form the basis of our **research methodology** – the way we approach research on a given topic and the particular methods we adopt. The methods we use often then determine the type of data we collect, which, in turn, influences the conclusions we draw. Thus, when we undertake research we make theoretical and methodological decisions about what questions to ask, what data to collect, and how to analyze the data collected. In doing so, we are making philosophical decisions that influence the nature and type of knowledge we produce and, ultimately, our understanding of "truth."

To illustrate, let us consider, for example, the question of poverty (see Box 13.01). There are many approaches one could take to study poverty (see Kitchin and Tate 2000). One could simply gather facts about poverty and then present them for interpretation by the reader of our research report. A second approach could be to try to explain the causes of poverty by collecting and scientifically testing data related to poverty. Both of these approaches, of course, raise questions about what the "facts" of poverty are and what the relevant data to collect would be. Another approach could be to observe how individuals interact with one another to create social conditions that maintain poverty – e.g., cycles of crime, low self-esteem, etc.

Box 13.01 Philosophies of science and the study of poverty

Empiricism
Facts about poverty would be collected and presented for interpretation by the reader (e.g., indices of poverty such as social welfare recipient, housing tenure, etc.).

Positivism
Poverty is explained through testing a hypothesis by collecting and scientifically testing data related to poverty (e.g., statistically testing whether poverty is a function of educational attainment).

Humanism
Poverty is understood by trying to gain insight into how poor people think about poverty and the world they live in (e.g., interview poor people on what it feels like to be poor, why they think they are poor, how they see themselves in relation to the rest of society).

Realism
Poverty is understood by trying to determine its root causes through an examination of the mechanisms underlying how society operates (e.g., examine whether poverty exists because of the uneven development of modernization).

Marxism
Poverty is explained through the examination of how poor people are exploited for capital gain (e.g., are poor people poor because it is in the interests of capital to retain unskilled, low-wage jobs rather than distribute fully corporate profit?).

Feminism
Poverty is understood by trying to adopt more emancipatory and empowering approaches that allow poor people to express experience and knowledge; seeks to understand how men and women differentially experience poverty as a result of the way patriarchy operates in society.

Postmodernism
Poverty is understood by trying to deconstruct and read the various ways poverty is constructed and reproduced in society (e.g., examine the ways poor people are excluded from society through unequal power relations).

Source: Adapted from Kitchin and Tate 2000: 20–2.

Or, one could try to understand how poor people think about poverty and what it feels like to be poor. Finally, one could try to explain poverty through examining the ways larger social and economic structures function to create and perpetuate inequalities. Each of these approaches is informed by different philosophical assumptions about the nature of the poverty and how we know about it. As you can imagine, the study results from each of these approaches would likely be quite different, with each study shedding light on poverty in different ways. When our research findings are used to form public policy, however, how we approach a topic, the ontological and epistemological assumptions we make, and the methods we adopt to collect and analyze data become especially critical. The policy "solutions" we formulate stem directly from how we define and investigate the "problem." Defining policy problems in different ways can result in very different policy solutions. Thus, theory is important.

Research Paradigms in Geography

We refer to the different approaches to research that stem from different philosophies of science as **research paradigms**. Research paradigms include a **theoretical framework** (a set of ideas, general propositions or analytical tools for understanding, explaining, and/or making predictions about a given phenomenon) and a related **methodology** (a set of methods or procedures for investigating a particular topic of study). Because geography is such a broad discipline encompassing both physical and social phenomena, geographic research has been influenced by several different scientific philosophies, and thus our theoretical frameworks and methodologies are quite diverse. Within human–environment geography, three sets of research paradigms have been prominent in the 20th century. These are: spatial science, humanistic geography, and critical theories.

Spatial Science

For much of its history geographic research was rooted in empiricism – a theory that states that knowledge only comes from evidence gathered through sense experience (observation) – and was largely *idiographic* (fact-gathering), or descriptive, in nature. It tended to examine patterns and processes primarily on a regional basis in order to understand particular places. The effort was to apply intuitive understanding to studying the unique and particular rather than search for any generalized explanations (Cloke et al. 2004). Then, in the early 1950s, a number of geographers began to argue that geography should become a *nomothetic* (law-producing) science; that is, geographic research should become more scientific in its method and should seek to identify underlying universal *laws* that *explain* spatial patterns and processes (Kitchin 2006). The argument was that up to that point in time geographic research produced empiricist descriptions that did not differentiate between causal correlations and accidental associations. In order to be sure that the patterns observed are in fact caused by a particular phenomenon, they should be tested scientifically. Thus, what ensued was the quantitative revolution in geography and the birth of **spatial science**.

Spatial science dominated geographic research in the 1960s and most of the 1970s and generated a search for law-like statements of order and regularity that could be applied to spatial patterns and processes (Cloke et al. 2004). In searching for orderly causal processes, geographic inquiry became a matter of asking: (1) How do objects and practices vary and/or move across the earth's surface? and (2) Why do these variations take the spatial forms that they do? Spatial science method primarily utilizes **quantitative data** (numbers or empirical facts that can be quantified) and generally employs **deductive reasoning** in which general theories or hypotheses are set forth and then tested through statistical analysis (Shaw et al. 2010). Ontologically, spatial science assumes that space and time are measurable and that objects of analysis (phenomena) are discrete (separate, distinct). The epistemological foundations of spatial science are objectivity and generality; that is, that general laws do govern the spatial organization of human behavior, and these can be discerned and predicted through objective observation and statistical analysis.

Humanistic Geography

Humanistic geography emerged in the 1970s and 1980s as a reaction against both the empirically based regional and landscape studies of the 1930s, 1940s, and 1950s and the positivist-inspired spatial science of the 1960s and 1970s. Humanistic geographers found the mathematical models of spatial science, which understood human actions as rational decisions based on perfect information in homogeneous and featureless spaces, overly simplistic, reductionist, and removed from the chaotic and complex worlds of everyday life. The earlier regional geography was also unsatisfactory because it tended to look for the sources of geographic variation in the ways of life of different social *groups* rather than focus on *individual* actors (see Entrikin and Tepple 2006). The interest of humanist geographers was to open up geography to "more realistic conceptions of humans as geographical agents who draw on their experiences, attitudes, and beliefs as well as their moral and aesthetic judgment, in making decisions that shape their environments" (Entrikin and Tepple 2006: 31).

Humanist geographers emphasize the human subject in a search for meaning and understanding. They argue that "[m]eaning is not something to be found in objects, but must be understood in relation to subjects. [P]lace, region, and landscape are not simply spatial categories for organizing objects and events in the world, but rather processes in the ongoing dynamic of humans making the earth their home and creating worlds out of nature" (Entrikin and Tepple 2006: 31). According to this view, people should be studied free of any preconceived theories about how they act, and that, for people, the world exists only as a mental construction, created in acts of intentionality (Kitchin and Tate 2000). This implies a preference for **inductive reasoning** in which the research effort is to produce primarily **qualitative data** (unstructured words, pictures, sounds) that is *interpreted* (as opposed to "analyzed") and then construct a theory for understanding the phenomena observed. The goal of humanistic geographic research is to reconstruct the worlds of individuals, their actions, and the meaning of the phenomena in those worlds to understand behavior. In contrast to scientific approaches, which treat phenomena as external objects that can be studied objectively, humanistic philosophies recognize subjectivity and demand that we reflect on our own consciousness of things in our experience to come to a deeper understanding of the world.

Critical Theories

Around the same time that humanistic geographers were beginning to influence geographic research, other groups of geographers were also advocating new approaches to research. Collectively we may loosely refer to this group of new approaches as **critical theories**, so named because the proponents argued for a more critical approach to scientific inquiry and in so doing challenged the ontological and epistemological bases of geographic research, especially spatial science, as well as the purpose of research. These geographers attacked the underlying positivist basis of spatial science for ignoring the various ways in which social structures and social relations influenced spatial patterns and processes and for reinscribing how reality *seemed* to be rather than considering how it *might* be different under different social conditions (Kitchin and Tate 2000). Critical theorists asserted that geographic research should serve the political purpose of revealing structures of domination, exploitation and oppression within society with the objective of effecting positive social change.

Critical theory encompasses a wide range of theoretical perspectives, among which we discuss just a few that have influenced human–environment geography. **Critical realism** emerged as a critique of spatial science and its assumption that one can discern causality from spatial variation as well as a critique of humanism, which places too much emphasis on the individual. Critical realists are concerned with investigating the underlying mechanisms and structures of social relations and identifying the "building blocks" of reality (Kitchin and Tate 2000). Although they recognize the need for seeing individuals as part of wider worlds, they argue that things happen in the world to cause it to change and these events can only be understood as having been produced by deeper structural forces and their causal mechanisms (Shaw et al. 2010). This is similar to the structural approach of **Marxist** geography, which focuses specifically on the economic and political constraints imposed upon spatial patterns and seeks to identify how social relations vary over space and time in order to reproduce and sustain capitalist modes of production and consumption (Kitchin and Tate 2000).

Feminist approaches and critiques began to gain influence in the discipline in the 1980s and have expanded significantly since the mid-1990s. Feminist geographers have critiqued geographic inquiry in two main ways. First, they argue that geographic research has ignored the lives of women and the role of patriarchy in society. Second, they have criticized the way research is conducted and the nature of the knowledge that is produced. The argument is essentially that because geographic knowledge has been predominantly produced by men who have ignored women's experiences and perspectives, it represents men's views of the world. However, such knowledge has been presented as some kind of universal truth. Thus, feminist geographers have adopted an epistemology that challenges conventional ways of knowing by questioning the concept of "truth" and validating other sources of knowledge such as subjective experience. In doing so, feminist geography turns our attention to the power relations and *non-neutrality* of research (Women in Geography Study Group 1997) and attempts to expand our understanding of the world by consciously incorporating attention to gender and gender relations in its methodologies. Similar critiques about the partiality of knowledge and the underlying power relations that generate it have come from **postcolonial theory**, which takes conventional geographic research to task for its bias toward Western ways of thinking and producing knowledge that marginalizes the perspectives and experiences of non-white, non-European, and non-elite peoples.

Postcolonial theory is part of a larger philosophical approach known as **postmodernism**, which began to influence the study of culture and society in the 1980s. **Modernism** is concerned with the search for a unified, grand theory of society and seeks to reveal **universal truths** and meaning, and both positivism and Marxism are examples of modernist approaches to knowledge. In contrast, postmodernism is based on the idea that there is no one answer nor any one discourse that is superior to another, and no one's voice should be excluded from dialogue (Dear 1988, cited by Kitchin and Tate 2000). Postmodernists argue that modernist approaches have failed to adequately account for differences in society. Rather than seeking "truth," postmodernism offers "readings" and "interpretations" in place of "observations" and "findings." Its analytical method consists of deconstructing (tearing apart) culture and social practices. For example, in the example of poverty cited above, postmodernism tries to deconstruct and read the various ways poverty is constructed and reproduced in society and how poor people are excluded from society (Kitchin and Tate 2000).

Drawing on the critiques emerging from various critical theories, **political ecology** is an approach toward studying human–environment interactions that has gained a lot of ground within geography over the past couple of decades (and informs a lot of the perspective of this book!). Political ecology is the study of the relationships between political, economic, and social factors and environmental issues and changes. It differs from other approaches to ecological studies by politicizing environmental issues and phenomena. That is, it integrates ecological studies with analysis of political and economic structures and processes in its examination of topics such as land degradation and marginalization, environmental conflict, conservation and control, and environmental identities and social movements (Robbins 2004). Through incorporating attention to differential power relations, issues of scale and the influence of discourse in influencing development policies, political ecology has offered more complex and nuanced analyses of environmental issues that go beyond simplistic explanations of population pressure and overuse.

Now that we have discussed research paradigms and methodologies, we are ready to discuss the actual collection of geographic data and its analysis. Each of the research paradigm shifts discussed above has introduced new methods into geographic research and broadened both the scope of study and our knowledge and understanding of the world. For example, spatial science expanded the horizons of geographic knowledge from the local and particular to an analysis of general patterns and processes. Humanist geographers incorporated attention to the individual and started using methods that provided structure to the tasks of meeting people, interacting with them on an everyday basis, talking to them in depth and "listening to their voices" (Cloke et al. 2004). Critical theories again broadened the scope and scale of our analytical tools by revealing complex structures and processes at work that underlie geographical phenomena and provided a means for understanding linkages between local and global processes and difference in experience within societies. The result is that geographers have at hand a number of tools for collecting and analyzing qualitative and quantitative data and for integrating these to produce rich geographies of human–environment interactions. At the center of such geographic research lies the method of **field research**, also known as **fieldwork**.

Collection and Analysis of Geographic Data: Approaches to Field Research

Field research has a long history in geography, and the early years of the 20th century in particular were marked by a keen interest in fieldwork and field-based data – that is, traveling to distant places to observe and collect data as opposed to conducting laboratory experiments. By the 1920s, fieldwork experience and knowledge of field methods was considered a mandatory basic attribute of all geographers. This sentiment was related in part to a major thrust toward large-scale land use mapping that emerged after World War I and required extensive field study. To the regional geography of the 1930s, field research was considered the principal source of information and knowledge. Much of this effort was devoted to understanding the interrelations between humans and their physical environment, the understanding of which necessitated on-site field studies. An interest in foreign places after World War II further encouraged field research through the 1950s.

In the 1960s, however, as the focus of geographic research shifted toward spatial science, geographic methods shifted away from collecting **primary data** (data collected at first hand through observation, interviewing, or collection of samples) from the field to utilizing non-field-based **secondary data** sources (pre-existing data, or data collected from someone else such as census data), and large-scale survey methods became more common. This was made possible by computer technology, and over time geography departments began hiring technique-oriented geographers in place of foreign area specialists.

In the latter 1980s, however, interest in field research was revived by the emergence of **new regional geography** and its use of critical theory. Whereas the regional geography of the 1920s and 1930s tended to focus on the uniqueness of events in particular places, the new regional geography recognized that unique local events are usually tied to wider structural processes. Understanding the interplay between local particularities and "global" (or national) structural processes became the focus of research within this subfield of geography as well as in the study of human–environment interactions undertaken by political ecologists. Fieldwork once again became central to geographic research.

Within the subfield of human–environment geography, a lot of fieldwork is geared toward investigating how humans understand their relationship to the non-human world and toward measuring and evaluating the impact of human activities on the environment (Robbins 2010). This requires collecting data from and about physical environments as well as human social, cultural, economic, and political processes as they interact with those environments. In the discussion that follows we review different types of data that human–environment geographers collect and the methods they use for collecting these types of data from the field.

Collecting Data from and about the Physical Environment

The data from and about physical environments that geographers are generally most interested in pertains to landforms, climate, and vegetation. The objective is to investigate the form and function of physical systems, the factors that influence their variability over space and time, and how they are impacted by human activities. The methods used to study physical landscapes (landforms, vegetation) and climate and the data geographers collect often have the objective of helping us understand change over time, but increasingly attention is being directed to the prediction of change that might occur in the future (Crozier et al. 2010). Thus, data collection techniques are generally of two types: historical (paleological) and modern survey. Core sediment collection, pollen fossil and tree-ring analysis, soil sampling and particle size analysis, archival data on temperature and precipitation, field experiments, and surveying of current conditions such as the distribution of plants are all methods used to collect data about physical environments. Human–environment geographers are particularly interested in the interactions between human activities and physical environments. For example, Dr. Grissino-Mayer (Box 13.02) describes below how he uses tree-ring analysis to study environmental history. Dr. WinklerPrins (Box 13.03) then discusses her research on how human activities contribute to soil formation.

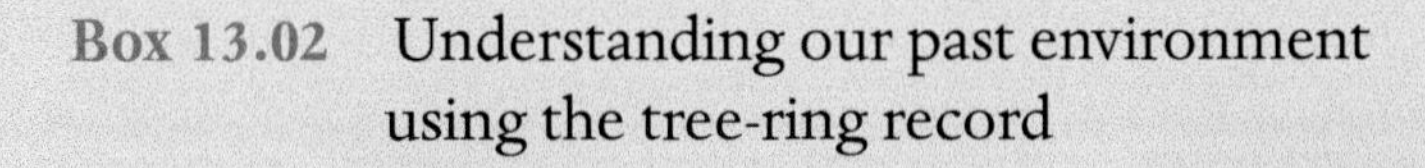

Box 13.02 Understanding our past environment using the tree-ring record

Figure 13.01 Dr. Henri Grissino-Mayer stands next to a contorted Douglas fir tree on the lava flows of New Mexico. The tree is well over 400 years old. *Source*: Henri Grissino-Mayer. Used with permission.

If anyone had told me 25 years ago that I would be spending the rest of my life inspecting tree rings for information on our past environment, I would have suggested they see a doctor. I first heard of **dendrochronology** back in 1983 when I was an undergraduate at the University of Georgia. Dendrochronology is the science that uses data gathered from tree rings to improve our understanding of past environmental processes, such as climate change and wildfires, even past volcanic eruptions, earthquakes, and air pollution. A series of quirky events during my Master's program allowed me to pursue this topic and I immediately became hooked on its level of precision, its power and breadth for interpreting the past environment, and its eloquence in design despite it being a highly quantitative discipline. My training as a physical geographer soon narrowed to specific skill sets in biogeography, the discipline within geography where dendrochronology has found a firm foothold in the past 20 years. Dendrochronology became my passion, and it remains so today.

The dating of tree rings was originally developed in the early part of the 20th century by Andrew Ellicott Douglass, who discovered that tree rings vary in their widths from year to year because climate in any one region differs from year to year. This creates a unique pattern over time in the tree-ring record, as unique as a fingerprint or a DNA sequence, and these patterns within the tree rings can be tied together across geographic areas. Even the tree rings in long-dead pieces of wood can be absolutely dated to the exact year in which they were formed by simply matching the outer sequence of their tree-ring patterns with the inner sequence on tree rings that are already dated. This is the technique by which old historic structures and even prehistoric pueblos in the American Southwest can now be absolutely dated with annual precision.

My interest in the use of dendrochronology first focused on the reconstruction of climate back in time using tree-ring data. The premise is simple. As we all know, one can look at the tree rings on a stump, count them, and determine how old the tree was. Dendrochronology takes this one step further, focusing on the variability in the widths of those annual rings. If weather conditions were favorable in a given year, the tree will grow a wide ring, while poor growing conditions (for example, drought) will cause a narrow ring to be formed. By calibrating historic climate data of the 20th century with tree-ring data from the same century, we develop an equation that basically says, "A tree ring with this width means this much rainfall fell in the year." Because we then have tree-ring data for many centuries in the past (using old-growth trees, logs from historic structures, and even plain wood lying on the forest floor), we can extend this reconstruction to well before climate data records were kept. This technique, for example, has given us insights into just how wild the climate of the 20th (and now the 21st) century was. In fact, tree-ring data have shown us that the decade of the 1990s was the hottest decade in the last 600 years!

In New Mexico near the city Grants, just west of Albuquerque, lies a lava badlands known as "El Malpais." Several years ago, my research team and I found these ancient lava flows to be covered with open woodlands of Douglas fir and ponderosa pine trees. How they grow here is anyone's guess. Coring these trees, we were shocked to learn how old they were. One Douglas fir that is still living is over 1290 years old! We've also found Rocky Mountain juniper trees here that are nearly 2000 years old. Using old wood we found lying on the lava, I was able to develop a continuous tree-ring chronology back to 136 BC and to reconstruct annual rainfall for west-central New Mexico for the last 2129 years. This reconstruction shows that century-scale drought and periods of wetness are cyclical in the Southwest. In fact, the period from 1800 to the present has been anomalously wet. If the Southwest ever returns to the dry conditions experienced during the "Mega-Drought" from about 1560 to 1610, human populations could be in major trouble.

I also use tree-ring data to address another major issue, one whose ferocity is witnessed each year in many portions of the United States. Wildfires appear to be growing in their intensity and spatial extent, but before we evaluate just how different wildfire activity is today, we must know about wildfire activity in the past. Enter tree rings! Many wildfires actually do not kill a tree. Instead, they leave their mark in the trunk of a tree as an obvious scar, and then the tree will attempt to grow over the injury, thus preserving a record of that wildfire event. Some ponderosa pine trees in the Southeast can record up to 30 or more wildfires this way in their tree-ring record. By dating the centuries-long record of wildfires found in scarred trees (which are pervasive throughout the US), we can gain a sense of the historical record of fires before we started putting them out in earnest in the early and mid-20th century.

A final application illustrates the versatility of tree-ring data to help provide a more accurate record of human history, as well. Because tree rings can be dated to their exact year, I am often asked to date the tree rings in the logs and timbers of historic structures. The outermost ring that formed before the tree was cut down can be interpreted as the construction date of that house, barn, or log cabin. This dating technique has helped many historians, historical agencies, and architects determine exactly when a structure was built. Often, the year of construction is not as long ago as previously reported. For example, the Abraham Lincoln Birthplace log cabin in Hodgenville, Kentucky, was built from trees cut in the 1840s and 1850s, yet President Lincoln was born in 1809. Tree rings therefore help us correct historical inaccuracies!

Source: Contributed by Henri Grissino-Mayer, PhD, Department of Geography, University of Tennessee.

Box 13.03 Doing human–environment geography research in the Brazilian Amazon

As a graduate student I was drawn to the Brazilian Amazon for its environmental issues, and started doing research there about 20 years ago with a focus on smallholder agriculture, local soil knowledge, and environmental history. I had lived in Brazil as a child and had a familiarity with Portuguese, which helped me get started. Some fortuitous connections via faculty and alumni from my graduate department helped secure a research location and the vital institutional collaboration needed for international work. For my dissertation I spent close to a year living and working on the floodplain of the Amazon River near the city of Santarém, in the state of Pará, Brazil, understanding floodplain livelihoods and soils.

From this research on the rural floodplain evolved research on urban home gardens as I had observed the importance of relatives (social connections) in the city to the residents of the floodplain. I "migrated" to the city alongside the people with whom I had conducted research on the floodplain, paralleling a common trend in the region (which is today 70% urban). By combining an analysis of the flows of people and products, collected through the use of semi-structured interviews, with an inventory of garden plants, we (my field assistant and I) were able to demonstrate the critical importance of the production of home gardens for urban survival.

During our work on the flows of products in home gardens, we noticed a distinct soil-management technique that created a product locally called *terra queimada* (TQ, or burnt earth), which is used as a soil conditioner. This observation led to my current research on the formation of Amazonian Dark Earths (ADEs) as the technique creates organic char, a key ingredient in ADEs. ADEs are anthrosols, soils formed under the influence of human action in the past, that are very dark in appearance and highly fertile. They occur as relatively small patches throughout the Amazon Basin and are actively sought out by current residents since the general background soil is not as fertile. The existence of ADEs challenges many assumptions about the Amazon environment. Throughout the last two decades there has been much research examining the soils themselves and their current use, but what remains poorly understood is how human activities contributed to their formation in the past. Given the severe depopulation of the region following European contact (estimated at 90% of the population), there are few groups of people with livelihoods that are similar today to what they were in the past. The people with whom I had been doing research on the floodplain and in the city are of mixed ancestry and are known as Caboclos, Amazonia's indigenous peasantry. Some of their resource-management strategies derive from Amerindian ways, and the soil management we witnessed in home gardens is likely such a strategy.

The research question for my current project is: does the use of TQ lead to the formation of ADEs? The methods I am using to try to answer this are a combination of soil fertility analysis and semi-structured interviews. Doing any type of research on and with soils in Brazil requires a local collaborator (foreigners are currently not permitted to handle Brazilian soils), a local lab where the analysis can be done, and the necessary research clearance and visa. My collaborator is Dr. Newton Falcão, a soil fertility expert from INPA, Brazil's national institute for Amazonian research. He has long been working on ADE fertility and on finding ways to re-create this fertile soil. He was the one who took the soil samples and supervised their analysis in his labs. I was in charge of the semi-structured interviews with garden managers, the ones making and using TQ, and conducting the analysis of this qualitative data.

We selected 40 home gardens in Santarém; most were ones that I had worked in before; others we obtained through snowball sampling. In each home garden we sampled the soil in two locations, one being the pile of TQ, and one the background soil of the garden (as best as that could be determined). The soil was sampled at three depths, 0–10 cm, 10–20 cm, and 20–30 cm. Two repetitions of each sample were taken. While Dr. Falcão and his soil technician were sampling the soil, I would conduct a semi-structured interview with the home garden manager. The objective of the interviews was to ask about the process of creating TQ and to try to understand the source of that knowledge. My field assistant assisted me with this interview and an American graduate student took at GPS point of our sample location.

Once all the sampling was completed, the soil samples were sent to the soil fertility lab at INPA, where they were analyzed for levels of N, P, K, Ca, Mg, Al, C, organic matter, pH, and a variety of micronutrients. The interview data were manually coded and collated and a map has been created of all of our sample locations using GIS.

Our results to date indicate that TQ has many of the chemical properties documented in ADEs. We also found that the background soil in the home gardens, although much less fertile than the TQ material, was more fertile than the typical soils in the region. This indicates that the use of TQ over

Figure 13.02 Dr. WinklerPrins with her research team in Brazil. *Source*: A.M.G.A. WinklerPrins. Used with permission.

time likely does lead to ADE formation. Our project was small in scale and low in number due to various logistical challenges, and a much larger study needs to be mounted to be able to conclude this definitively.

The results of the interviews indicate that there are people who deliberately make TQ to fertilize their garden. Others do it as a way to get rid of household waste. This runs parallel to findings from archaeology and from ethnographic research work with Amerindians which demonstrates that there is a range of ADEs formed through human action, some very deliberate charring to fertilize crops in fields and some incidental ADE formation as a consequence of long-term inhabitation and waste disposal. Truly understanding how and why TQ is made necessitates much more time in the field, including participant observation. Home garden soil management is a part of daily practice and few people can articulate in an interview why they do what they do. Their behavior needs to be observed to be better understood.

Source: Contributed by A.M.G.A. WinklerPrins, PhD, Department of Geography, Michigan State University.

Collecting Data from and about People

Geographers collect and utilize data from and about people using a number of different methods. Primary sources of data might come from surveys, interviews, participant observation, focus group discussions, mental maps, and oral histories. Secondary sources of data include documents such as newspapers, diaries, census reports, and other types of texts. The types of data generated for human geographic research can include both quantitative and qualitative data. Quantitative data are structured empirical facts that can be quantified and analyzed using statistical methods. Examples include facts such as household incomes, number of residents, number of migrants, age of marriage, and so on. In contrast qualitative data are unstructured and usually consist of words, images, or sounds that reflect attitudes, beliefs, values, historical events, etc. Although the different research paradigms discussed above tend to favor either quantitative or qualitative data analysis, increasingly geographers combine data collection/analysis methods in what is called a **mixed method** approach. For example, survey data might be used to identify general patterns within or across populations and are then supplemented with in-depth interviews or focus group discussions with select individuals to gain deeper insight into what factors might generate general patterns observed. Or, in the example of Dr. WinklerPrins' research described in Box 13.03 above, soil sampling and analysis is combined with semi-structured interviews to understand the effect of soil management technique on soil type formation. Dr. Jennifer Mandel's research in post-earthquake Haiti, described below, is another example of how different sources of data are utilized in a mixed method approach that combines quantitative

Box 13.04 Providing critical information in post-earthquake Haiti

It's funny how life comes full circle. I started graduate school in a Masters of Public Affairs program intending to work on housing issues in developing countries. At the end of my first year, I had my first consultancy with USAID in Côte d'Ivoire, in which I conducted two studies, one on the Ivoirian government's decentralization process and one evaluating a squatter settlement redevelopment project. I then decided to continue my interest in conducting socially important research and applied to PhD programs in geography. I ended up in geography because that was the field of most of the literature I was reading for my ongoing squatter settlement research. As a PhD student, my focus shifted to women's livelihood strategies in Benin, still emphasizing practical development applications. Throughout graduate school and during seven years as a college professor, I occasionally took consulting jobs focused on development project monitoring and evaluation.

When an opportunity to do research for a non-governmental organization (NGO) in Washington, DC, presented itself, I grabbed it. A year and half later, I decided to set up shop as a full-time consultant. Although I mainly do monitoring and evaluation, I often have opportunities to do other types of research such as in my current job, where I do research in post-earthquake Haiti to support the disaster recovery efforts of an organization called Internews-Haiti.

Internews is an NGO that empowers local media worldwide to provide quality, independent news and information. In a crisis context such as Haiti's following the January 12, 2010 earthquake, this is critically important. People need news and information because it provides access to humanitarian assistance that will ensure people's survival. News and information about recovery and reconstruction processes are also key to giving people a sense of security and stability. Internews provides this through a 15-minute daily radio program called *Enfomasyon Nou Dwe Konnen-ENDK* (News You Can Use). It has also provided radio stations resources to facilitate their reconstruction/recovery and training to enhance local journalists' capacity in investigative reporting, especially around humanitarian assistance.

To support these activities Internews included a research department as part of its Haiti project. While they understood this would be key to the project's success, they really didn't know what they needed by way of research. So I spent my first two weeks in Haiti learning all about humanitarian assistance communications and Internews' work. I then designed the research program that I thought could help the project. We have three objectives: (1) track audience information needs; (2) assess the pre- and post-earthquake status and assistance needs of local media outlets; and (3) monitor and evaluate Internews activities. The audience assessment serves three main purposes: (1) it informs ENDK programming so that it responds to Haitians' information needs; (2) it provides the humanitarian community information about Haitian's information needs so they can also appropriately respond; (3) it facilitates monitoring ENDK's success.

To accomplish this, I work with nine Haitian researchers on several different research activities – some that serve multiple purposes. It's important to work with local staff because they speak the local language and understand the culture. We integrate both quantitative and qualitative methodologies so that we can answer questions about what is going on and how and why it is happening.

We started the audience assessment with 24 focus group discussions (12 each with men and women in different earthquake-affected areas that were also differentiated based on how their homes were impacted – completely destroyed, damaged, or intact). The purpose was to understand their access to and use of the media pre- and post-earthquake and their current information needs. This provided a base line for evaluating Internews-Haiti's success in meeting audience information needs and information critical to developing the survey. For example, the focus group discussions provided radio names that we then used to pre-code survey question answers.

We analyzed the focus group discussion data using *N-vivo* software to identify broad patterns in people's responses. The analysis will be compared to the end-line focus group discussion that will be conducted at a later date. The objective will be to assess how people's access to and use of the media and their information needs have changed since the beginning of the project. It will also help in determining if ENDK has successfully provided people with news they can use. Initial results of the comparative analysis indicate that it has.

Following the focus group discussions, we began a rolling audience survey in the same neighborhoods to track how people's information needs changed over time. Every two weeks, four team members conduct face-to-face survey interviews in the same places. They either work in a commercial district randomly stopping adults or go door-to-door in neighborhoods and camps requesting surveys at every third door. The survey data are entered into an SPSS spreadsheet by another Haitian researcher the following day. My assistant, one of the nine Haitian staff, does a random accuracy check of all entered data and analyzes it using SPSS. Interestingly, very simple analyses, frequencies, and cross-tabs are all that is needed for our research. The analysis looks for differences in people's responses along four different dimensions: gender, age, camp / non-camp living, and geographic area. The results are published weekly and reported to the ENDK staff and the humanitarian assistance community.

Between March and July 2010 health issues were the number one requested information theme by survey participants. However, because survey data are limited, it's impossible to know which people really want information. Working with several other NGOs, we integrated questions about health

Figure 13.03 Dr. Mandel and the research staff of InternewsHaiti. *Source*: Jennifer Mandel. Used with permission.

information needs into many focus group discussions, including the end-line audience assessment. I will integrate and analyze these data. Then we'll publish a brief report summarizing the results. That ENDK and the humanitarian community have satisfied this information need will hopefully be evident in later survey results when a new issue should become the most important to Haitians.

The media assessment involved visits and in-depth interviews with radio station management. Using highly structured interviews, another researcher discussed each radio station's pre- and post-earthquake status, including their funding sources, program format, staff, and equipment. To coordinate assistance efforts, the results were made available to many NGOs working to support the media.

The results of both the audience and media assessments will be integrated in a report analyzing changes in the media landscape as a function of the earthquake. Another report will detail Internews' impact assisting the media recovery, training journalists, and, most importantly, addressing Haitians' need for news and information.

Source: Contributed by Jennifer L. Mandel, PhD, Director of Research and Evaluation, Internews-Haiti.

and qualitative analysis. Which method(s) are utilized and which types of data are to be generated depend on the purpose of the research, the particular research questions one is investigating, and the kinds of practical constraints one faces.

Using GIS and Remotely Sensed Data – "Peopling the Pixels"

Although the majority of research methods used by human–environment geographers are the same as those used by geographers working in other subfields of the discipline, human–environment geographers are specifically interested in understanding the multiple processes through which humans use, alter, or relate to non-humans and vice versa (Robbins 2010).

This research focus creates a methodological challenge because such ecological interactions are enormously complex and fluid, and the direction of causality is not always clear. It requires the amassing of a wide range of data from both the physical environment and human societies, making the management and analysis of such data a potentially Herculean task. It raises questions about what kind of evidence is required and what counts as credible evidence (Robbins 2010), and the issue of scale also becomes problematic. For example, surveying vegetative cover or building stock over a relatively small territory at one moment in time might be fairly easily accomplished. But, how do we capture broad-scale seasonal patterns that encompass hard-to-reach places?

To overcome these issues, geographers have increasingly turned to the use of **remote sensing** and **geographic information systems** (GIS). A GIS is a computer-based system designed to capture, store, manipulate, analyze, manage, and present **geographically referenced data**, that is, data that are given a location on Earth's surface. Geographic information systems are able to store large quantities of different types data in a single database. Typically a GIS allows a researcher to arrange information about a place on a series of maps that are overlaid on one another (see Figure 13.04). Once so arranged, information on the different layers

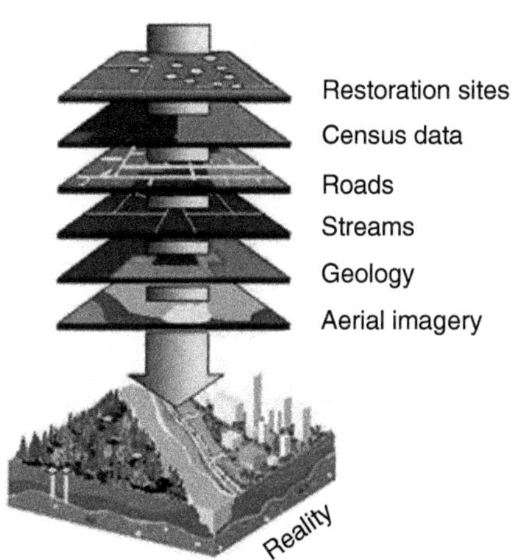

Figure 13.04 A geographic information system.

can be compared and analyzed, thereby facilitating the identification of spatial patterns and processes. Geographic information systems are used by scholars as well as industry to perform a number of different functions related to activities such as natural resource mapping, urban and regional planning, emergency management, habitat restoration, transportation services, and environmental modeling. One of the sources of data used in a GIS comes from remote sensing.

Remote sensing of the environment involves collecting data about features on Earth's surface and in the atmosphere without being in direct contact with those features (Stow 2010), or the collection of data about an object from a distance. Geographers usually accomplish this with the use of mechanical devices called remote sensors that are mounted on helicopters, planes, and satellites and measure the transmission of electromagnetic energy from reflecting and radiating surfaces (Pidwirny et al. 2010). Aerial photographs and ultraviolet, infrared, and radar imagery are examples of the form remotely sensed data may take (see Figure 13.04).

Remote sensing may be used for *reconnaissance* (to obtain a qualitative perspective of an environment or landscape), to *inventory* land surface features such as natural resources, to *map* phenomena such as land use, road networks, or insect damage, and to *monitor* landscape change (see Stow 2010). The applications of remote sensing for analyzing physical environments are apparent, but increasingly geographers are attempting to incorporate remote sensing methods into social science, or the "human" side of the human–environment equation – an effort some have referred to as "peopling the pixels" (Liverman et al. 1998). The contribution remote sensing makes to human–environment geography is that it helps provide important information about the biophysical contexts in which humans live, work, and play. "People live their lives in contexts, and the nature of those contexts structures the way they live" (Rindfuss and Stern 1998). Remote sensing data may also be used in **triangulation** (cross-checking) with other types of data to assess, and possibly alleviate, real-world problems such as the threat of famine, as described by Dr. John Unruh in Box 13.05.

Box 13.05 Working with the USAID famine early warning system in Ethiopia

I worked for the USAID-funded Famine Early Warning System (FEWS) in Ethiopia from 1998 to 2000 after earning my PhD in geography from the University of Arizona. At the time, FEWS sought to provide analysis-based early warning of extreme food shortage (which could potentially lead to famine) in Ethiopia and select other countries in Africa. Through a combination of specific monitoring techniques, early warning could be provided up to three months prior to the onset of severe food shortage. The overall purpose of FEWS was to provide the time needed for shipments of food aid to be sourced and loaded in donor countries, shipped to destination countries, offloaded and then transported to often remote locations before the onset of extreme food shortages or famine. More specifically, the objectives were to: (1) provide food aid in sufficient quantities to those in need; (2) terminate occurrences of extreme food shortage in specific locations before such shortage spread to much wider areas thus causing famine – in other words prevent the jump from extreme food shortage (geographically constrained) to full-onset famine (geographically widespread); and (3) to save on the overall cost of food aid delivery – with ocean shipments able to transport much larger volumes of food aid at much less cost, albeit more slowly, than air shipments of food aid.

Our approach to famine early warning in Ethiopia was based on monitoring physical and human geography hunger-related measures. The primary physical geography technique utilized remote sensing imagery from satellites to determine the photosynthetic vigor of plant communities. In this case, satellite imagery was provided every 10 days on spectral information given off by plant photosynthesis, which indicated if plant communities were stressed (usually due to too little or too much water), compared to a long-term average. While such information could not separate crops from weeds, bushes, and trees in Ethiopia due to their highly mixed arrangement on the landscape, if the general vegetation was doing poorly, then rain-fed crops and grazing were likely doing poorly as well. By taking into account the timing of the agricultural calendar, comparisons with the long-term average vegetation vigor for specific periods of the year and specific locations, a general estimate could be made of the coming harvest (good, average, poor, extremely poor). However, this approach had to be used with other indicators because the satellite only "saw" the vegetation canopy it first came into contact with (the uppermost canopy). Thus it had difficulty assessing the vegetation layer below the uppermost canopy, occasionally causing problems. For example if the bush layer was green, but the grass layer below was dead, then the satellite would only see the "good"-status vegetation and not the brown grass underneath. If the local livelihood was cattle pastoralism (cattle graze on grass, camels browse on bushes) then the local livelihood would be in trouble despite the satellite indicating that the area was in good shape. The approach to handling this limitation is to also use human geography indicators.

A primary human geography early warning indicator used together with remote sensing was market price monitoring. In this case, the prices of key commodities were monitored in important markets around the country, which revealed what was going on in larger "market sheds" (geographic areas from which people traveled to attend a specific market). Most important in this regard were the prices of grains and livestock relative to each other.

Figure 13.05 Vegetative cover over Ethiopia; darker areas (excepting lakes) indicate more surface biomass. *Source*: USGS/EDC. Reprinted by permission of USGS.

Thus when the price of livestock (usually cattle or camels) dropped dramatically and at the same time grain prices rose dramatically, then we were concerned that severe food shortages were looming.

An additional indicator was the collection and compilation of reports from a wide variety of NGOs, government agencies, district officials, aid agencies, and others who traveled through areas of concern. When such actors reported problems, we would verify these and compare them with other indicators. Like remote sensing and market price monitoring, these regular reports allowed us to monitor very large areas.

We also monitored a variety of livelihood-specific indicators in certain areas. For example, when farmers began selling large amounts of charcoal at the roadside (much larger than usual), this could be an indicator of livelihood stress as people who are no longer able to produce food begin to engage in extractive activities. Furthermore, when children were taken out of school, sent to live or work with relatives, or began to suffer from sickness in significant numbers (above average) this could be an indication of larger-scale extreme food shortages. Use of wild foods, selling of livelihood assets, eating fewer meals than normal, and migration were also indicators of possible population-wide problems associated with food shortage. Finally, field assessments comprising visits to specific locations thought to be the initial pockets of extreme food shortage were important in order to move quickly with analyses regarding severity and spatial extent of food shortage.

Most important in the above-noted "tools" of early warning were their agreement with each other. If all indicators agreed that the situation departed from normal or average, then the certainty of the early warning was higher than if the indicators did not agree. Disagreement between indicators usually required us to visit the locations of concern to ascertain why such disagreement was occurring, and what the actual situation was.

An important issue was the limitation of only three months of early warning. Ideally a year or more would have been more useful. However, the problem was that for periods greater than three months, the indicators being monitored became lost in the variability of the average functioning of the local human and physical geography, such that actual indications of "deviation from average" could not be discerned.

The Ethiopian government became quite adept at using early warning information to address its food security problems. It currently has a number of pre-positioned food aid warehouses distributed in famine-prone areas of the country. As such, when a local extreme food shortage has been identified, the food aid response can be quick and appropriate in scale, dealing with a food shortage before it spreads to neighboring areas.

Source: Contributed by John Unruh, PhD, Department of Geography, McGill University.

Understanding the "Big Picture" of Human–Environment Interactions

Now that you have some idea about the kinds of research methods that geographers use and the type of research we undertake, we turn to the question of analysis and geographic knowledge generation. How do geographers arrive at a theoretical understanding of human–environment interactions? How do we make sense of our analyses in a broader context? What do our research observations, findings, and interpretations mean in terms of the "big picture" of human–environment interactions? As you might imagine, there are many ways this can be done. To illustrate one way, let us return to the story that opened this chapter.

The event described in the opening story is a protest march and demonstration that was organized by the National Fishermen's Forum and a collection of fishworker unions based in the Indian state of Kerala, which lies along India's southwest coast. Fisherfolk from several fishing communities in Kerala gathered in the state capital that day to express their discontent with the central government's new policy allowing foreign fishing fleets into India's territorial waters – a policy that these traditional artisanal fisherfolk were concerned would adversely impact their fishing efforts and ability to earn a livelihood. Since the 1970s artisanal fishing communities had been observing a decline in their harvests as the forces of national development and globalization began transforming the fisheries sector in profound ways. Now it seemed that a new threat – in the form of foreign ships – further threatened their right to the sea. As it turns out, the demonstration described above was the latest event in what has been an organized, sustained resistance movement on the part of traditional fisherfolk to a series of environmental and economic dislocations that have unfolded over the past four decades.

One framework geographers use to understand the causes and consequences of ecological crisis, such as the one faced by Kerala's fisherfolk, relies on the analysis of **commodity chains** and **livelihoods**. In this section we explain how a commodity chain analysis reveals the links between production in one place and consumption in another which, combined with livelihoods analysis, indicates how broad-scale economic, political, and environmental transformations impact people's ability to survive. The focus of our explanation will be the global fish food commodity chain and fisherfolk livelihoods in Kerala, India.

A commodity chain is a network of labor and production processes beginning with the extraction or production of raw materials and ending with the delivery of a finished commodity (product) with **added value** at each point along the way. One of the outcomes of economic globalization is that the distance products travel from their point of production to final consumption has expanded considerably, and this is true of our food. By tracing commodity chains we can gain an understanding of the ways in which sites of consumption are linked to and interact with sites of production. To illustrate, let us take the example of frozen shrimp produced in India. The frozen shrimp commodity chain begins with the fisherman who harvests the shrimp. Once harvested, the shrimp are then sold via auction to agents who work for companies that process the shrimp for export. In between the fishermen and these agents are middlemen who have loaned money to the fishermen in exchange for the right to sell their harvests for a (hefty) portion of the sales. The shrimp then go to sheds for processing, which consists of peeling, freezing, and packaging for export. The processing of shrimp is performed by women working for very low wages in very poor conditions. Once they are packaged, the Indian exporter sells the shrimp to importers in the US who then distribute them to retail outlets where they are purchased and eaten by consumers.

What is significant about the frozen shrimp commodity chain is that the emergence of consumer markets for luxury seafood products such as shrimp in countries such as the United States in the early 1960s has influenced the nature and direction of fisheries development in countries such as India (see Chapter 12) over the past 50 years. Prior to the 1960s India had pursued an integrated rural development program centered on increasing production of fish for domestic consumption and improving the welfare of fisherfolk communities, which were among the poorest of the poor in India. This development approach included a technological component intended to improve the productive capacity of traditional boats targeting domestically consumed species of fish like sardine and mackerel. (Shrimp at the time were very cheap and so abundant that the excess was used as fertilizer.) With the emergence of new consumer markets in the West, development policy quickly changed in response. Mechanized trawlers designed specifically to target shrimp were introduced, which completely transformed India's fisheries both economically as well as ecologically. First, trawlers are quite expensive so only a small portion of fisherfolk households could afford them. The remainder continued to use traditional craft and gear or became laborers on the new boats. Many trawlers are actually owned by "outsiders" and new forms of investment have reconfigured production relations within the sector. Second, trawl nets are quite destructive to marine habitats. They destroy nesting grounds and capture a lot of other marine species that are then discarded as waste. The effect of this is to reduce stocks of other fish species that could be harvested and sold locally by traditional fishermen, and the disposal of this "bycatch" causes further environmental damage to marine systems. Third, the rush to cash in on "pink gold" has also led to overfishing, which threatens the long-term viability of the fisheries sector. Finally, systems for distribution (or marketing) of fish and seafood have also undergone transformation.

Previously fish and shrimp were sold to petty traders on the shore (most of whom are women from fishing households) who then sold it directly to consumers. The emergence of global consumer markets and the expansion of fish and shrimp production to chase these markets have reconfigured the commodity chain and the structure of distribution relations. The commodity chain has been extended as new actors and outlets entered the sector, which means that fish traders are connected to the commodity chain in new ways. Second, the structure of market networks has become more complex and stratified in ways that have benefited some groups of fish traders and disadvantaged others (see Hapke 2001).

To understand the diversity of ways in which fishers and fish traders have been impacted by globalization at the household and individual levels, we can turn to an analysis of livelihoods.

"Livelihood" refers to the way one makes a living or the means of supporting one's existence or securing the necessities of life. According to Ellis (2000), "livelihoods" refers to "the *assets* (natural, physical, human, financial and social capital), the *activities* [strategies of use], and the *access* to these (mediated by institutions and social relations) that together determine the living gained by the individual or household" (2000: 10). According to this understanding, individual and household livelihoods are shaped by local and distant institutions, social relations, and economic opportunities. Assets and access (opportunities) interact to define the possible livelihood strategies (activities) available to individuals or households, which may change from season to season or from year to year as assets are built up and eroded and as access to resources and opportunities change "due to shifting norms and events in the social and institutional context surrounding ... livelihoods" (Ellis 2000:10). **Livelihoods analysis**, then, refers to a range of methods for measuring household and individual assets, activities, and access in a given ecological setting and how these change or are impacted by broad-scale economic and environmental transformations. While commodity chains analysis reveals general patterns of change, directing analytical focus to the level of the household or individual through an analysis of livelihoods allows us to ask questions such as: To whom and under what circumstances do global economic trends provide access and opportunity? For whom and under what circumstances do such trends create constraints? How do these differ for men and women? How do individuals and households forge strategies to adapt to and cope with macro-level trends and processes and institutional structures?

In the case of Indian fishing communities, livelihoods would be influenced by factors such as current fish stocks, boat ownership/non-ownership, access to harbors and other landing sites, household size and composition, education and skill levels, family ideologies and work identities, and entrepreneurial initiative. For example, households that are able to expand their asset base are not only able to withstand the shocks of ecological crises, they may even benefit directly from new market trends and opportunities. The willingness of households to deploy women in paid work, the extent of their mobility, and their level of skill shapes the particular strategies households adopt, which then shapes future options for livelihood strategies. The extent to which men responsibly contribute their earnings to household budgets determines women's livelihood options (see Hapke 2008). A close examination of fisherfolk livelihoods in the context of ecological change reveals that different households have been impacted in very different ways that at times contradict one another. The implications of such contradictions for collective action, state policy, and the local politics of development are profound.

Geographic Research, Politics, and Social Change

As indicated in an earlier section of this chapter, geographic research is undertaken for a number of different purposes (*exploration*, *description*, *explanation*, *understanding*, or *prediction*). Just as our purpose for conducting research varies so does the relationship a researcher has to her/his research subjects. We might refer to this relationship as the **politics of research**. As David Smith explains, during the quantitative revolution (in the 1960s), "it was widely believed that geography could be value free or neutral. Measuring the characteristics of spatial organization in numerical form ... could be conveyed as purely 'scientific', as could the adoption of mathematical models purporting to explain what was observed" (Smith 2010: 412). The critiques of spatial science by humanism and critical theories that emerged in the late 1960s and 1970s and the reorientation of research focus toward the conditions of human life that ensued ultimately generated a desire to conduct research that was "socially relevant." Issues such as poverty, hunger, crime, the condition of ethnic minorities, and environmental degradation received attention, which inevitably raised questions about values as well as the possible beneficiaries of research. The radical geography movement associated with several schools of critical theory (Marxism, feminism, postcolonialism) shifted the intended beneficiaries of research away from academics toward the public at large, and, in its research on inequality and social justice, toward poor and disenfranchised people in particular (Smith 2010). This gave rise to another purpose for conducting research,- namely, **advocacy**, or the use of research to serve a particular cause, or social change.

The emergence of advocacy as an objective of research and the concern that local populations benefit from research findings have been linked to a methodological shift in which research is conducted with local people in a strongly collaborative and participatory manner. This has come to be called **participatory research** (PR), or, when there is a clear advocacy component, **participatory action research** (PAR). This type of research reworks the relationship between research scholars and subjects from one of "extraction," in which "experts" enter a community and take away data to write reports back home, to one of collaboration, in which research is conducted by local people for local people. In PR and PAR local participants are directly involved in identifying research problems, developing research tools (e.g., survey questions), collecting data, analyzing results, and enacting an action plan on the basis of these results. Such involvement serves another research purpose, that of **empowerment**.

The influence of participatory research approaches in the larger research community has been profound. Participatory approaches have been incorporated into a wide range of scientific disciplines, including international development, plant breeding, technology development, psychology, and education. A number of institutions and organizations devoted to promoting PR/PAR have been born. Numerous publications about participatory research methods have become available. And even large research funding institutions such as the National Science Foundation in the United States have made social relevance/impact and local research capacity-building mandatory components of the projects they fund. Finally, participatory (action) research has expanded the realm of field research beyond academia to other contexts. Most notably the last three decades have witnessed an expansion of field research undertaken by **non-governmental organizations** working in local communities. Dr. Mandel's research for InternewsHaiti described above is a perfect example of an NGO conducting field research in collaboration with local individuals to solve local environmental problems.

Chapter Summary

This chapter has explored the various methods geographers adopt to conduct research. We began with a brief discussion of the theoretical frameworks and scientific philosophies that have influenced geographic research. We then outlined different approaches for collecting and analyzing data, including different approaches to fieldwork and the use of remotely sensed data in geographic research. To illustrate how geographic research is used to understand real-world situations, we discussed commodity chains analysis and livelihoods analysis. Finally, we concluded the chapter with a brief discussion of the relationship between geographic research and social change and the use of participatory research methods.

Critical Questions

1. What is the relationship of philosophy to research? Why are philosophical questions important for geographic research? How do ontology and epistemology inform the way we approach geographic research?
2. What are the major research paradigms that have influenced geographic research in modern times? In what ways has each expanded geographic knowledge?
3. What is the difference between quantitative data and qualitative data? What are the strengths and weaknesses of each type of data with respect to geographic knowledge generation?
4. What are some of the methods human–environment geographers use to conduct research?

Key Vocabulary

added value
advocacy
commodity chains
critical realism
critical theories
deductive reasoning
dendrochronology
empowerment
epistemology
feminism/feminist approaches
field research/fieldwork
geographic information systems
geographically referenced data
humanism
inductive reasoning
livelihoods
livelihoods analysis
Marxism
methodology/research methodology
mixed method
modernism
new regional geography
non-governmental organization
ontology

participatory (action) research
political ecology
politics of research
positivism
postcolonialism
postmodernism
primary data
qualitative data
quantitative data
remote sensing
research methodology
research paradigm
secondary data
spatial science
theoretical framework
triangulation
universal truths

References

Cloke, P., Cook, I., Crang, P., Goodwin, M., Painter, J., and Philo, C. (2004) *Practising Human Geography* (Thousand Oaks, CA: Sage Publications).

Crozier, M.J., Hardenbicker, U., and Gomez, B. (2010) Physical landscapes. In B. Gomez and J.P. Jones III (eds.), *Research Methods in Geography: A Critical Introduction*, pp. 93–115 (Malden, MA: Wiley-Blackwell).

Dear, M. (1988) The postmodern challenge: reconstructing human geography. *Transactions of the Institute of British Geographers*, 13, pp. 262–74.

Ellis, F. (2000) *Rural Livelihoods and Diversity in Developing Countries* (Oxford: Oxford University Press).

Graham, E. (2005) Philosophies underlying human geography research. In R. Flowerdew and D. Martin (eds.), *Methods in Human Geography: A Guide for Students Doing a Research Project*, 2nd edn. (Harlow, Middx: Pearson Education).

Entrikin, J.N. and Tepple, J.H. (2006) Humanism and democratic place-making. In S. Aitken and G. Valentine (eds.), *Approaches to Human Geography* (Thousand Oaks, CA: Sage Publications).

Hapke, H.M. (2001) Petty traders, gender and economic transformation in an Indian fishery. *Economic Geography*, 77(3), pp. 225–49.

Hapke, H.M. (2008) *Gendered Livelihoods in the Global Fish Food Regime*. Paper presented at the International Geographical Union Commission on the Dynamics of Economic Spaces, Barcelona, Spain.

Herod, A. and Parker, K.C. (2010) Operational decisions. In B. Gomez and J.P. Jones III (eds.), *Research Methods in Geography: A Critical Introduction* (Malden, MA: Wiley-Blackwell.

Kitchin, R. (2006) Positivistic geographies and spatial science. In S. Aitken and G. Valentine (eds.), *Approaches to Human Geography* (Thousand Oaks, CA: Sage Publications).

Kitchin, R. and Tate, N.J. 2000. *Conducting Research in Human Geography: Theory, Methodology and Practice* (New York: Prentice Hall).

Liverman, D., Moran, E.F. Rindfuss, R.R. and Stern, P.C. (eds.) (1998) *People and Pixels: Linking Remote Sensing and Social Science* (Washington, DC: National Academy Press).

Pidwirny, M., Hassan, G., Hussein, G., and Banks, A.C. (2010) Remote sensing. In C.J. Cleveland (ed.), *Encyclopedia of Earth* (Washington, DC: Environmental Information Coalition, National Council for Science and the Environment). http://www.eoearth.org/article/Remote_sensing (accessed July 7, 2011).

Rindfuss, R.R. and Stern, P.C. (1998) Linking remote sensing and social science: the need and challenges. In D. Liverman et al. (eds.), *People and Pixels Linking Remote Sensing and Social Science* (Washington, DC: National Academy Press).

Robbins, P. (2004) *Political Ecology: A Critical Introduction* (Oxford: Blackwell Publishing).

Robbins, P. (2010) Human–environment field study. In B. Gomez and J.P. Jones III (eds.), *Research Methods in Geography: A Critical Introduction*, pp. 241–56 (Malden, MA: Wiley-Blackwell).

Shaw, I.G.R., Dixon, D.P., and Jones, J.P. III (2010) Theorizing our world. In B. Gomez and J.P. Jones III (eds.), *Research Methods in Geography: A Critical Introduction* (Malden, MA: Wiley-Blackwell).

Smith, D.M. (2010) The politics and ethics of research. In B. Gomez and J.P. Jones III (eds.), *Research Methods in Geography: A Critical Introduction*, pp. 411–23 (Malden, MA: Wiley-Blackwell).

Stow, D.A. (2010) Remote sensing. In B. Gomez and J.P. Jones III (eds.), *Research Methods in Geography: A Critical Introduction*, pp. 155–72 (Malden, MA: Wiley-Blackwell)

Women in Geography Study Group (1997). *Feminist Geographies: Explorations in Diversity and Difference* (Harlow, Essex: Longman).

14

Conclusion

Making a Difference

Icebreaker: Three Human–Environment Geographers who Made a Difference

Geography is a small but rapidly growing discipline with a long and distinguished history of studying human–environment questions. Consider the brief stories of three human–environment geographers who have changed the way we understand the world and – directly or indirectly – influenced policy and programs.

An Introduction to Human–Environment Geography: Local Dynamics and Global Processes,
First Edition. William G. Moseley, Eric Perramond, Holly M. Hapke and Paul Laris.

Published 2014 by John Wiley & Sons, Ltd.

The British geographer **Piers Blaikie** *published an influential book in 1985 entitled* The Political Economy of Soil Erosion. *Up until this point, many environmental problems (soil erosion in this instance) had been studied as biophysical processes in relation to local management practices. What Blaikie expertly combined was an understanding of soil erosion with an analysis of broader-scale political and economic processes. He showed, for example, that erosion produced by farming on steep slopes often had as much, or more, to do with poor farmers being pushed off of good land into marginal areas by powerful actors than it did with the actual farming practices of the poor farmers in question. His book, and the approach he described, eventually led to development of an empirically and theoretically rich interdisciplinary field known as political ecology. His work also changed development policy and practice in many areas of the world.*

Diana Liverman *exemplifies several aspects of what human–environment geographers have to offer to agricultural questions. Trained in both human and physical geography, Liverman has a long-standing interest in the human dimensions of global change. Born in Ghana, and educated in the UK, Canada, and the US, she initially was interested in the potential and limitations of predicting climate impacts using both crop-simulation models and the first generation of global models that allowed for the assessment of climate change impacts. However, as it became clear to her that the scientific community's knowledge of climate impacts in the developing world was insufficient for modeling, and that some of the most interesting questions were about how people and places became vulnerable to climate change, much of her work came to focus on the vulnerability to drought of farmers in the drylands of Mexico. By studying small and large farmers in the Sonora and Puebla states of Mexico, Liverman was able to quantify the impacts of land tenure and technology on vulnerability to drought. Here she found that those with access to irrigation have lower drought-related crop losses, and farmers on communally held ejido land are more at risk to drought than large private farms. Of course these two factors (technology and land tenure) are correlated in Mexico as the large private farms are more likely to have irrigation than communal (or ejido) land. Furthermore, private landowners were more likely to have access to higher-quality land which has a bearing on crop losses during low rainfall years. Liverman's work in this area was important because it showed that patterns of crop loss could depart from levels of rainfall because of differences in agricultural vulnerability between households. She also has considered the influence of politics and economics on farming and ranching decisions in the face of changing climatic conditions. Liverman chaired the US National Academy of Sciences Committee on the Human Dimensions of Global Environmental Change, and has sat on advisory committees for the National Oceanic and Atmospheric Administration (NOAA), NASA, and the Inter American Institute for Global Change.*

Contrary to the conventional wisdom that slaves were a mere labor force in the US South, geographical scholarship has shed light on the powerful role that Africans had in developing the agricultural and economic potential of the region. Geographer **Judith Carney***'s research reshaped contemporary thinking about Africans in the Americas. Her seminal work,* Black Rice: The African Origins of Rice Cultivation in the Americas *(2001), is part of a growing literature on the Atlantic world, a global region centered on the Atlantic Basin. Carney effectively demonstrated how African understanding of tidal mangrove rice cultivation in West Africa, and especially the knowledge of women on this topic, was vital* *for the establishment of highly productive rice plantations in the Carolinas in the 17th and 18th centuries. What is interesting about Carney's career is that she worked in Africa first (e.g., Carney 1993), and only later became interested in the US Southeast. It was her detailed understanding of rice cultivation in West Africa that allowed her to recognize the agency of Africans in the historical development of rice farming in the Carolinas. Without a previous background in African agriculture, Carney's exploration of the African connections to the historical development of rice farming systems in the Carolinas could never have been made as forcefully and as convincingly.*

Chapter Objectives

1 To review some of the main themes articulated in this book.
2 To show that the way we understand the world, through theories and models, influences how we act in the world.
3 To explore examples of how geographical concepts and ideas may be applied in the real world.
4 To consider the value of bridging academic and applied work by being an engaged scholar or a thoughtful activist.

Introduction

This chapter reviews some of the main themes articulated in this book. It then makes an argument for how the many of these ideas may be employed in the real world to understand environment-related problems and bring about change. We articulate three broad take-home messages for the reader. First, the way we understand the world has important implications for how we act in the world. Second, human–environment geographic research need not necessarily be an activity isolated from "real world" policy and change. There are many examples wherein the actual process of research (particularly action-research) is a vital component of the change process. Third, there are strands of human–environment geography which are applied or practical in nature and there are a number of geographers who have worked directly on policy issues. Twenty-first-century geography students are well positioned to bridge the worlds of theory and practice by being thoughtful practitioners and engaged scholars.

A Brief Review of Major Themes from the Book

This book had two broad objectives: (1) to convey geography's theoretically rich tradition and unique approach to environmental issues, and (2) to offer a text that would be accessible to introductory students, many from allied environmental fields who were encountering geography for the first time, others in geography for whom this was their first course on human–environment themes. You will recall that this book is divided into four parts. The first part was a broad overview of the basic information needed to understand human–environment geography, moving from the geographic perspective, to environmental politics, to some basic physical geography and ecology. The second section explored a sampling of geography's rich theoretical traditions in the realm of human–environment geography, namely cultural and political ecology, environmental history, hazards geography and human vulnerability, and environmental justice. The third part was more thematic in nature, covering: climate, atmosphere and energy; the population–consumption–technology nexus; agriculture and food systems; biodiversity, wildlife, and protected areas; and water resources and fishing livelihoods. The final part (which includes this chapter) is meant to connect the book's material to the real world by showing the student how geographers undertake fieldwork and collect and analyze data; and to make suggestions for using the concepts in this text to understand environment-related problems and influence change.

While this book had many themes which are too numerous to delineate here, some of the broad conceptual ideas which we hope you retain about human–environment geography include the following.

1. Geography is a broad discipline that essentially seeks to understand and study the spatial organization of human activity and of people's relationships with their environment. It is also about recognizing the interdependence among places and regions, without losing sight of the individuality and uniqueness of specific places.
2. Fundamental geographic perspectives that often characterize the discipline's approach to human–environment questions include: scale-sensitive analysis, attention to spatial patterns of resource use, a conception of the human–environment system as a single unit (rather than two separate parts), and a cognizance of the connections between places and regions.
3. In contemporary Western society, nature has often been constructed as a place without humans. Once we understand that humans, like other animals, modify and are influenced by their environments, then we appreciate that we are a part of the environment and it is problematic to conceptually separate ourselves from it. The construction of nature as a place without humans has led to a myriad of problems, including the destruction of some rural livelihoods to create parks and insufficient care and attention for more humanized landscapes.
4. Humans modify their landscapes, and these changed environments, in turn, influence human behavior.
5. The globalized nature of the contemporary economy means that production and consumption rarely occur in the same place. This is often problematic because there is little to no environmental feedback for consumers to heed when making choices.

6 There is no such thing as merely acting in the world. Action is always influenced by one's own theories and models of how the world works. The persistence of some theories or models may be influenced by the interests of powerful entities.

The remainder of this chapter explores how some of the major themes in this text may be applied in the real world to understand environment-related problems and influence change.

Theory, Scholarship, and the "Real World"

Many a student has grown weary of theory and pined to simply "act" to make a difference. It is not unusual to want to go out and work in the real world rather than muse about abstract and seemingly irrelevant theory. One of the central arguments of this book, however, is that theory does matter and we need to be cognizant of this. Whether we wish to believe it or not, there is no such thing as simply "doing," as all action is guided by some theoretical understanding or mental model of the world. A **model** is a simplification of the world and **theory** is an explanation for how the world works, often involving generalizations or simplifications that cut across contexts. In order to make sense of the world, all of us operate with theories and models in our heads, whether we wish to explicitly acknowledge this or not.

In their work on northern Côte d'Ivoire, Tom Bassett and Koli Bi Zuéli (2000) encountered a strong regional narrative of deforestation which had been enshrined in that country's National Environmental Action Plan. For decades, the country's forestry officials, development banks, and international consultants had argued that declining forest cover was the primary problem which must be combated with afforestation efforts. In other words, these officials and development actors were, knowingly or unknowingly, operating with a certain theory of regional environmental change which was influencing their perception and diagnosis of the problem.

Bassett and Zuéli questioned this theory, likely because they had been informed by other work in the region on environmental counter-narratives (discussed in Chapter 4), and found that the cover of woody species in the area studied was on the rise. The real problem was a degradation of range lands, a situation exacerbated by expanding tree cover in part due to shifting fire regimes. This led them to conclude that environmental narratives could be counterproductive because they overshadow real solutions to real problems. In other words, environmental narratives may lead to the misdiagnosis of the problem – leading to inappropriate (and sometimes counterproductive) solutions.

As such, it is terribly important to recognize that the way we understand the world may influence how we see the world. If we are cognizant of the influence of models and theories on the way we see the world, then we can sometimes get outside of these views, or work with a different set of theoretical assumptions, to ask an alternative set of questions which may yield new insights. Ultimately, understanding what really happened is important for determining appropriate future action.

Geographic Research and Social Change

Human–environment geographic research need not necessarily be an activity isolated from "real world" policy and change. There are many examples wherein the actual process of research (particularly action-research) is a vital component of the change process. As discussed in Chapter 13, fieldwork may be undertaken in a variety of ways with different degrees of engagement with the local community.

In the traditional research paradigm, the research question is defined by the research, and usually relates to some key debate occurring in the scientific literature (rather than a question of local import). The researcher then caries out his or her fieldwork to answer the question. Local people may be interviewed, but little effort is expended to explain the goal of the research project and the results or findings may never be shared with the community. This more traditional approach to research is sometimes referred to as **extractive research** because the research is essentially extracting information from local people and giving very little in return.

While extractive traditional research is one extreme, the most participatory form of research would be **participatory action research** (PAR). Practitioners of PAR typically start with a series of meetings with the local community to identify their chief concerns, problems or questions. Once a research question has been identified which is a priority for the community, the researcher works in an open and collaborative fashion with local people to answer that question. Participatory research techniques (such as cognitive or mental maps, transect walks, proportional piling, and Venn diagrams) are often employed, which allow both literate and non-literate community members to fully participate in the research process. At the end of the field research process, all findings are shared with the local community.

The reality is that many researchers do not operate at either extreme, but somewhere in the middle. Because of institutional constraints, many researchers may not have the time to undertake full-blown PAR, or they may need to work on questions that have broader import than those most pressing for one particular community. That said, even more traditional research can be modified in a number of ways to make it meaningful, beneficial, and empowering for local community members. For example, because of bias and prejudice, many rural people, minority groups, women, or less powerful individuals have never been asked in a respectful manner to share their understanding and insight on a particular situation. When this is done well, and research results are shared back with the community for comment, this can often be a very empowering experience (particularly if the insights of the community are validated). While it may seem trivial, such empowerment can itself be a catalyst for change.

Making a Difference

Human–environment geographers contribute to solving real-world questions and problems in two broadly different ways. First, by producing socially relevant research, geographers at universities and research institutes create the understanding that is necessary for informed action. See Box 14.01 for examples of geographic research and insights that have been applied to development questions in one region of the world, Africa. We would also invite you to explore the most recent (2010) National Research Council (NRC) report on geography, entitled *Understanding the Changing Planet: Strategic Directions for the Geographical Sciences*, which outlines a number of key, socially relevant research themes for geographers (many of which are human–environment related).

The second broad way that geographers become involved in solving real-world questions is by using their insights in applied professions. Outside of research institutes and universities, human–environment geographers work in business, for government and international organizations, and in the non-profit sector. Some apply their technical skills as remote sensors, cartographers, and GIS analysts. Others use their understanding of the physical environment to work in natural resource management agencies. Still others use their understanding of natural disasters, or of development to work for units of the United Nations, World Bank, or non-governmental organizations (NGOs). Many human–environment geographers work in multidisciplinary teams with other natural and social scientists. Geographers are often valued for their unique perspectives as well as their ability to easily work with insights from others disciplines and to bridge the social/natural science divide.

Box 14.01 Applying the insights of human–environment geography to development: three African examples

This section provides just a few examples of how geographers have contributed and are contributing to development in Africa. Geographers, with their tradition of fieldwork, study of coupled human–environment systems, spatial analysis tools, area studies tradition, and attention to the connections between places and across scales, have used their conceptual models and unique mix of skills to further the theory and practice of development in this region. We outline three examples of where geographic concepts and techniques have played crucial roles: famine early warning, gendered use of natural resources, and local knowledge.

First, the conceptual and technical development of **famine early warning systems**[1] is a good case in point (with the FEWS example discussed in Chapter 13 being one example of this). Interest in famine early warning systems developed in the wake of major droughts which racked the African Sahel (a semi-arid region just south of the Sahara Desert) in the early 1970s and mid-1980s (Moseley and Logan 2001). Such systems, often run by national governments, international agencies (such as the UN Food and Agriculture Organization) or non-governmental organizations, seek to identify those populations or groups that are likely to experience a food shortfall in a given year, and the magnitude of that shortfall. Hutchinson (2001) has described how geographers have been at the forefront of developing these systems. While not a specialist on Africa, Dando (1980) was one of the first to recognize that famine could occur in areas where there was plenty of food available on the market, yet people could not afford to purchase this food (although Amartya Sen [1981] was the first to really bring this fact to the attention of the world). Up until this time, famine scholars had been heavily influenced by Malthusian thinking (see Chapter 9) and saw hunger as a supply problem, i.e., an absolute lack of food on the market. The Sahel offered a classic case of this problem as food was being exported from the region during a major drought and famine in the early 1970s (Franke and Chasin 1980). Working in northern Nigeria, Watts (1983) produced pioneering work on coping strategies in which he described the rationality of attempts by the rural poor to deal with food shortfalls, such as reducing the number of meals eaten, collecting wild foods, and selling livestock. There was often a clear progression of such strategies, and therefore they could be monitored as predictive signals in the run-up to a famine. Beyond these important early conceptual contributions, geographers have gone on, and continue, to make important technical contributions combining remote sensing, geographic information systems (GIS), and social science surveys to predict and map those areas in Africa that are most vulnerable to hunger in any given year (Hutchinson 1998).

Second, up until the 1980s many development programs in Africa run by international agencies, bilateral donors, and NGOs could be faulted for largely looking at communities as undifferentiated wholes and, in practical terms, largely working with men. In the 1980s and 1990s, a group of geographers (Carney 1993; Rocheleau et al. 1997; Schroeder 1997) began to analyze natural resource management and development projects in Africa from the perspective of gender. Simple tools, like **cognitive or mental maps**, were used by these geographers and others with focus groups within communities to better understand a subgroup's relationship with the landscape and various natural resources (see Figure 14.01). What they showed is that men and women in Africa often interact with the natural environment differently because of their societal roles. So, for example, because many African women are responsible for collecting firewood, fetching water, and cooking, they are much more aware of water scarcity and increasing distances to collect wood.

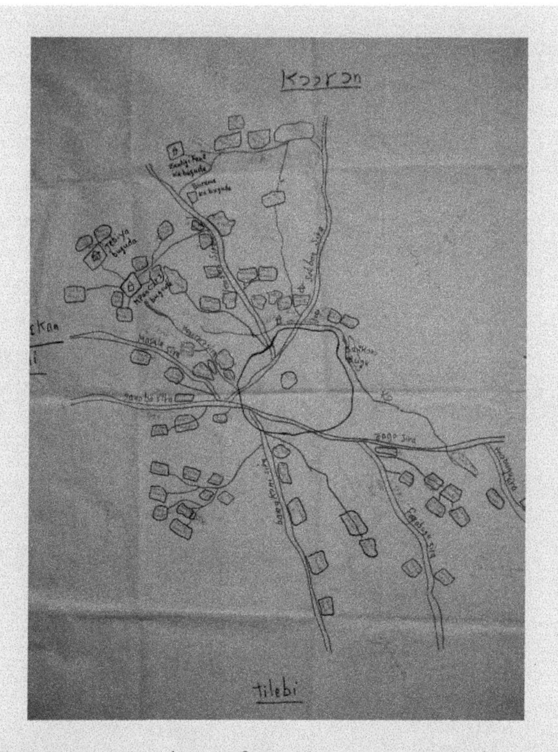

Figure 14.01 Cognitive or mental map of a rural community in Mali. *Source*: Photo by William G. Moseley. Used with permission.

This insight has two important implications. First, men and women in Africa may be impacted differently by environmental change because of their distinct social roles and responsibilities. Secondly, development organizations may only understand part of a community's development challenges if they solely speak to men. Schroeder's (1997) study of development work in the Gambia is a classic case of outsiders unwittingly subverting the position of women by acting on the priorities of men. The work of geographers on gender and development helped spawn a new generation of development approaches that were more sensitive to the needs of women in Africa, sometimes referred to under the banner of **women in development** (WID) or simply "gender."

Finally, appreciation for **local or indigenous knowledge** (see Chapter 4) of agricultural, forestry, and range management is another realm where geographers have made vital contributions to African development thinking and practice. Until the 1980s, conventional, modernist development thinking suggested that local natural resource management practices were primitive and backward. Development, therefore, meant replacing these local approaches with industrial or Western methods. Richards (1985) and Bayliss-Smith (1982), for example, demonstrated that traditional practices such as agroforestry and intercropping (see Chapter 10) actually made local farming systems more efficient in terms of energy consumed per unit of output. Others have demonstrated the ecological and economic rationality of traditional approaches to herd and pasture management (Turner and Hiernaux 2002) as well as forestry (Fairhead and Leach 1996). Geographers' approach to local knowledge and resource management has often been distinguished by its joint focus on both human and biophysical systems as well as its embrace of hybrid research employing mixed methods (Batterbury et al. 1997). These methods have ranged from soil sampling and remote sensing, to interviews, to a whole host of **participatory rural appraisal (PRA)** techniques, such as mental maps, transects, or proportional piling (see Figure 14.02) (see Chapter 13).

Figure 14.02 Women in rural Zimbabwe undertaking a proportional piling exercise on household wealth in their community. *Source*: W.G. Moseley. Used with permission.

Thoughtful Practitioners and Engaged Scholars

Part of breaking down the barrier between the academy and practitioners is to realize that both worlds need each other. As discussed in an earlier section of this chapter, there is no such thing as merely acting in the world. Action is always influenced by one's own theories and models of how the world works. As such, practitioners must be mindful of the theories and models which may be influencing the decisions they make regarding programs and policies.

In a similar way, those geographers who are academics and researchers cannot afford to completely shut themselves off from the world around them. This may mean working on questions that are of import to society or weighing in on societal questions to which one's research has some connection (or both). Those academics who are proactive about engaging with the public on important social questions are sometimes referred to as **public intellectuals**.

Public intellectuals are often academics or writers with particular expertise who regularly address issues of broad interest in the media. Posner (2003: 2) has defined public intellectuals as "intellectuals who opine to an educated public on questions of, or inflected by, political or ideological concern." As such, these are essentially individuals who take the time to address an important public debate when they have a valuable and well-supported perspective to add to the conversation. Public intellectuals may be distinguished from social commentators (such as newspaper columnists) who regularly write on a broad range of issues on which they may or may not have any particular expertise; as well as from thought leaders who are generators of new ideas in a particular field. Our assertion, which is shared by many, is that society benefits when its public discourse is enlivened and enriched by a diverse and informed set of perspectives and that public intellectuals are key actors in this process.

From the perspective of a geographer, Alexander Murphy has argued (2006: 3), "that our understanding of issues and problems will be impoverished if geographical perspectives are not part of the mix." Others assert that academics have an ethical obligation to engage in **public scholarship**[2] where appropriate (Said 1995). Kathryn Mitchell (2006: 205), paraphrasing Karl Marx, asserts "that the point of scholarship ... is not just to interpret the world but to change it." One way to participate in the discussion of an important public policy question is to write an opinion piece (or op-ed) for your local, regional, or national news outlet (see Box 14.02). The same rules would generally apply for print as well as electronic media.

Box 14.02 Civic engagement 101: publishing a newspaper opinion article

The following are some general rules to consider when trying to publish a newspaper opinion article.

First, one must be timely. Writing something quickly, or having something ready to go, when the issue is on the front pages is critical. Many issues have a three- to four-day news cycle, which means you have about 24–36 hours to submit something for consideration for publication.

Second, you need to have some expertise to offer – and ideally you will have undertaken research on the issue at hand. Sometimes it requires a creative mind to make connections between your research and the issue under debate.

Third, writing for a popular newspaper audience demands a different style of writing than that employed for scholarly publications. Above all, one must be concise. Going above 800 words is an absolute sin, and ideally one should be in the 650- to 750-word range. One should also write as simply and clearly as possible. An op-ed page editor once wrote the following in response to a submission: "Love the topic. But the writing is too academic by a mile for my purposes." Short sentences and very short paragraphs are quite common. It is also typical to begin with a provocative or catchy opening sentence or two that conveys the main point of the article. The conclusion should emphasize concrete policy recommendations. Just as you would do with a journal article submission, it is helpful to send your submission to the editorial editor with a cover email succinctly explaining why the issue you are discussing is important and why you are qualified to comment.

Fourth, many colleges and universities have press offices that are willing to work with faculty (and sometimes students) to help them prepare and submit op-eds to the press. This assistance is most helpful when people in the press office actually know the editorial page editor (as this may increase the odds of having your piece read by an editor). Staff in the press office may also be able to provide feedback on your editorial, allowing you to make vital improvements.

Fifth, closely read the instructions for editorial contributors for each newspaper. Some papers require exclusive consideration while others do not. While we would all like to publish an editorial in the *New York Times* (USA), *The Guardian* (UK), or *The Australian*, the probability of this occurring is extremely small. Given the importance of timing (as described in the first point above), the week-long period that the top papers ask to consider an editorial submission may effectively eliminate the possibility of it being published in another paper if you are declined. As such, unless you are a big-name academic known to the general public, it may be risky to submit your most time-sensitive articles to the top two or three papers. However, for articles with a several weeks-long shelf life (sometimes referred to as "evergreen"), it certainly does not hurt to try the big papers first, and then move on to other papers if you are not successful at the highest levels.

Sixth, carefully consider the best newspaper outlet for your editorial. Are you writing about a local or a national issue (as the latter would suggest that you target a newspaper with a more national circulation)? What is the political orientation of the paper – liberal or conservative? – as this may influence whether or not they publish your submission. Finally, newspapers with larger circulations generally tend to be more desirable as they research larger audiences. That said, there are newspapers with smaller circulations that are influential because of their target audience, their number of online visitors, or their ability to place your article in other papers.

Source: These points are largely based on Moseley 2010.

Chapter Summary

This chapter had two main aims. The first, and arguably of lesser import, was to review some of the main themes in the book. The second goal was to encourage students to explore how the human–environment perspective may be used to better understand problems in the real world and to provide examples of how this view has actually made a difference. Clearly the way we understand the world has huge implications for the way we act in it. Human–environment geography research is very diverse, with some scientists working on abstract questions and others seeking to tackle queries with direct policy relevance. In either case, the process of engaging in research can often be transformative in and of itself (for researchers and participants). Twenty-first-century geography students are well positioned to bridge the worlds of theory and practice by being thoughtful practitioners and engaged scholars.

Critical Questions

1. In your opinion, what were some of the most interesting themes and perspectives covered in this whirlwind tour of human–environment geography?
2. Why is it important to be cognizant of the theories and mental models which may influence what we "see" in the world?
3. How can field research be empowering or disempowering for local participants?
4. How have geographical insights changed the way we look at some issues?
5. Do you believe that academic geographers have an ethical obligation to engage in public scholarship where appropriate? Why, or why not?

Key Vocabulary

cognitive or mental maps
Diana Liverman
extractive research
famine early warning systems (FEWS)
Judith Carney
local or indigenous knowledge
model
op-ed
participatory action research (PAR)
participatory rural appraisal (PRA)
Piers Blaikie
public intellectuals
public scholarship
theory
women in development (WID)

Notes

1. Systems used to predict the potential onset of famine in order to deliver humanitarian assistance in time to avoid such a disaster.
2. Traditional public scholarship is scholarly writing which is accessible to a broad yet well-educated audience. This is different than academic writing, which is intended for a specialist audience.

References

Bayliss-Smith, T. (1982) *The Ecology of Agricultural Systems* (New York: Cambridge University Press).

Bassett, T.J. and Koli Bi Zuéli (2000) Environmental discourses and the Ivorian Savanna. *Annals of the Association of American Geographers*, 90, pp. 67–95.

Batterbury, S., Forsyth, T., and Thomson, K. (1997) Environmental transformations in developing countries: hybrid research and democratic policy. *The Geographical Journal*, 163(2), pp. 126–32.

Blaikie, P.M. (1985) *The Political Economy of Soil Erosion in Developing Countries* (London: Longman).

Carney, J. (1993) Converting the wetlands, engendering the environment: the intersection of gender with agrarian change in the Gambia. *Economic Geography*, 69(4), pp. 329–48.

Carney, J. (2001) *Black Rice: The African Origins of Rice Cultivation in the Americas* (Cambridge, MA: Harvard University Press).

Dando, W.A. (1980) *The Geography of Famine* (London: Edward Arnold).

Fairhead, J. and Leach, M. (1996) *Misreading the African Landscape: Society and Ecology in a Forest-Savanna Mosaic* (Cambridge: Cambridge University Press).

Franke, R. and Chasin, B. (1980) *Seeds of Famine: Ecological Destruction and the Development Dilemma in the West African Sahel* (Montclair, NJ: Rowman & Littlefield).

Hutchinson, C.F. (1998) Social science and remote sensing in famine early warning. In D. Liverman, E. Moran, R. Rindfuss, and P. Stern (eds.), *People and Pixels: Linking Remote Sensing and Social Science*, pp. 189–96 (Washington, DC: National Academy Press).

Hutchinson, C.F. (2001) Famine and famine early warning: some contributions by geographers. *Yearbook of the Association of Pacific Coast Geographers*, 63, pp. 139–44.

Mitchell, K. (2006) What's left? Writing from left field. *Antipode*, 38(2), pp. 205–12.

Moseley, W.G. (2010) Engaging the public imagination: geographers in the op-ed pages. *Geographical Review*, 100(1), pp. 109–21.

Moseley, W.G. and Logan, B.I. (2001) Conceptualizing hunger dynamics: a critical examination of two famine early warning methodologies in Zimbabwe. *Applied Geography*, 21(3), pp. 223–48.

Murphy, A. (2006) Enhancing geography's role in public debate. *Annals of the Association of American Geographers*, 96(1), pp. 1–13.

National Research Council (NRC) (2010) *Understanding the Changing Planet: Strategic Directions for the Geographical Sciences* (Washington, DC: National Academy of Sciences of the USA).

Posner, R.A. (2003) *Public Intellectuals: A Study of Decline* (Cambridge, MA: Harvard University Press).

Richards, P. (1985) *Indigenous Agricultural Revolution: Ecology and Food Production in West Africa* (London: Westview Press).

Rocheleau, D.E., Thomas-Slayter, B., and Wangari, E. (eds.) (1997) *Feminist Political Ecology: Global Perspectives and Local Experience* (New York: Routledge).

Said, E.W. (1995) *On Defiance and Taking Positions. Beyond the Academy: A Scholar's Obligations*. American Council of Learned Societies Occasional Paper 31, http://www.acls.org/op31said.htm#said.

Schroeder, R. (1997) Reclaiming land in the Gambia: gendered property rights and environmental intervention. *Annals of the Association of American Geographers*, 87(3), pp. 487–508.

Sen, A. (1981) *Poverty and Famines* (Oxford: Clarendon Press).

Turner, M.D. and Hiernaux, P. (2002) The use of herders' accounts to map livestock activities across agropastoral landscapes in semi-arid Africa. *Landscape Ecology*, 17(5), pp. 367–85.

Watts, M. (1983) *Silent Violence: Food, Famine and Peasantry in Northern Nigeria* (Berkeley, CA: University of California Press).

Index

An Introduction to Human–Environment Geography: Local Dynamics and Global Processes, First Edition. William G. Moseley, Eric Perramond, Holly M. Hapke and Paul Laris.

Published 2014 by John Wiley & Sons, Ltd.